Russian Research Center Studies, 69

NEWTON AND RUSSIA

THE EARLY INFLUENCE, 1698–1796

Frontispiece. Newton, by Sir Godfrey Kneller, painted four years after Newton's supposed meetings with Peter the Great. From the National Portrait Gallery, London.

VALENTIN BOSS

NEWTON AND RUSSIA

The Early Influence, 1698–1796

Harvard University Press   Cambridge, Massachusetts   1972

To the memory of my father

Чудный закон
Природы крылся.
Но Бог всесильный рек: родись Невтон!
Исчезла тьма и свет явился.

One of several Russian versions of Pope's famous lines on Newton, this anonymous translation is from *Drug iunoshestva* (1811).

PREFACE

I wish to record my particular gratitude to Professor Richard Pipes for his encouragement, kindness, and interest. Without his loan of a volume of the *Obzor fondov Akademii nauk,* the line of inquiry leading to the inventory of J. D. Bruce's library would not have been made. Both he and Professor Robert L. Wolff have been good enough to read parts of the manuscript at various stages. The generosity of the Russian Research Center at Harvard helped me to complete the present work, while the assistance of the Inter-University Committee of Indiana enabled me to spend several months in Soviet libraries.

Likewise, I am grateful to the custodians of various Soviet collections, in particular that of the XVIIIth-century section of the Library of the Academy of Sciences of the USSR in Leningrad whose kind cooperation was so essential in tracing Bruce's books. I also wish to thank the librarians of the Archive of Ancient Acts (ZGADA) in Moscow, and of the rare manuscript section of the Saltykov-Shchedrin Library in Leningrad. Dr. V. L. Chenakal, director of the Lomonosov Museum on the banks of the Neva, spent many valuable hours showing me around this monument to Russian science. The St. Petersburg Academy's inaugural letter to Newton is reproduced by courtesy of the Royal Society; the portrait of Newton by Kneller (frontispiece), by courtesy of the Earl of Portsmouth and Lady Rupert Nevill; the portrait of Peter the Great, also by Kneller, by courtesy of The Queen's Collection, London; the portrait of Bruce, by courtesy of the Hermitage, Leningrad; Newton's own drawing of his telescope by courtesy of the Royal Society; Lomonosov's sketches of the "Newton-Lomonosov" telescope and of the "Tubus Nyctopticus-N." by courtesy of the USSR Academy of Sciences; the Newtonian speculum made by

vii

Bruce and the older telescope in his collection by courtesy of the Lomonosov Museum in Leningrad.

My thanks are also due to the late Dr. Gordon M. Trim, President of Babson College in Wellesley, Massachusetts, for access to its Newton collection, as well as to its helpful curator, Mrs. Virginia Harrison, for permission to reproduce Prince Menshikov's unpublished letter to Newton and two drafts of Newton's reply, as well as figures 4–9, 14, 20–22, 30, 33, and 47. Prince Antiokh Cantemir's portrait is reproduced from Barkov's edition of his poems, published in 1762, which may be found in the Kilgour Collection of the Houghton Library at Harvard. The portrait of Francesco Algarotti is taken from the *Opere del Conte Algarotti,* edizione novissima, Volume I (Venice, 1791); the diagram illustrating Newton's "experimentum crucis" with light from the *Dialloghi sopra l'ottica Neutoniana* is from the same edition of Algarotti's *Opere.* Title pages from *Evklidovy elementy iz dvenatzati neftonovykh knig vybrannyia* (1739) and Kurganov's *Pismovnik* are reproduced from editions in the Library of Congress, as also the extracts from the *Kniga mirozreniia ili mnenie o nebesnozemnykh globusakh;* Bruce's "Ex Libris" and figure 13, from the Library of the USSR Academy of Sciences in Leningrad; all other title pages, from works in the British Museum.

I am also indebted to P. S. Laurie of the Solar Department at the Royal Greenwich Observatory for information concerning Peter the Great's visits to the Royal Observatory, as well as to the Deputy Master of the Royal Mint for access to Newton MSS. Portions of Chapters 3 and 4 previously appeared in the *Archives internationales d'Histoire des Sciences* (Nos. 60–61); my appreciation is due the editors for permission to use some of the same material, as well as to Professors I. B. Cohen and the late Alexandre Koyré, who read it prior to publication. Parts of the "Bibliographic Note" and Chapter 2 were communicated in a paper ("N'iuton i Rossiia") given at the XIIIth International Congress of the History of Science in Moscow. Finally, I would like to thank Miss Rita Howe of Harvard University Press for her unfailing good judgment in editing the manuscript.

In transliterating from the Russian, proper names as always present a special problem. Antiokh Cantemir's name, for example, is often written with a "K" in deference to the Cyrillic practice, but since he always spelled his name with a "C" when writing it in languages other than Russian, it seemed more considerate to follow the poet's own predilection. On the other hand, in the case of names already Anglicized through frequent mention in the eighteenth century—for example, Leonard rather than *Leonhard* Euler, John rather than *Johann* Bernoullii—there seemed no point in allowing the autographs to dictate present usage. With Russified names, such as Brius (Bruce) or Rikhman (Richmann), I have followed the original spelling. There are also instances where custom defeats logic: Lomonosov, for example, usually wrote his name with a double "s" and a "w" at the end when writing in Western languages (in deference to German usage). This is sometimes observed in English, but I have preferred to follow the simpler Russian form. In the eighteenth century Russian spelling was often eccentric, especially when dealing with Western names, titles, and so forth. I have made no attempt to make it seem more consistent than contemporary practice allowed. I have applied the same principle in citations from French, German, Italian, and other sources of the period.

The present study concludes with Newton's apotheosis in the Catherinian era; the most productive period of Newtonian scholarship in Russia, however, began more than a century later—this will be the subject of a companion study .

CONTENTS

Preface   vii
Illustrations   xiii
Abbreviations   xvi
Introduction   1

PART I   THE PETRINE ERA

1   Did Peter the Great Meet Newton?   9
    Newton and Bruce   15

2   Flamsteed, Halley, Newton, and the "Collonel"   19
    Bruce and "Ivan Kolsun"   29

3   Russia's First Newtonian   33
    J. D. Bruce and the Compilation of His Library   33
    The *Principia* and Related Commentaries   35
    The "Fluxionary" Method and the Mathematical Writings   42

4   Prince Menshikov's Exchange of Letters with Newton   45

5   "Worlds within Worlds"   50
    The First Russian Explanation of Gravity   50
    Planetarians   57

6   Bruce as Translator and Instrument Maker   61
    The Origin of the Translation and Its Reception   61
    Newton's Optical Discoveries   68
    The Newtonian Telescope   71
    Bruce's Newtonian Telescope   73

Contents

7   Newton and the Naval Academy   78
    Newton and Andrew Ferquharson   78
    "Domkino," alias George Peter Domcke   84
    'S Gravesande's *Introductio ad Philosophiam Newtonianam*   89

PART II   NEWTON AND THE ST. PETERSBURG ACADEMY OF SCIENCES

8   The Foundation of the Academy and the Death of Newton   93

9   The Principal Objections to the Cartesian Theory   97

10   Newtonian Physics at the New Academy   102

11   The Opening Controversy over Gravitation   106

12   Bilfinger and James Jurin   112

13   "Il Propagatore del Newtonianismo"   116

14   The Shape of the Earth   128

15   Newton's Experiments with Light, and the Controversy
       over "vis viva"   138

16   Electricity and Action at a Distance   152

17   Newton, Lomonosov, and Christian Wolff   165
    Lomonosov and Wolff   165
    from The *Principia*   170
    Empiricism versus Rationalism   176

18   Lomonosov and the *Opticks*   185

19   A New Kind of Newtonian Telescope?   200
    The Newton-Lomonosov Telescope   200
    "Tubus Nyctopticus modo Lom.N."   207

xii

Contents

20   Newton in the Catherinian Era   211

Conclusion   231

Appendix I   Draft of Newton's Letter to Menshikov   241

Appendix II   Gravitation and Language   243

Bibliographic Note   247

Bibliography   257

Index   287

Illustrations

(*frontispiece*)
Newton, by Sir Godfrey Kneller.

(*preceding Part I*)
1. Peter the Great, by Kneller.
2. Jacob Daniel Bruce, by an unknown artist.
3. Title page from Newton's *Principia* (Amsterdam, 1714).
4. Title page and frontispiece from William Whiston's *New Theory of the Earth* (London, 1696).
5. Title page from William Whiston's *Praelectiones Astronomicae* (London, 1707).
6. Title page from Jacques Rohault's *System of Natural Philosophy* (London, 1735).
7. Title page from John Clarke's commentary on the *Principia* (London, 1730).
8. Title page and facing page from William Whiston's exposition of the Newtonian system (London, 1716).
9. a. A bound copy of the first edition of Newton's *Principia*.
   b. The second edition of Newton's *Principia* (Cambridge, Eng., 1713).
10. Title page from Bruce's personal copy of Newton's *Universal Arithmetick* (London, 1728).
11. A note on "Specifick Gravity" from the flyleaf of Newton's *Universal Arithmetick*.
12. Bruce's "Ex libris."
13. Bruce's signature in Newton's *Universal Arithmetic*.
14. Title page and frontispiece from Newton's *Method of Fluxions* (London, 1736).
15. John Flamsteed's *Historiæ Cœlestis Britannica* (London, 1725)—title page and frontispiece (portrait of Flamsteed).
16. Title page from Flamsteed's *Atlas Cœlestis* (London, 1729).
17. Menshikov's letter to Newton.
18. Menshikov, a contemporary engraving.
19. Draft of Newton's letter to Menshikov, 1714.
20. Title page from Newton's *Opticks* (London, 1704).
21. Title page from Newton's *Optical Lectures* (London, 1728).
22. Title page from Newton's *A Treatise of the System of the World* (London, 1728).
23. The Newtonian speculum made by Bruce.
24. The inscription on the back (convex) side of Bruce's Newtonian speculum.
25. One of the older, pre-Newtonian refracting telescopes in Bruce's collection.

(*preceding Part II*)
26. Title page from *Kniga mirozreniia ili mnenie o nebesnozemnykh globusakh* (St. Petersburg, 1717; Moscow, 1724).

27. First page from *Kniga mirozreniia ili mnenie o nebesnozemnykh globusakh.*
28. The first explanation, in Russian, of Newtonian gravitation "Takovoezh Gospodin Isak Niuton vnov' iz'iasnil, kako ot sikh prichin eklipticheskiia krugi planet, svoe proiskhozhdenie imeiut . . ."
29. The Cartesian theory of "vortices"—from Descartes's *Principia Philosophiae.*
30. Title page from *Evklidovy elementy iz dvenatzati neftonovykh knig vybrannyia* (Moscow, 1739).
31. Title page from George Peter Domcke's *Philosophiae Mathematicae Newtonianae illustratae* (London, 1730).
32. The St. Petersburg Academy's inaugural letter to Newton.
33. Title page from the first Newtonian textbook to be used in the St. Petersburg Academy.
34. a. Antiokh Cantemir.
    b. Francesco Algarotti.
    c. Lord Hervey's lines to Algarotti, inserted in Algarotti's *Neutonianismo per le Dame* (Naples, 1739).
35. Title page from David Gregory's *Astronomiae Physicae & Geometricae Elementa* (Oxford, Eng., 1702).
36. Title page from John Wallis' *Operum Mathematicorum Volumen Tertium* (Oxford, Eng., 1699).
37. Title page from Francesco Algarotti's *Neutonianismo.*
38. Sketch illustrating Newton's "experimentum crucis" from Francesco Algarotti's *Neutonianismo,* with Cantemir's commentary (from his *Fourth Ode*).

(preceding Chapter 17)
39. Mikhail Lomonosov.
40. Drawing of Newton's reflecting telescope.
41. Newton's own drawing of a telescope and its parts.
42. Lomonosov's drawing of the "Newton-Lomonosov" telescope, showing the inclination of the speculum and the position of the eyepiece.
43. Lomonosov's drawing of the "Newton-Lomonosov" telescope, indicating the path of light.
44. Lomonosov's initial sketch of the "Newton-Lomonosov" telescope with the eyepiece in the same position as in Newton's reflector telescope.
45. The "Tubus Nyctopticus modo Lomonosov-Newton," indicating a return to the refracting principle.
46. Title page from N. G. Kurganov's *Pismovnik,* the first edition (St. Petersburg, 1769).

(in Appendix I)
47. Rough draft of Newton's letter to Menshikov.

NEWTON AND RUSSIA

THE EARLY INFLUENCE, 1698–1796

Abbreviations

Babson: A Descriptive Catalogue of the Grace K. Babson Collection of the Works of Sir Isaac Newton and the Material Relating to Him in the Babson Institute Library, Babson Park, Mass. (New York: Herbert Reichner, 1950).

Babson II: A supplement to the Catalogue of the Grace K. Babson Collection of the Works of Sir Isaac Newton . . . (Babson Park, Mass.: Babson Institute, 1955).

Correspondence: The Correspondence of Isaac Newton, ed. H. W. Turnbull *et al.,* Vols. I–IV (Cambridge, Eng.: the University Press for the Royal Society, 1959–1967).

Gray: G. J. Gray, *A Bibliography of the Works of Sir Isaac Newton, Together with a List of Books Illustrating His Works,* 2nd ed. (Cambridge, Eng.: Bowes and Bowes, 1907; facsimile reprint by Dawsons of Pall Mall, London, 1966).

MIAN: [Imperatorskaia Akademiia nauk], *Materialy dlia istorii Imperatorskoi Akademii Nauk 1716–1750,* 10 vols. (St. Petersburg, 1885–1901).

Opticks: Fourth English edition (1730) of Newton's work.

Polnoe sobranie: M. V. Lomonosov, *Polnoe sobranie sochinenii,* 10 vols. (Moscow and Leningrad: Akademiia nauk, 1950–1957).

Principia: English translation of Newton's work by Andrew Motte.

RGO: Royal Greenwich Observatory.

SK: Svodnyi katalog russkoi knigi grazhdanskoi pechati XVIII veka 1725–1800, ed. I. P. Kondakov *et al.,* 5 vols. (Moscow: Izdanie Gosudarstvennoi Biblioteki SSSR imeni V. I. Lenina, 1962–1967).

TIIE: Trudy Instituta istorii i estestvoznaniia

TIIE-T: Trudy Instituta istorii estestvoznaniia i tekhniki

INTRODUCTION

In the decades treated in the present volume, Russia underwent a transformation in outlook more rapid and lasting in its significance than any other country in Europe. At the end of the seventeenth century the heliocentric theory had hardly been heard of outside Kiev and Moscow; and St. Petersburg, the cradle of Russian science, had not even been built. Yet by the time Newton died in 1727, the protracted process of intellectual development that lay between the rise of the Schoolmen and the triumph of Cartesianism had either been circumvented or passed. Newton's revolutionary doctrines had taken root in the St. Petersburg Academy, and, with astonishing speed, a nation long identified with backwardness, superstition, and ignorance vaulted into the modern era.

Yet, remarkably enough, the role of Newton's ideas in this dramatic development has not previously been examined. Any reader of even the least familiar studies of the period might well conclude that the "incomparable Mr. Newton," the idol of Voltaire and the "philosophes," whose accomplishments in physics, optics, astronomy, and mathematics dominated science for generations to come, played no part in eighteenth-century Russian thought. Indeed, it has even been suggested (by an early biographer of Evgenii Bolkhovitinov) that his ideas were unknown in Russian society before the end of the eighteenth century.[1] No work on the period has seen fit to question this sweeping assumption in any detail.[2]

1. See S. Shmurlo, *Mitropolit Evgenii kak uchenyi. Rannie gody zhizni, 1767–1804* (St. Petersburg, 1888), p. 185.
2. See, however, the essay by T. P. Kravets, "N'iuton i izuchenie ego trudov v Rossii," in *Ot N'iutona do Vavilova[,] ocherki i vospominaniia* (Leningrad: Izdatel'stvo "Nauka," 1967), pp. 7–20, which, though devoted mainly to the nineteenth and the second half

1

In reality, Newton's influence in Russia began, as we hope to show, almost a hundred years earlier. His major works found their way there long before Newton's death, and there is reason to believe that Newton himself may have directly contributed to having the *Principia* sent to Russia, where his scientific views were known at least a quarter of a century before the opening of the St. Petersburg Academy. This may seem all the more surprising when we recall that the Copernican system had yet to be heard of in Russia when Newton was born in 1643; the cosmology that then retained almost canonical authority was derived from a work written in the sixth century.[3]

The present study may owe its genesis to an error committed in the nineteenth century when Zabelin, one of the more erudite Russian historians of the time, published the inventory of a library belonging to J. D. Bruce, one of Russia's first mathematicians and natural philosophers. His books comprised the earliest known private collection of "modern" scientific literature in Russia. Later it passed to the St. Petersburg Academy of Sciences where, owing to the misguided efforts of the eighteenth-century Orientalist, Theophil Bayer, it was dispersed. The inventory would, nonetheless, have retained considerable historical value had Zabelin deciphered some of its entries, most of which are either garbled or incomplete, but this he failed to do; besides, he was careless enough to publish only half the inventory—a fact of which his readers remained long unaware. Indeed, when Ivask came out with his bibliography in 1911, he unwittingly helped to perpetuate Zabelin's oversight.[4]

The full inventory attracted interest again in the 1930's, when the USSR Academy of Sciences set itself the task of deciphering it, but the project was apparently abandoned. My own partial transcription showed the inventory to be more extraordinary than even Zabelin had

---

of the eighteenth centuries, traces Newton's influence in Russia to the opening of the St. Petersburg Academy. See the "Bibliographic Note."

3. *The Christian Topography of Cosmas Indikopleustès*, an Egyptian monk, whose work was translated into Slavonic at an early and as yet unestablished date. His influence in Russia is described by T. I. Rainov in his *Nauka v Rossii XI–XVII vekov* (Moscow and Leningrad: Akademiia nauk, 1940) which is the only full-length work on pre-Petrine science, though there are interesting chapters on Russian medieval astronomy and cosmology in B. E. Raikov, *Ocherki po istorii geliotsentricheskogo mirovozzreniia v Rossii*, 2nd ed. (Moscow and Leningrad: Akademiia nauk, 1947). On Cosmas Indikopleustès see Wanda Wolska, *La topographie chrétienne de Cosmas Indikopleustès* . . . (Paris: Presses universitaires de France, 1962).

4. See U. G. Ivask, "Chastnye biblioteki v Rossii: opyt bibliograficheskogo ukazaniia," *Russkii Bibliofil* (St. Petersburg), (No. 3, 1911), p. 74.

vaguely suspected.[5] Not only does it include all of Newton's major writings, but most of the better-known commentaries as well. How did Bruce acquire them? The evidence supplied by his library seemed more tantalizing still in light of the knowledge that Bruce remained in England for further training in mathematics with "Ivan Kolsun," after which he returned to Russia to found one of the first nonclassical schools in the world. He was to play an important part in Peter's reforms at a time when the dissemination of Newton's ideas was still being held in check in more advanced societies.

How long they took to gain ground is well known. For example, John Winthrop, a professor of mathematics and natural philosophy at Harvard, who was probably the first Newtonian in the western hemisphere, acquired a copy of the *Principia* (the third edition) only in 1739. Yet Bruce, so the new evidence suggested, had become acquainted with it almost four decades earlier. On further investigation, it transpired that Newton's name was familiar in the Tsar's entourage. One contemporary English account has Peter expounding the Newtonian system to "his Lords around him," while Bruce, it turned out, was personally engaged in translating into Russian a book in which the theory of universal gravitation was explained to his countrymen for the first time. As to the Newtonian works in Bruce's library, most survived Bayer's reorganization, as well as the natural calamities to which the St. Petersburg Academy fell victim during the eighteenth century. This my subsequent journey to Leningrad happily confirmed.

One problem, however, remained. The inventory, for all its revealing associations, failed to establish a direct connection with Newton. By a stroke of good fortune, publication of the fourth volume of the Royal Society's *Correspondence of Isaac Newton (1694–1709)* provided it. Here "Colonell Bruce" does indeed make more than one appearance. Though his identity is not revealed by the editor, the context leaves no room for doubt. Bruce, it transpires, was personally involved in a misunderstanding between Flamsteed and Newton over the latter's lunar theory, and the *Correspondence* further makes clear that Flamsteed was intimately acquainted with the mysterious colonel and apparently met him regularly, as did Edmond Halley, who may have contemplated an invitation to go to Russia with Bruce. More evidence concerning the "Colonell" will no doubt appear with

5. See V. J. Boss, "Russia's First Newtonian: Newton and J. D. Bruce," in *Archives internationales d'Histoire des Sciences* (Nos. 60–61, 1962), pp. 241 ff.

time, but there could now at least be a certain finality about the historical origins of Newton's influence in Russia.

The first part of the present volume deals largely with the introduction of Newton's doctrines into Russia during the reign of Peter the Great, whose reforms more or less coincided in time with the last three decades of Newton's life. Part II continues the narrative from the foundation of the Imperial Academy of Sciences in 1725/26, when Newton was still president of the Royal Society, to the death four decades later of Lomonosov, the most remarkable scientist and poet Russia produced in the course of the eighteenth century. His interest in Newton's work extended beyond the *Principia* and the *Opticks* to the telescope Newton invented (which Lomonosov tried to improve). Together with Leonard Euler, Lomonosov was to be the most formidable critic of Newtonianism in St. Petersburg, as well as one of its last outstanding foes.

Though the controversies over Newton's theories within the St. Petersburg Academy in that period have not been described before, some of the disputants (such as Bilfinger, Daniel Bernoulli, Jacob Hermann, and Franz Ulrich Theodor Aepinus) are well known through the contributions they made to natural philosophy while outside Russia. Conversely, the part they played in the history of the Academy of Sciences is not always adequately described in non-Russian sources, and my rule in such cases has been to limit the discussion as far as possible to their work in St. Petersburg. Thus, in elucidating the effect of James Jurin's polemical excursions, it seemed wisest to follow only that part of the controversy that actually affected members of the Russian Academy. Similarly, it seemed superfluous to delineate the familiar controversy over Newton's lunar theory since the St. Petersburg Academy's involvement in it, culminating with its award of a special prize to Clairaut, was purely formal.

By the end of the eighteenth century the influence of Newton's doctrines came to be felt far beyond the confines of so-called natural philosophy. Literature, poetry, and even everyday speech—all had felt their impact. This has so far been wholly neglected. In the last chapter some attempt is made to show how Newton's ideas were popularized during the reign of Catherine II. Their effect on the thought, science, and language of the period proved to be just as significant as elsewhere in Europe, though subject to the peculiar conditions that gave eighteenth-century Russian society its unique character.

When Newton made his great discoveries, Russia still lay outside the orbit of Western secular culture. Indeed, the first secular works to be printed in Russian in any quantity did not appear until the very last years of the seventeenth century. Mathematical learning, as opposed to rudimentary practice) was virtually nonexistent in Muscovy when the *Principia* was first published in 1687. Nor were scientific instruments—apart from a few bismars, telescopes, clog calendars, and *shchety* (Russian abaci)—apparently put to use until the Petrine era.[6] The novelty of Newton's doctrines could not, therefore, be measured against previous scientific theories. They represented a new vision of the universe which threatened the continued survival of a long-established way of life. What happened in Russia at the beginning of the eighteenth century was therefore in marked contrast to the process of "westernisation" described by Dr. Needham in his brilliant study of science and civilization in China.

Writing of the Jesuit missions there during the seventeenth century, he notes that "their aim was naturally to support and commend the 'Western' religion by the prestige of the science from the West which accompanied it."[7] The new sciences might be true. What mattered for the missionaries, however, was that it originated in Christendom. The implicit logic was that only Christendom could have produced it. "But the Chinese," continues Dr. Needham, "were acute enough to see through all this from the very beginning. The Jesuits might insist that Renaissance natural science was primarily 'Western,' but the Chinese understood that it was primarily 'new.'"

The Russian response took quite a different form. From the earliest attacks of the Old Believers on books connected with the Newtonian enlightenment, it is clear that the latter appeared to be both "Western" and "new." And "Western" religion was just as much an anathema as "Western" science. Ivan Pososhkov would assume as a matter of course that Copernicus and Martin Luther were somehow diabolically related, and that Lutherans were responsible for dis-

---

6. Even eighteenth-century scientific instruments exhibited in Leningrad and Moscow museums are, for the most part, of English, French, or German origin. The absence of detailed museum catalogues (an affliction by no means confined to the USSR) makes it difficult to judge to what degree instruments were used before that time. Nor is enough known about eighteenth-century Russian instrument makers, a subject explored by Dr. V. L. Chenakal, director of the Lomonosov Museum. See his *Russkie priborostroiteli pervoi poloviny XVIII v.* (Leningrad: Akademiia nauk, 1953).

7. Joseph Needham, F.R.S., *Science and Civilisation in China* (Cambridge, Eng.: Cambridge University Press, 1959), III, 449.

seminating heretical notions such as the heliocentric theory.[8] The historical significance of this confusion cannot be overestimated. It meant that in Russia the "new science" was often regarded as an alternate faith. This in fact is what it eventually came to represent once the revolutionary implications of the Petrine reforms were realized. Having crushed the autonomy of the Church, the new secular state that Peter built became the official protector of science, and Newton its undeclared patron saint.

8. I. T. Pososhkov, *Zaveshchanie otecheskoe,* new ed., ed. E. M. Prilezhaev (St. Petersburg, 1893), pp. 128–129.

PART I

The Petrine Era

60

1. Peter the Great, by Kneller, painted shortly before Peter's visit to London. From the Queen's Collection, Kensington Palace, copyright reserved.

2. Jacob Daniel Bruce (1670–1735), by an unknown artist. From the Hermitage.

PHILOSOPHIÆ

NATURALIS

PRINCIPIA

MATHEMATICA.

AUCTORE

ISAACO NEWTONO,

EQUITE AURATO.

EDITIO ULTIMA

AUCTIOR ET EMENDATIOR.

AMSTÆLODAMI

SUMPTIBUS SOCIETATIS,

MDCCXIV.

---

3. Title page from Newton's *Principia* (Amsterdam, 1714). This edition passed from Bruce into the library of Peter the Great.

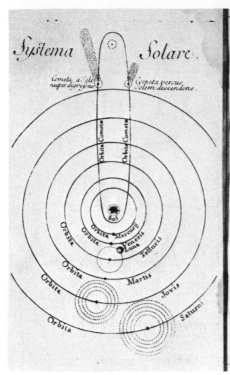

4. Title page and frontispiece from William Whiston's *New Theory of the Earth* (London, 1696).

5. Title page from Whiston's *Praelectiones Astronomicae* (London, 1707). Bruce had the English translation of this rare work.

ROHAULT's
SYSTEM
OF
Natural Philosophy,

ILLUSTRATED WITH

Dr. SAMUEL CLARKE's Notes

Taken mostly out of

Sir ISAAC NEWTON's Philosophy.

VOL. I.

Done into ENGLISH
By JOHN CLARKE, D.D. Dean of *Sarum*.

The THIRD EDITION.

LONDON,
Printed for JAMES, JOHN, and PAUL KNAPTON,
at the *Crown* in *Ludgate-Street.*
MDCCXXXV.

---

6. Title page from Jacques Rohault's *System of Natural Philosophy* (London, 1735), with Samuel Clarke's notes. This Cartesian work introduced the Newtonian system to Cambridge.

A
DEMONSTRATION
Of some of the
PRINCIPAL SECTIONS
OF
Sir ISAAC NEWTON's
PRINCIPLES
OF
Natural Philosophy.
IN WHICH
His peculiar METHOD of treating that useful Subject, is explained, and applied to some of the chief Phænomena of the *System of the World.*

By JOHN CLARKE, D.D.
Dean of *Sarum.*

LONDON:
Printed for JAMES and JOHN KNAPTON,
at the *Crown* in St. *Paul's Church-yard.*
MDCCXXX.

---

7. Title page from John Clarke's commentary on the *Principia* (London, 1730).

BOOKS *writ by* W. WHISTON, M.A.

I. THE Elements of *Eu.lid*, with select Theorems out of *Archimedes*. By the Learned *Andrew Tacquet*. To which are added, Practical Corollaries, shewing the Uses of many of the Propositions. The whole abridg'd, and in this Third Edition publish'd in *English*.

II. Astronomical Lectures read in the Publick Schools at *Cambridge*. Whereunto is added, a Collection of Astronomical Tables; being those of Mr. *Flamstead*, corrected; Dr. *Halley*; Monsieur *Cassini*; and Mr. *Street*. For the Use of Young Students in the University. And now done into *English.*

III. An Account of a Surprizing Meteor seen in the Air, *March* the 16th, 1715. at Night. Containing, 1. A Description of this Meteor, from the Author's own Observations. 2. Some Historical Accounts of the like Meteors Before; with Extracts from such Letters and Accounts as the Author has receiv'd. 3. The principal Phænomena of this Meteor. 4. Conjectures for their Solution. 5. Reasons why our Solutions are so imperfect. 6. Inferences and Observations from the Premises. The Second Edition.

IV. An Humble and Serious Address to the Princes and States of *Europe*, to Admit, or at least openly to Tolerate the Christian Religion in their Dominions. Containing 1. A Demonstration, that none of them do, properly speaking, Admit or openly Tolerate the Christian Religion in their Dominions at this Day. 2. The true Occasions why it is not Admitted or openly Tolerated by them. 3. Some Reasons, why they ought to Admit, or at least openly Tolerate this Religion. 4. An earnest Address to the several *European* Princes and States, grounded on the Premises, for the Admission, or at least the open Toleration of the same Christian Religion in their Dominions.

All Printed for W. TAYLOR, *at the* Ship *in* Pater-Noster-Row.

Sir *Isaac Newton's*
MATHEMATICK
PHILOSOPHY
More easily DEMONSTRATED;
WITH
Dr. *Halley's* ACCOUNT of
COMETS Illustrated.

Being Forty LECTURES Read in the Publick Schools at *Cambridge.*

By WILLIAM WHISTON, M.A.
Mr. LUCAS's, Professor of the Mathematicks in that University.

For the Use of the Young Students there.

In this *English* Edition the Whole is Corrected and Improved by the AUTHOR.

LONDON:
Printed for J. SENEX at the *Globe* in *Salisbury-Court*; and W. TAYLOR at the *Ship* in *Pater-Noster-Row.* 1716.

---

8. Title page and facing page from William Whiston's exposition of the Newtonian system (London, 1716).

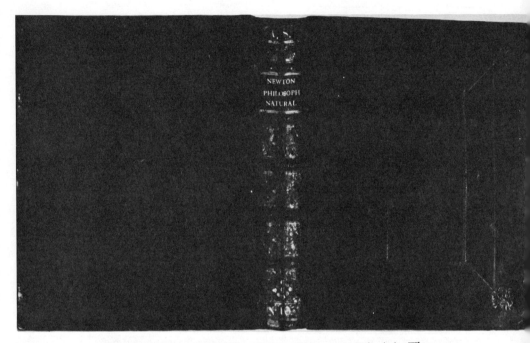

9a. A bound copy of the first edition of Newton's *Principia*. The abbreviated title on the spine may account for the erroneous Russian transcription in Bruce's inventory (see p. 40).

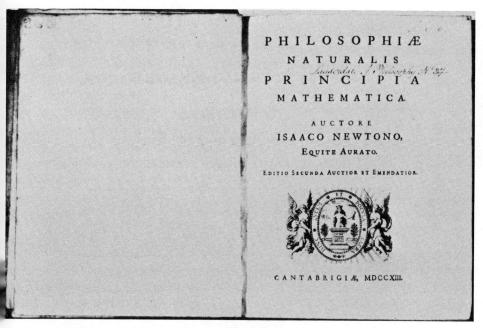

9b. The second edition of Newton's *Principia* (Cambridge, Eng., 1713).

UNIVERSAL
ARITHMETICK:
OR, A
TREATISE
OF
ARITHMETICAL
COMPOSITION and RESOLUTION.

To which is added,

Dr. HALLEY's Method of finding the
Roots of EQUATIONS Arithmetically.

Written in LATIN by Sir ISAAC NEWTON, *and Translated by the late Mr.* RALPHSON, *and Revised and Corrected by Mr.* CUNN.

The Second Edition, very much Corrected.

LONDON:
Printed for J. SENEX, in *Fleet-street*, W. and J. INNYS, near
S. *Paul's*, J. OSBORNE and T. LONGMAN in *Pater-noster-row*.

M. DCC. XXVIII.

10. Title page from Bruce's personal copy of Newton's *Universal Arithmetick* (London, 1728).

The Specifick Gravity of the Mixture, & the Specifick Gravity of both the things mixt are given a how to find the proportion of the things mixed pag 78

MAT.

11. A note on "Specifick Gravity" from the flyleaf of Newton's *Universal Arithmetick,* in Bruce's handwriting.

12. Bruce's "Ex libris."

Univerſal Arithmetick:

OR, A

TREATISE

OF

ARITHMETICAL

COMPOSITION and RESOLUTION.

13. Bruce's signature in Newton's *Universal Arithmetick,* in
the same hand as the notes on the flyleaf (Figure 11).

Sensibiles sensibilium velocitatum mensuræ. vid. pag. 273.

Τὰ ϰοινὰ ϰαινῶς, τὰ ϰαινὰ ϰοινῶς.

THE
METHOD of FLUXIONS
AND
INFINITE SERIES;
WITH ITS
Application to the Geometry of CURVE-LINES.

By the INVENTOR
Sir ISAAC NEWTON, K.t.
Late President of the Royal Society.

Translated from the AUTHOR'*s* LATIN ORIGINAL
not yet made publick.

To which is subjoin'd,
A PERPETUAL COMMENT upon the whole Work,

Consisting of
ANNOTATIONS, ILLUSTRATIONS, and SUPPLEMENTS,

In order to make this Treatise
A compleat Institution for the use of LEARNERS.

By *JOHN COLSON,* M.A. and F.R.S.
Master of Sir *Joseph Williamson*'s free Mathematical-School at *Rochester.*

LONDON:
Printed by HENRY WOODFALL;
And Sold by JOHN NOURSE, at the *Lamb* without *Temple-Bar.*
M.DCC.XXXVI.

14. Title page and frontispiece from Newton's *Method of Fluxions* (London, 1736).

JOHANNES FLAMSTEEDIUS Derbiensis
Astronomiæ Professor Regius. Anno Ætatis 74 Obijt
Decem 31 1719

HISTORIÆ COELESTIS
BRITANNICÆ
VOLUMEN PRIMUM.

Complectens

STELLARUM FIXARUM

Nec non

PLANETARUM OMNIUM

OBSERVATIONES

Sextante, Micrometro, &c. peractas.

Quibus subjuncta sunt

PLANETARUM LOCA

ab iifdem OBSERVATIONIBUS deducta.

Obfervante *JOANNE FLAMSTEEDIO*, A.R.

In OBSERVATORIO Regio

GRENOVICENSI

CONTINUA SERIE

Ab Anno 1675, ad Annum 1689.

LONDINI: Typis H. MEERE. M.DCC.XXV.

15. John Flamsteed's *Historiæ Cælestis Britannicæ* (London, 1725)—title page and frontispiece (portrait of Flamsteed).

ATLAS

COELESTIS.

By the late Reverend

Mr. *JOHN FLAMSTEED*,

REGIUS PROFESSOR of ASTRONOMY at *Greenwich*.

LONDON, Printed in the Year M.DCC.XXIX.

16. Title page from Flamsteed's *Atlas Cœlestis* (London, 1728), a copy of which was in Bruce's possession.

17. Menshikov's letter to Newton.

18. Menshikov (1672–1729), a contemporary engraving.

Potentissimo et maxime colendo, Domino, D.no Alexandro Menzicoff,
Romani et Russi Imperij Principi, D.no de Oranienbourgh.
Czaranæ Majestati Primo a Concilijs, Equitum Stratego,
Devictarum Provinciarum Dynasti, Ordinis Eliphantis, &
Albæ a Nigræ Aquilæ Equiti &c.

Isaacus Newton salutem.

Cum Societati Regiæ dudum innotuerit Imperatorem
vestrum, Czariensem Majestatem, Artes et Scientias in Regnis
suis quam maxime promovere, eumqz ministerio vestro, non
solum in rebus bellicis et civilibus administrandis, sed etiam in
bonis literis et scientijs propagandis, apprime adjuvari: maximo
omnes suffusi fuimus gaudio, quando Mercatoris Angli nobis
significarunt, Excellentiam vestram pro humanitate summa,
& singulari suo in scientijs affectu, & erga Gentem nostram
amore, in corpus Societatis nostræ cooptari, se dignari. Atque
eo quidem tempore calibus nostris finem, pro more, imposi-
turi eramus. Sed hoc audito semel adhuc convenimus ut
Excellentiam vestram suffragijs nostris eligeremus: id quod

fecimus

fecimus unanimi consensu. Et jam, ut primum cœtus
nostros renovare licuit, Electionem Diplomate sub
Sigillo nostro communi ratam fecimus. Societas
autem Secretario suo in mandatis dedit, ut transmisso
ad vos Diplomate, Electionem vobis notam facret.
Vale.

Datam Londini
xxv Octob. Anno
MDCCXIV.

19. **Draft of Newton's letter to Menshikov, 1714.**

OPTICKS:

OR, A

TREATISE

OF THE

REFLEXIONS, REFRACTIONS, INFLEXIONS and COLOURS

OF

LIGHT.

ALSO

Two TREATISES

OF THE

SPECIES and MAGNITUDE

OF

Curvilinear Figures.

LONDON,

Printed for Sam. Smith, and Benj. Walford, Printers to the Royal Society, at the *Prince's Arms* in St. *Paul*'s Church-yard. MDCCIV.

20. Title page from Newton's *Opticks* (London, 1704). This edition had no author indicated on the title page.

OPTICAL
LECTURES

Read in the

PUBLICK SCHOOLS

OF THE

Univerſity of *CAMBRIDGE,*

Anno Domini, 1669.

By the late Sir *ISAAC NEWTON,*
Then Lucaſian Profeſſor of the Mathematicks.

Never before Printed.

Tranſlated into *Engliſh* out of the Original *Latin.*

LONDON:

Printed for FRANCIS FAYRAM, at the
South Entrance of the *Royal Exchange.*

M. DCC. XXVIII.

21. Title page from Newton's *Optical Lectures* (London, 1728).

A

TREATISE

OF THE

SYSTEM

OF THE

WORLD.

BY

Sir *ISAAC NEWTON*.

Tranflated into ENGLISH.

LONDON:

Printed for F. FAYRAM at the South Entrance under the *Royal Exchange*.
M DCC XXVIII.

22. Title page from Newton's *A Treatise of the System of the World* (London, 1728).

23. The Newtonian speculum made by Bruce.

24. The inscription on the back (convex) side of Bruce's Newtonian speculum.

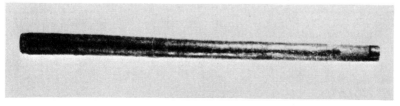

25. One of the older, pre-Newtonian refracting telescopes in Bruce's collection.

1

Did Peter the Great Meet Newton?

Peter the Great's visit to England in 1698 aroused great public interest, and that interest has not diminished with time. Even contemporaries were aware that his journey abroad, the first ever undertaken by a Russian tsar, represented a startling new departure in the affairs of his isolated country. The people of Russia soon felt the practical consequences of that journey when Peter on his return built a fleet modeled on the Royal Navy, founded a new capital in St. Petersburg, and, for the first time in the history of his country, established a system of scientific and mathematical education.

Other vast reforms followed. Their effects are still being felt today, and the debate over their significance has not yet ceased. But if any episode in Peter the Great's reign captures their symbolical meaning, it is his supposed meeting with Isaac Newton. For in the Russian context Peter's transformation of his country represented an intellectual upheaval corresponding in time and dimension to the so-called "scientific revolution" culminating with the Newtonian synthesis. Lomonosov, who was both a scientist and a historian, saw this parallel when he adapted Pope's famous couplet about Newton to read: "God said 'Let Peter be!' and there was Light."

Did Peter the Great really meet the author of the *Principia?* Naturally enough this question has long interested Soviet scholars, particularly the late Dr. Andreev, who went to some pains to demonstrate that this "historic" meeting (so he believed) took place.[1] It allegedly

1. A. I. Andreev, "Peter v Anglii," in *Petr Velikii, Sbornik statei,* ed. Andreev, Vol. I (Moscow and Leningrad: Akademiia nauk, 1947), 86.

occurred in the course of Peter the Great's sojourn in London, when Newton was charged with reforming the coinage. The state of the currency was then causing grave anxiety in England, and Newton's friend Charles Montague, who became Chancellor of the Exchequer in 1695, offered him the post of Warden of the Mint.[2] Newton accepted and his appointment was confirmed by royal warrant dated April 13, 1696.

The position was not then a sinecure. As in Russia, where the prevalent practice of clipping similarly debased the coinage, the problem of maintaining the value of the currency was further compounded in England by the impossibility of preventing gold coins issued by the Royal Mint from being immediately withdrawn and either melted or exported. When Newton took up his post, the true value of a coin was determined by the amount of metal in it, and a careful investigation showed that the value of the average shilling was not more than six pence. On him fell the responsibility for the later part of the recoinage (with which John Locke as well as Montague and Sommers, the Lord Keeper, were also associated). In 1698 the recoinage was still in process, and Newton moved down from Cambridge to live in London where he spent most of his time at the Mint. It is this circumstance that makes his meeting with the Tsar possible.

When Peter the Great planned his first journey to Europe, he did not originally intend to visit London. Initially, the Great Embassy's route took the Tsar through several German principalities. He made his debut in European society in the town of Koppenburg, memorable chiefly for the fact that this was where he missed an opportunity to meet Leibniz.[3] But Peter's goal was Holland, and he arrived in Zaandam (then known as Saardam) on August 18, 1697. While there, he sat for Sir Godfrey Kneller, whose magnificent portrait of the young Tsar looking like "Mars in the uniform of the Preobrazhenskii Guards" (see fig. 2) was so much admired by Karamzin during his

2. The Warden was immediately under the chief officer, who was Master of the Mint. Newton remained associated with the Mint until his death, having become Master of it at the close of 1699. In his later days, the work was much lighter, and, during the last years of his life, Newton rarely went there. On the confusion in the currency before Newton's reform, see Chapter XXI of Macaulay's *History*.

3. See W. I. Guerrier, *Relations de Leibniz avec la Russie et Pierre le Grand d'après des papiers inédits de la Bibliothèque de Hanovre* (St. Petersburg, 1871). See also V. I. Chuchmarev, "G. V. Leibnits i russkaia kul'tura nachala 18 stoletiia," *Vestnik istorii mirovoi kul'tury* (No. 4, 1957), pp. 120–32.

visit to London later in the century.[4] Some weeks after his arrival in Zaandam Peter paid a call on William III at Utrecht and was invited to England.[5]

The ostensible purpose of the Tsar's journey was to find allies for Russia's drive against the Turks, but Peter's acceptance of William's invitation to come to Great Britain was apparently based not so much on diplomatic considerations as on his disenchantment with the instruction he had been receiving from Dutch ship builders. He is said to have observed that their methods were empirical rather than mathematical and that, consequently, they could not furnish him with the theoretical information that he sought. As Peter later indicated in his notebooks, what he wanted was to study the science (*nauka*) of shipbuilding, and it was his opinion that Dutch shipwrights lacked this knowledge.

It has also been suggested that Peter's departure from Holland was prompted by his disappointment with the capacity of Dutch natural philosophy to satisfy his curiosity,[6] though this in itself is hardly a convincing explanation. As Verenet's neglected study has shown, he had ample opportunity to meet Dutch scientists and mathematicians, some of whom belonged to the very first rank.[7] He met not only one of Holland's best astronomers, Christian Hartzoeker, who helped the Tsar with his astronomical observations, he also learned about the solid work of Dutch mathematicians such as Hans Gouda, Dirk Raven, and Hans Isbrandtsen Hoogzaat, and he met men of proven scientific accomplishment, such as Van Coehorn and Leewenhoek, who introduced him to the microscope.[8]

4. See Karamzin's *Pis'ma russkogo puteshestvennika*, in *Izbrannye sochineniia*, 2 vols. (Moscow and Leningrad: Izdatel'stvo "Khudozhestvennaia literatura," 1964), I, 552. (Karamzin wrongly states that the painting was done in London during the Tsar's stay there.)

5. The most exhaustive treatment of the Tsar's first journey is to be found in volume II of M. M. Bogoslovskii's unfinished work, *Petr I: materialy dlia biografii* (Moscow: Ogiz Gosudarstvennoe Sotsial'no-Ekonomicheskoe Izdatel'stvo, 1941). Early English biographies of the Tsar, such as John Barrow's *A Memoir of the Life of Peter the Great* (London, 1832) or John Banks's *The History of the Life and Reign of the Czar Peter the Great*, 2nd ed. (London, 1740), also contain much undocumented information.

6. As may be judged from the orders the Tsar issued to some of these shipwrights on his return to Russia, insisting that their work be supervised by English, Danish, or Venetian foremen. See N. G. Ustrialov, *Istoriia tsarstvovaniia Petra Velikogo* (St. Petersburg, 1858), III, 91.

7. George Verenet, *Pierre le Grand en Hollande et à Zaandam dans les années 1697 et 1717* (Utrecht, 1865).

8. *Ibid.*, p. 66. The Tsar's attempt to recruit Hartzoeker for service in Russia met with refusal.

It should be added that several of these meetings took place during a brief stay in Holland after Peter left London on his journey home. It is difficult to deny A. I. Andreev's suggestion that his scientific interests were to some extent aroused by his experiences in England.[9] He had been enthusiastic about "scientific attractions" even during his first weeks in Holland when he visited Jacob de Wilde's archaeological and anthropological museum, the collection of specimens and corpses assembled by Boerhaave at Leyden University, and the anatomical exhibit belonging to Frederic Ruysch, with whom Peter was on particularly cordial terms for he attended Ruysch's lectures and called on a hospital in his company to observe surgical operations.[10] But it may well be that in England Peter's concern with science grew more serious, as is duly reflected in some of the traditions passed down concerning his stay there.

Peter the Great's arrival in London early in 1698 elicited much curiosity in the newspapers of the day. The *Postman*, for example, after congratulating him on the nobility of his "design," went on to praise the Tsar for his discernment in taking William III as his "fit pattern." But such contemporary accounts are not informative, largely because Peter insisted on traveling incognito. He dressed as a commoner and shunned all ceremony. This is why several aspects of his stay in England are still so vague. No official record was kept of his activities in the countries he visited. Even the Tsar's celebrated tour of Oxford is not officially recorded in the university archives.[11] His life in Deptford, where Peter worked in the shipyards with his hands, has been described by startled contemporaries, and the extraordinary impression he conveyed has been preserved in various accounts by Englishmen who either met the Tsar or came in contact with members

9. Cf. A. I. Andreev, "Petr v Anglii," Andreev, ed. For a more recent but brief English survey of the Tsar's sojourn in England, see Ian Grey, "Peter the Great in England," in *History Today*, VI (April 1956), 225–234, which recapitulates some of the better known facts concerning Peter's contacts in the course of his stay in Deptford and elsewhere.

10. P. P. Pekarskii, *Nauka i literatura v Rossii pri Petre Velikom*. Vol. I: *Vvedenie v istoriiu prosveshcheniia v Rossii XVIII stoletiia* (St. Petersburg, 1862), p. 9.

11. University records contain no mention of Peter's visit, or of his having received an honorary degree, despite the legend which has drawn largely on Baron Heinrich von Huyssen (SK, I, 264, s.v. "Giuissen"), buttressed by a notation in the writings of John Henry. According to the Tsar's *Journal*, the trip to Oxford was made on April 8, and a notice in the *Postboy* of April 6, 1698, confirms this: " 'Tis said that the Czar of Muscovy has been (the close of last week) to see the University of Oxford, but he made a short stay there."

of his suite. The only document that even approximates an official record, however, is Peter's own cryptic diary, or *Journal*. It is this diary[12] that Andreev used to support his speculations about the Tsar's encounter(s) with Newton.

The *Journal* describes in the barest outline some of Peter the Great's preoccupations in 1698, but it does not provide any ordered résumé either of his appointments or of his callers. It refers with tantalizing brevity to Peter's visitation of the Royal Society, but few personal names of other than Russian provenance are mentioned. Nor is Newton cited anywhere in it by name. But it does record the fact that the Tsar visited the Tower (in which the Royal Mint was housed) on January 27 and April 13, 1698. Was Newton present on either of these two occasions? Andreev's conviction that an examination of Newton's unpublished papers at the Mint would provide an affirmative answer has proved overly sanguine.[13] For, like Peter the Great, Newton rarely made note of his appointments or social engagements.

Nor are his papers at the Mint devoted exclusively to the coinage of the realm. He often used paper on which he had already written for further drafting either on the Mint or on a completely different subject. Hence, the collection of manuscripts at the Mint includes a number of fragments on scientific, historical, and theological matters, as well as the controversy with Leibniz, but it does not contain any record of a meeting with Peter the Great.

There were, however, at least *three* more recorded occasions (besides the two noted in the *Journal*) on which the Tsar could have met Newton. At the time of Peter's sojourn in London, Newton was living on Jermyn Street. A year before the Tsar's arrival, his stepniece, Catherine Barton, was installed to run Newton's household. She was an intimate friend of Montague, the Lord Chancellor, and through her, if not through Montague himself, Newton would almost certainly have heard of the Tsar's interest in the Mint. According to Narcissus

12. The entire "diary" is included in Ustrialov, *Istoriia*, III, 602, 605. A French version was published in London in 1773 (*Journal de Pierre le Grand*).

13. Most of Newton's papers at the Mint, of which many tend to duplicate in one form or another material to be found in other collections, have remained unpublished. A preliminary inventory of the five bound volumes of MSS, in typescript, has been compiled by Sir John Craig. It contains about 900 items and may be seen, together with the papers themselves, at the Royal Mint, which moved to its five-acre site on Tower Hill, only a few hundred yards from the place it occupied inside the Tower in Newton's day, in 1810; it has recently been moved again to Llantrisant in the Vale of Glamorgan.

Luttrell's diary (not known to Andreev), the Chancellor of the Exchequer entertained Peter at the Tower on April 21.[14] Moreover, Sir John Craig, in his history of the London Mint published one year after Andreev's essay, reports that Montague was also present during Peter's visit on *February 3*.[15]

That visit, the only one recorded by Craig, is also omitted from the *Journal*. There is, finally, one more fresh piece of evidence in the form of a note to Newton—now in the archives of the University Library in Cambridge:

Most hond Sr
> The Czar intends to be here to morrow before 12 and I thought my self obliged to signify to You hee likewise expects to see You there. I have taken all possible care to have things in readiness & have not time to add more but that I am

Most hond Sr
Your most humble & obediant
Servant
J. Newton[16]

Fryday Feb 5th. 1697/8

Isaac Newton's family tree shows several relatives with the name of John Newton, but his identity in the given context is less important than the date of the note. February 5 fell on a Saturday. On Sunday Peter is known to have visited the shipyards in Deptford. Presumably, therefore, John Newton's note was written on Friday, February *4*, and the Tsar's meeting with Isaac Newton would then have occurred on Saturday, *February 5*.

Thus there were at least *five* occasions (not just two) on which Peter may have met Newton at the Mint—in addition to the one time when, according to the *Journal*, the Tsar visited the Royal Society on January 27. Its president then was Charles Montague. Could Peter's visit have served as an inducement for arranging an appointment with its most distinguished Fellow on February 3rd? And, in the light of Peter's impulsive and spontaneous nature, would it not have been

14. Narcissus Luttrell, *A Brief Historical Relation of State Affairs, September 1678 to April 1714*, 6 vols. (Oxford, Eng., 1857), IV, 372.
15. Sir John Craig, *History of the London Mint from* A.D. *287 to 1948*, (Cambridge, Eng.: the University Press, 1953), p. 195.
16. J. Newton to Newton, February 5, 1697/98, Cambridge University Library, Add. 3966.15, fol. 310; previously published for the Royal Society in *The Correspondence of Sir Issac Newton*, ed. J. F. Scott, Vol. IV, *1694–1709* (Cambridge, Eng.: the University Press, 1967), p. 265, no. 583.

characteristic of the Tsar to order another appointment with Newton on the following day?

We cannot be absolutely sure of the answer. Nonscientific matters were of little interest to Newton's colleagues, and they saw no reason to record them. Thus, only an oral tradition supports Charles II's famous "visit" to the Royal Society, and Peter the Great's presence there might be similarly dismissed as apocryphal were it not for more tangible (and so far neglected) evidence supplied by one of the Tsar's companions. According to the *Journal,* on April 13 Peter made his (fourth) visit to the Mint. But he did not go alone. He was accompanied by 'Iakov Brius': *Byl . . . s Iakovom Briusom v Ture, gde den'gi delaiut* (see fig. 2). It was largely through this man that Newton first came to be known in Russia.

NEWTON AND BRUCE

Bruce often crops up in the *Journal*—usually as "Iakushka," or "little Jacob," for Peter was fond of him, having known him almost from childhood. He began his dazzling career by joining the Tsar's regiment of boy soldiers as early as 1683; eventually his intellectual gifts won great respect from Peter and resulted in a friendship which lasted throughout his reign.

Though carrying a famous Scottish name, Jacob Daniel (Iakov Vilimovich) Bruce was in actual fact born in Moscow in 1670, scion of a family which had emigrated from Scotland in 1647. He is perhaps one of the most arresting, if neglected figures of the Petrine era, all too often inaccurately described in Western sources as "un Ecossais au service de la Russie" or as "an Englishman" and "a captain" in the service of the Tsar.[17] In making its way to Russia the Bruce family was by no means unique for several Scotsmen had distinguished themselves at the Russian court. The manner in which the Bruce family settled there is surely exceptional as we learn from the memoirs of Peter Henry Bruce, who served successively in the Prussian, Russian, and British armies. He narrates that two Bruces,

17. Cf. Kazimierz Waliszewski, *L'Héritage de Pierre le Grand, règne des femmes, gouvernment de favoris, 1725–1741* (Paris: Plon-Nourrit et cie., 1900), pp. 186, 368; Mairin Mitchell, F.R.C.S., *The Maritime History of Russia* (London: Sidgwick and Jackson, 1949), p. 67.

James Bruce and John Bruce, cousins and descendants of the family of Airth, in the county of Stirling (a branch of the family of Clackmannan) in Scotland, formed a resolution, during the troubles of Oliver Cromwell, to leave their native country, in order to push their fortunes abroad: and as there were some ships in the port of Leith ready to sail for the Baltic, they agreed to go together to that part of the world: but as there happened to be two shipmasters of the same name, by an odd mistake the cousins embarked in different vessels, the one bound to Prussia, the other, to Russia, by which accident they never again saw each other.[18]

James Bruce served as major general in the army of Tsar Feodor Alekseevich and died in 1680. His son William Bruce, who was born in Scotland, pursued the same career; after attaining the rank of colonel, he died courageously in battle in 1695 on Peter the Great's first Azov campaign against the Turks.[19] Jacob Daniel Bruce, William's son, is usually referred to in Russian sources as Iakov Vilimovich. Of his childhood little is known; though it appears that he was exceptionally well educated by the standards of the time and country. He became passionately interested in mathematics and natural philosophy at an early age—interests which remained with him all his life.[20]

At the time of his father's death, J. D. Bruce served as an army engineer; in 1696, after a voyage from Voronezh to Azov, he drew the first map of Russian possessions in Tartary and Asia Minor, a feat which earned him the rank of colonel.[21] During this period he belonged to a group which styled itself the "Society of Neptune," presided over by François Lefort, which included Peter himself among its members. It used to meet secretly in the Sukharev Tower in Moscow, and its members supposedly read scientific tracts and practiced the esoteric arts of natural philosophy, which at this time in Russia was widely regarded as heretical.[22]

18. *Memoirs of Peter Henry Bruce Esq., A military Officer in the services of Prussia, Russia, and Great Britain, containing an Account of his Travels in Germany, Russia, Tartary, Turkey, the West Indies, &c as also several very interesting private Anecdotes of the Czar, Peter I of Russia* (London, 1782), p. 2. Peter Henry Bruce was a descendant of the German branch of the family and was invited into the service of the Russian army by Jacob Daniel Bruce in 1710 (a nice instance of clannish feeling).
19. Cf. *Russkii biograficheskii slovar'* (St. Petersburg, 1908), III, 413.
20. The fullest (though incomplete) biography of Bruce was published by M. D. Khmyrov in the *Artilleriiskii Zhurnal for* 1866: "Vtoroi generalfel'dzekhmeister Iakov Vilimovich Bruce," pp. 81–136, 153–199, 249–291.
21. *Russkii biograficheskii slovar'*, III, 416.
22. See D. O. Sviatskii, "Chudesnost' i estestvennost' v nebesnykh iavleniiakh po predstavleniiam nashikh predkov," in *Mirovedenie*, XVI (No. 2, 1927), 91–102. On

On Peter the Great's European journey in 1697/98, Bruce was chosen as one of the Tsar's sixteen companions on his tour of inspection in England. Oxford University, The Royal Society, Hampton Court, Woolwich Arsenal, the House of Commons, the Royal Observatory at Greenwich, the Tower, the Royal Mint, the London theaters, Quaker meetings, the Deptford docks—these and the other institutions and features of English life mentioned in the *Journal* were presumably seen by Bruce as well as by the Tsar, Menshikov, and others in his suite, though Bruce's place in it was rather special. It would seem that the main stimulus to Peter's inquiries about scientific matters came from him. He had the advantage not only of understanding Russian needs, but of speaking English, German, and probably Dutch. It was thus to Bruce that the task of purchasing mathematical instruments and books on navigation and shipbuilding was entrusted. He was also instructed to look for mathematicians who might enter the Tsar's service,[23] though this did not by any means exhaust the broad responsibilities and duties entrusted to him.

The prescriptions for new habits in dress and social deportment, reform of the currency and fiscal practices, readjustment of the relationship between church and state, administrative reforms, the creation of a bureaucracy and far-reaching legal changes (enactment, for example, of the inheritance law of 1714)—all this was anticipated in one way or another by some of Peter's interests and preoccupations in 1697/98. While the reforms themselves may well have been hastened by military necessity, Bruce's activity in London serves to show quite clearly that even that early many of the ideas already existed in Peter's mind. Why would he have commissioned Bruce to study English laws on primogeniture in 1698 if he was not already thinking of enacting the law that he decreed a decade and a half later?[24] Bruce's concern with monetary problems falls within the same category. When Peter was faced with a drastic reform of the coinage later in his reign, he appointed Bruce to supervise the direction of the Tsar's 'Monetnyi Dvor.' In this position Bruce's task was to solve some of the same problems which Newton had earlier so successfully overcome at the Mint.

In this context Peter the Great's visits in 1698 to the Tower—

Lefort and his times, see Moritz Posselt, *Der General und Admiral Franz Lefort—sein Leben und seine Zeit* (Frankfurt am Main, 1866), I and II.

23. Bogoslovskii, *Petr I: materialy*, III, 302.

24. Andreev, "Petr v Anglii," p. 98.

"gde den'gi delaiut" (where they make money)—may seem far more revealing in retrospect than the brief entries in the *Journal* and Narcissus Luttrell's diary convey. If it was Peter's purpose to learn how Newton was accomplishing his reforms at the Mint, it would have been only natural for him to take with him to the Tower the one man among his companions who could understand the technical problems involved. As Master of the Mint, on March 3, 1711–1712; June 23, 1712; and September 21, 1717, Newton issued reports explaining his reforms of the English coinage. They appear to have found their way into Bruce's library after the latter returned to Russia.[25]

The occasions indicated in the *Journal,* by Narcissus Luttrell, Sir John Craig and by the note in the University Library in Cambridge are not, however, the sole opportunities Bruce would have had for meeting Newton. At some point during the Tsar's sojourn in England, the decision was taken that Bruce should stay on in London for the further study of navigation and mathematics. As a result of this decision, Bruce became acquainted with John Flamsteed, the astronomer who had furnished Newton with observations turned to account in the *Principia*. Through Flamsteed Bruce was to learn of the celebrated controversy that led in 1698 and 1699 to the bitter rupture of Flamsteed's long friendship with Newton.

25. Cf. MIAN:
726) Изъяснение содержания нѣкоторой пробы о поправленіи сребряной монеты.

"I would not Injure Mr. Halley either with Mr. Neuton (on whom I know he has dependance) nor the Collonel (by whom he may make some advantage) . . ."—John Flamsteed

2

Flamsteed, Halley, Newton, and the "Collonel"

The *Journal* records Peter the Great's visits to the Greenwich Observatory which largely owed its existence to the single-minded efforts of John Flamsteed, the first Astronomer Royal (see fig. 15). The observatory itself was designed by Christopher Wren, but the government failed to provide it with any instruments. On a pitiful salary Flamsteed was compelled to acquire or construct them himself, and the rest of his courageous life was dedicated to the gargantuan task of "rectifying the places of the fixed stars," the tables then in use being widely erroneous and out of date. In spite of ill health and continued "ill-usage," Flamsteed achieved amazing results, and in the year that Peter visited him he was laboriously compiling, with the aid of Abraham Sharp's mural arc, the observations which were later to appear in his masterpiece, the *Historia Cœlestis*. The Tsar, according to the record made in Flamsteed's own handwriting, came on two occasions: once on February 6, a Sunday, and again on March 9[1] On each occasion, as noted by Flamsteed, Peter made his observations with instruments he brought himself, but what appears to have impressed itself even more on Flamsteed's mind was the fact that on his first visit "His Most Serene Majesty, Emperor of Muscovy" was accompanied by one "Bruceus Parentibus Scotis Moscuae Natus."[2] In those days this must have been considered remarkable.

1. See Ia. *Observing book of the Mural Arc* (RGO MSS Item 6) February 6 (Sunday), 1698, in Flamsteed's handwriting: "Serenissima Majestas Petri Imperatoris Moscovitiae adearat vespere(?) Instrumenta in Observatorio habitu privato comitantibus Bruceo Legato militari progenie Scota, Moscoviae Nato: Wolfio et Stiles Angli . . ."
2. Cf. Ib. Fair copy of the above (RGO MSS Item 16) (same date): "Serenissimo

Perhaps Flamsteed saw Bruce again on April 6 when, according to the *Journal,* Peter made his third visit to the Royal Observatory.[3] And, by the end of the year, according to the English astronomer's correspondence, he and Bruce knew each other well.[4] The Tsar left London, but Bruce stayed. Flamsteed was preparing a "Table of Refractions" for Bruce,[5] and in a letter to him[6] he apparently complained of a certain "report" he believed Bruce had helped circulate concerning Newton's lunar theory. The report was untrue, and to prevent it "Injureing either Mr. Newton or me," Flamsteed unwittingly revealed the nature of the growing controversy between himself, Newton, and Edmond Halley.

Flamsteed's animosity was directed mainly against Halley, but it was later to include Newton as well, whom Flamsteed had known and admired even before 1674 when he went to Cambridge to receive his M.A. and attended Newton's Lucasian lectures.[7] In the following year, Charles II signed the warrant appointing Flamsteed "astronomical observator"; Flamsteed was later to supply Newton with observations which the Observatory of Greenwich could alone have provided. Newton visited it from time to time, and even after the publication of the *Principia* Flamsteed's cooperation was still of great consequence to him. To verify the equations Newton had deduced from the theory

---

Petrus Moscoviae Czaro observatorium primum visum venit lustratisque Instrumentis habitu privato abijt: aderant secum Bruceus Parentibus Scotis Moscuae Natus, legatus militaris, J. Wolfias et Stileus mercatores Angli."

The references to the Tsar in the observing book are noted in Flamsteed's own handwriting, according to which Peter made another visit to the Royal Observatory on March 8/9, e.g., *Observatory book* (RGO MSS Item 16): "Observante CZARO ipso Petro Serenissimo Moscoviae Magno Duco," and, in the margin, "Czari Petri observationes." Also: Fair copy (RGO MSS Item 16), March 8, 1698: "Observante Serenissimo Petro Moscoviae Czaro." The difference in dates of the March visit is accounted for by Peter's observation being made a quarter of an hour before midday, i.e., the astronomical day was reckoned from midday to midday, so although Peter made his observation a quarter of an hour before noon on Civil Day, March 9, the time was still March 8, 23^h45^m by astronomical reckoning.

3. This visit was not recorded by Flamsteed.

4. "Collonel[l] Bruce" makes his appearance in Volume IV of *Correspondence of Sir Isaac Newton,* ed. Scott, published in 1967, but he is not identified by the editor.

5. Flamsteed to Colson, October 10, 1698, *Correspondence,* IV, 285, no. 594.

6. *Ibid.,* p. 284: "Nor would I write (to the) Collonell, nor had to you . . ." Presumably this is not the only letter Flamsteed wrote to Bruce, but their correspondence has apparently been lost.

7. See *The Mathematical Papers of Isaac Newton.* Vol. III, *1670–1673,* ed. D. T. Whiteside, assisted by M. A. Hoskin and A. Prag (Cambridge, Eng.: the University Press, 1969), p. xix. Whiteside notes: "our search of the records yields the name of no one who can be shown to have attended [Newton's early Lucasian lectures] before John Flamsteed in 1674." See also *ibid.,* n. 38.

of gravity, it was essential for him to receive accurate lunar observations, which Flamsteed continued to give him. In return, Newton promised not to communicate them to anybody and much less to publish them without Flamsteed's consent:

And for my part I am of opinion that for your Observations to come abroad thus wth a *Theory wch you ushered into ye world* & wch by their means has been made exact would be much more [This word, notes the editor of the *Correspondence,* is indistinctly written, but seems to be what Newton intended.] for their advantage & your reputation then to keep them private till you dye or publish them wthout such a Theory to recommend them. For such a Theory will be a demonstration of their exactness & make you readily acknowledged ye Exactest Observer that has hitherto appeared in ye world. But if you publish them without such a Theory to recommend them, they will only be thrown into ye heap of ye Observations of former Astronomers till somebody shall arise that by perfecting ye Theory of ye Moon shall discover your Observations to be exacter than the rest. But when that shall be God knows: I fear not in your life time if I should dye before tis done. For I find this Theory so very intricate & the Theory of Gravity so necessary to it, that I am satisfied it will never be perfected but by somebody who understands ye Theory of gravity as well or better then I do. But whether you will let me publish them or not may be considered hereafter. *I only assure you at present that without your consent I will neither publish them nor communicate them to any body whilst you live, nor after your death without an honourable acknowledgement of their Author.*[8]

After this letter was written, the delicate relationship between Newton and Flamsteed was marred by the intrusion of Edmond Halley, who had also begun to investigate irregularities in the moon's orbit. Flamsteed took a deep dislike to the younger man, and, according to David Gregory, Newton told him in December 1698 that the real reason for Flamsteed's hostility was that Flamsteed's lunar tables "were first made and computed by Edmond Halley." Indeed, Flamsteed "published them without the knowledge of Halley, and . . . this theft was the origin of the eternal quarrels between Halley and Flamsteed"[9]—in which Newton was to take Halley's side.

This, then, is the background to those passages in a letter Flamsteed wrote in October 1698 complaining that "Collonell [*sic*] Bruce" (and

8. Newton to Flamsteed, February 16, 1694/5 *Correspondence,* IV, 87–88.
9. Noted by David Gregory in the margin of his annotations to the *Principia,* as cited by Rigaud and quoted in translation from the Latin by David Brewster in his *Memoirs of the Life, Writings, and Discoveries of Sir Isaac Newton,* (Edinburgh, 1855), II, 165–166.

his friend Colson) had said that "Mr. Newton had perfected the Theory of ye Moon *from Mr. Halley's Observations* and imparted it to him with Leave to publish it . . ." This, wrote Flamsteed angrily could not be true, for

Newton has seen a Synopsis of 152 Observed places of the Moon wth her calculated places, and the Ellements of the Calculation all done by my own hand, and knows I Imparted them wth as many more as made them above 200 to Mr Newton and that their is a very Fair correspondence keept between us for this purpose: and some have been given him very lately. But to clear you wholly and take of all Occasions of your Injureing either Mr Newton or me by Spreading this or the like false storyes for the future, I must Acquaint you that Mr Newton assures me he has not imparted his Lunar Theory to Mr Halley (so that all he knowes of it must be onely collected from discourse he has had with him) *nor made use of one of Mr Halleys Observations in rectyfying of it* . . . I would not Injure Mr Halley either with Mr Neuton (on whom I know he has dependance) nor the Collonel (by whom he may make some advantage): therefore when I found Mr Newton concerned at the report (wch I gave in near as few words as I have wrote it) I added no more, but that I wondered why or by whom it should spread. Nor would I write (to the) Collonel, nor had to you but that I find my servant discorsed to you on his own head and omitted w(ha)t I cheafly injoyned him wch I have marked before wth a Line underneath . . .[10]

The reference to Halley's relations with "the Collonel (by whom he may make some advantage)" probably alludes to Halley's desire, which he is said for a while to have seriously entertained, to accept a suitable post in Russia. During Peter the Great's sojourn in London, the two conversed on scientific subjects, and the Tsar is said to have been much taken by Halley,[11] who was at that time looking for em-

10. Flamsteed to Colson, October 10, 1698, *Correspondence*, IV, 284, no. 594.
11. Andreev suspected very strongly that Halley, who succeeded Flamsteed as Astronomer Royal on the latter's death in 1719, may also have met with Peter at Greenwich. This assumption is based largely on undocumented information in John Barrow's *A Memoir of the Life of Peter the Great* (London, 1832). Brewster's biography of Newton, and John Timbs's *London and Westminster: City and Suburb* (London, 1868), which indicate that Halley was in England at the time of Peter's sojourn. Presumably the source for the story is Halley's obituary in *The Gentleman's Magazine and Historical Chronicle*, XVII (November 1747), 506, though according to E. F. McPike's *Hevelius, Flamsteed, and Halley, Three Contemporary Astronomers and Their Mutual Relations* (London: Taylor and Francis, 1937), Halley had already left the country. According to the *Dictionary of National Biography*, as well as other sources, this is incorrect; Halley left Portsmouth on the *Paramour Pink* at the end of November 1698. Owing to insubordination among the crew, he was compelled to return from Barbados in the following June, but he set out again in September 1699, and was back in England in September 1700.

ployment. His brilliant gifts, which even Flamsteed acknowledged, were already widely recognized, but his advancement was arrested by suspicions concerning his religious beliefs. Thus, Bishop Stillingfleet had refused to recommend Halley to the Savilian Chair of Geometry in Oxford on the grounds of his materialist views:[12] and on his appointment to the Mint, Newton had come to Halley's rescue by securing for him the office of deputy controller of the Mint in Chester. In the course of 1698, however, the five country mints were discontinued, and Halley was again seeking employment. Newton once more tried to help his friend by offering him a position worth ten shillings a week to teach applied mathematics for two hours a day to engineers and officers of the army.[13] Halley, it seems, turned this down and may well have considered an invitation to go to Russia instead, which Bruce would certainly have been empowered to extend. According to Halley's obituary, the invitation was actually issued, but he did not accept it for, at the same time that Flamsteed was casting aspersions on Halley's integrity and religious orthodoxy, Halley's fortunes took a dramatic turn for the better. At his own request, he was appointed by the King to command the *Paramour Pink* which sailed a month later, at the end of November 1698. The purpose of Halley's voyage was to study the variations of the compass in different parts of the globe, and after sailing beyond Chile and Peru he returned to England in triumph.

Meanwhile, in his absence, Flamsteed's relations with Newton rapidly deteriorated. "On Sunday the 4th December, 1698, in the time of evening service,"[14] Newton visited the Greenwich Observatory to obtain twelve computed places on the moon, which Flamsteed had corrected for him. After Christmas, Flamsteed sent him a correction of the time in one of the observations, and, having afterward realized that the results needed to be further improved, he returned Newton's visit on December 30 or 31 to inform him of this. According to Flamsteed, Newton now broke his promise by being unusually "reserved" in not disclosing to him the particulars of his lunar theory.

At this point Newton's swiftly deteriorating trust for Flamsteed

12. Nonetheless, Halley was appointed Savilian Professor on the death of Wallis in 1703, the same year in which he was chosen secretary to the Royal Society. See S. J. Rigaud's *Defence of Halley against the Charge of Religious Infidelity* (Oxford, 1844); the matter is also discussed by Brewster in *Memoirs of . . . Sir Isaac Newton*, II, ch. xviii.

13. Brewster, *Memoirs of . . . Sir Isaac Newton*, II, 196.

14. *Ibid.*, p. 202.

was exacerbated by the unfortunate intervention of two of the greatest mathematicians of the period, John Wallis and David Gregory. Wallis was at that very moment preparing for the press the third volume of his *Opera Mathematica* (see fig. 36). On November 8, 1698, he had written to Flamsteed to obtain information concerning the latter's alleged "discovery" of the parallax of the polestar. Wallis was eager to include Flamsteed's observations in his forthcoming volume and urged him to send them as soon as possible: "... [I]t will be to the Reputation of our Nation, to be the first that have been able to make-out that Parallax."[15] Flamsteed immediately replied, but in doing so referred to Newton's lunar theory and the aid which he, Flamsteed, had been giving him:

I had become closely associated with Mr. Newton, at that time learned Professor of Mathematics at the University of Cambridge, to whom I had given 150 places of the Moon, deduced from my observations previously made, and at the time of these observations, her places as computed from my tables, and I had promised him similar ones for the future as I obtained them, together with the elements of my calculations in due order, for the improvement of the Horroccian theory of the moon, in which matter I hope he will have the success he expects.[16]

Wallis acknowledged this letter on December 10, 1698, and seeing in it "nothing ... but what is fit to be published," allowed David Gregory to read it, who then acquainted Newton with its contents. Newton found Flamsteed's use of his name deeply offensive and directed Gregory to request Wallis to suppress the injurious lines. Wallis then wrote posthaste to Flamsteed asking permission to omit or modify the offending passage on the ground that Gregory had asked him "not to print any Paragraph of your letter which speakes of your giving to Mr. Newton Observations of the Moon."[17] Flamsteed reacted by writing a long letter to Newton explaining the inoffensive nature of his communication to Wallis, but he had not kept a copy of his original letter to him, which means that all he could do was to cite the passage above (in Wallis' Latin translation.) We have therefore no way of knowing whether the English original that elicited such anger on Newton's part

15. Wallis to Flamsteed, November 8, 1698, *Correspondence, IV*, 287, no. 596.

16. Flamsteed to Newton [citing his earlier letter to Wallis], January 2, 1698/99, *ibid.*, p. 293, no. 600. See also Brewster, *Memoirs of ... Sir Isaac Newton*, II, 202–203, where the translation of this passage is from the text published by Baily. Except for the final sentence, where Brewster's version seems more apt, I have used the translation offered in the *Correspondence*.

17. Wallis to Flamsteed, December 28, 1698, *Correspondence, IV*, 288–289, no. 598.

when Gregory showed it to him corresponded exactly to Wallis' rendition of it.

Flamsteed was to claim that his original letter was wholly innocent of the construction put on it by Newton, and that Newton himself had been intentionally misled by David Gregory. Gregory's friendship with Edmond Halley prompted Flamsteed to suspect that Gregory bore him ill will, and he closed his letter to Newton with an admonition to "enquire what company Dr Gregory kept yt you my not be deceaved in his Character."[18] Newton was not appeased by Flamsteed's explanation, and on January 6, 1699, he replied in a celebrated letter[19] which reached a new level of acrimony in his relations with the astronomer:

Upon hearing occasionally that you had sent to Dr Wallis about ye Parallax of ye fixt starrs to be printed & that you had mentioned me therein with respect to ye Theory of ye Moon I was concerned to be publickly brought upon ye stage about what perhaps will never be fitted for ye publick & thereby the world [When Mr Halley boast tis done & given him as a secret tells ye Society so & forreigners see Mr Colsons letter to me.] *put into an expectation of what perhaps they are never like to have.* I do not love to be printed upon every occasion much less to be dunned & teezed by forreigners about Mathematical things or to be thought by our own people to be [Was Mr Newton a trifler when he read Mathematicks for a sallery at Cambridge. Surely ye Astronomy is of some good use tho his place be more beneficiall.] *trifling* away my time about them when I should be about ye Kings business. And therefore I desired Dr Gregory to write to Dr Wallis against printing that clause wch related to that Theory & mentioned me about it. [I know what I have to doe without teling.] You may let the world know if you please how well you are stored with observations of all sorts & what calculations you have made towards rectifying the Theories of ye heavenly motions: But there may [where persons thinke too well of themselves to acknowledge they are beholden to those who have furnisht them with ye feathers they pride themselves in when they have great fr(iends) & c.] *be cases* wherein your friends should not be published without their leave. And therefore I hope you will so order the matter that I may not on this occasion be brought upon the stage . . .

As the editor of the *Correspondence* points out, "it is difficult to understand Newton's attitude" as revealed in this letter: nor is it certain whom exactly Newton had in mind on this occasion when he

18. Flamsteed to Newton, January 2, 1698/99, *ibid.,* p. 295.
19. Newton to Flamsteed, January 6, 1698/99, *ibid.,* pp. 296–297, no. 601. The notes that appear in brackets were added by Flamsteed.

referred to "forreigners" who "teezed him about Mathematical things." Flamsteed, in the note he added referring to this passage, wrote "forreigners see Mr Colsons letter to me" (see above), which implies that he, at any rate, was thinking of "Collonel Bruce." For it was Bruce who, together with John Colson, had some weeks earlier announced to Flamsteed's servant that Newton had completed his lunar theory. Flamsteed's humiliation on hearing such "rumours" is easy to understand: for while Bruce and Colson supposedly knew that Newton had "perfected" his theory, Newton in his encounters with Flamsteed refused (so the latter believed) to discuss the matter with him.

Flamsteed must have readily explained this unjustified recalcitrance on Newton's part by recourse to rumors from the same source. Halley, so Bruce and Colson had reported, was about to publish Newton's lunar theory. To add insult to injury, Halley, so they also reported, was claiming that he and not Flamsteed had supplied Newton with the observations making that theory possible. It is to Flamsteed's credit that, in spite of this morass of misunderstanding in which rumors, rivalry, and personal animosity played so large a part, he continued to supply Newton with observations, but after this their friendship was irretrievably sundered. Perhaps Flamsteed's relationship with Bruce was also momentarily affected, for in his letter to Colson he adds:

I intend to be in London, god willing, on fryday next but the days being short I shall not have leasure to See the Collonel at your house. I shall be at Garways betwixt one and two If you come down hither in the mean time let it not be on Wednesday for I have company that day: at any other you shall be wellcome to

Your Friend & Servant
J. F. MR.[20]

Garways or Garraways was a noted coffeehouse in Change Alley, Cornhill, "a celebrated place for sandwiches, sherry, pale ale, and punch" which is often mentioned in the Hooke *Diary*. Presumably Bruce also frequented it. Wallis finally published the third volume of his mathematical works in 1699—the volume containing the letter of Flamsteed that had caused so much offense to Newton—and Bruce acquired it.[21] More significantly, long after he was back in Russia he also acquired Flamsteed's monumental *Historia Cœlestis Britannica*

20. Flamsteed to Colson, October 10, 1698, *ibid.*, p. 285, no. 594.
21. See Chapter 3, below, p. 44 and n. 24.

(see fig. 15), as well as his *Atlas Cœlestis* (see fig. 16). Owing to difficulties with engravers and lack of funds, the latter did not appear until 1729, ten years after Flamsteed's death. That it appeared at all was due to the zeal of his widow and his assistant, Crosthwait, whose labors in editing his master's works extended for over ten years and entailed the sacrifice of his own prospects in life.

The publication of the *Historia Cœlestis* was even more involved, making the low point in the depressing story of its author's relations with Newton and Halley to which the altercations of 1698/99 were the real prologue. Flamsteed's work first came out in 1712, in a so-called "pirated" edition, for which Halley was largely responsible. It then consisted of one folio volume, in two books, the first containing the catalogue and sextant observations; the second, observations Flamsteed made with the aid of Sharp's mural arc which covered the period Bruce lived in London. Halley, who wrote its ambiguous preface, was said to have boasted in Child's coffeehouse of his pains in correcting Flamsteed's errors, and Flamsteed retaliated by calling Halley "a lazy and malicious thief." The funds necessary for publishing the work were acquired through Newton's recommendation, but in return Flamsteed was more or less cajoled into giving up control over its production, which passed to a committee of the Royal Society that included David Gregory and Newton. This arrangement was bound to lead to further bitterness. Convinced that his work had been intentionally sabotaged and that he, as its author, was robbed of the credit due to him, Flamsteed resolved to bring out a corrected and improved edition of the *Historia Cœlestis;* in 1716 he was compelled to resort to legal proceedings for the recovery from Newton of several volumes of the original entries of his observations. Newton's anger is reflected in the second edition of the *Principia,* where several passages in which he had in 1687 acknowledged his gratitude to his former friend are omitted.

Flamsteed did not live to see the new and improved three-volume edition of the *Historia Cœlestis Britannica* appear in print as it was published only in 1725, but it is this version and not the earlier one (which Flamsteed had partially committed to the flames) that Bruce acquired.[22] Presumably he would have known something of its check-

---

22. See MIAN, V, nos. 132 and 32:
История зелестис на латинском языке Ягана Фламштента в трех томах [and] Атлас небесной Ягана Фламфета [*sic*].

ered history, for it was the controversy over the lunar theory in 1698 that contributed both to the difficult progress of Flamsteed's magnificent opus through the press and to its author's passionate quarrel with Newton later. The first volume of the *Historia Cœlestis* contained the observations of Gascoigne and Crabtree for 1638–1643; those made by Flamsteed for 1668–1674, and the sextant observations at Greenwich in 1676–1689 which were spared from the flames along with the edition of 1712. The second volume continued his observations with the mural arc from 1689 until his death; the third opened with an introduction on the progress of astronomy from the earliest times, which Bruce may have found particularly valuable because of the account with which it closes, describing the instruments and methods employed at the Royal Observatory. It also contained the catalogues of Ptolemy, Ulugh Beigh, Tycho Brahe, the Landgrave of Hesse, and Hevelius—with whose works Bruce was well acquainted.[23] Finally, there appeared the "British Catalogue" of 2,935 stars observed at Greenwich, in addition to the stars Halley had observed in the southern hemisphere during his voyage to the Pacific.

All this would have been of considerable interest to Bruce. Though neither the *Atlas* nor the *Historia Cœlestis* were likely to have been the first works from Flamsteed's pen to fall into his hands, he possessed two of Flamsteed's earliest productions, published in 1672 and 1673, one of which furnished Newton with data on the diameters of planets used in the third book of the *Principia*.[24] It is possible that these early works of Flamsteed were acquired by Bruce while he was still in

23. *Ibid.*, nos. 42–46. The most important of these from Bruce's point of view would no doubt have been the *Machina Cœlestis* (1673–1679), in which Hevelius describes the mounting of his telescopes and the erection of his famous observatory, and the *Selenographia* (1647), which is described in the inventory of Bruce's books as "Solenograpiia." This monumental opus is one of the most beautiful works to be published in the seventeenth century, the first to carry detailed plates of the moon.

24. See Rare Book Room, USSR Academy of Sciences, Leningrad, $\frac{1\text{V Bd}}{3}$ ann, and $\frac{1\text{V Bd}}{3}$ acc.6. Neither of these items appear to be entered in any recognizable form in the inventory of Bruce's library (on the way it was compiled, see Chapter 3), but it is likely that both in fact belonged to him. Flamsteed's *De Temporis Æquatione Diatriba. Numeri ad Lunae Theoriam Horoccianem* was published as a supplement (1673) to the *Opera Posthuma* (1672) of Jeremiah Horrocks, which may account for its omission from the inventory, Flamsteed made the discovery that Horrocks's lunar theory was the only one that did not contradict the varying dimensions of the moon he had observed (before Horrocks' theory was communicated to him by Towneley). At the joint request of Newton and Oldenburg, Flamsteed prepared the latter for publication, with the addition of his own more comprehensive explanation and numerical elements. The work itself was edited by John Wallis.

London. Certainly, his enthusiasm for astronomy, which he shared with Peter the Great, could only have been stimulated by his contacts with Flamsteed.

Unfortunately, the correspondence between Flamsteed and Bruce does not appear to have survived, but it is not difficult to surmise why he and the "Colonell" would have found their relationship mutually profitable. Bruce's intimacy with the Tsar was presumably well known in English circles. Besides, he disposed of funds which Peter had left behind for Bruce's further education, as well as the acquisition of scientific instruments and books. Flamsteed, on the other hand, was perpetually short of money. On his government salary of a hundred pounds a year, cut down by taxation to ninety, he was supposed not only to run the Royal Observatory, but to pay for the assistance of a "surly, silly labourer" (who had to be available for moving the sextant) and to instruct two boys from Christ's Hospital. On top of this, there was his large expenditure in procuring skilled aid and improved instruments. This is why Flamsteed was obliged to take private pupils, and between 1676 and 1709 he had about 140 of them, many of the highest rank. Bruce may not formally have belonged to their number, for he had an official instructor in mathematics, who, it transpires, was also a friend of both Flamsteed and Newton. It is to him we must now turn.

BRUCE AND "IVAN KOLSUN"

Peter the Great's record of expenditures in England refers to a certain "Ivan Kolsun," who was paid 48 guineas on April 17, 1698, for "training Jacob Bruce over a period of six months, as arranged by contract, including board and lodging . . ."[25] Another entry refers to a slightly larger transaction: "the sum of 50 guineas paid at the behest of the Great Sovereign to Jacob Bruce, who learnt the mathematical trade from Ivan Kolsun . . ."[26] Kolsun's name, sometimes spelled "Konston" or "Konton," recurs elsewhere. He is the recipient of 150 guineas paid by the Tsar for "a small yacht . . . made of cypress wood." We are also told in the *Journal* that Peter the Great had dinner with

25. Quoted by M. M. Bogoslovskii, "Pervoe zagranichnoe puteshestvie", in *Petr I[:] Materialy dlia Biografii,* ed. V. Lebedev, Part II ([Moscow]: Ogiz, 1941; reprinted by Mouton in cooperation with Europe printing, Vaduz, at The Hague, 1969), p. 378.
26. *Ibid.,* p. 378.

him on February 7, 1698[27], only two days after the Tsar's third supposed meeting with Newton.

Following Bogoslovskii, it has been assumed that this man was
John Colson (b. 1671) who, after early experience at sea, opened a
mathematical boarding school in Wapping. It has further been suggested that Bruce probably attended this school and that his relations
with the English mathematician were probably close for when he left
London—in all likelihood at the beginning of 1699—he took with
him a letter from Colson, addressed to Peter the Great, on the subject
of the Tsar's "State science."[28]

This John Colson died in 1709. Another John Colson living in
London, also a mathematician, later was called to Cambridge to take
up the Lucasian chair of mathematics earlier occupied by Newton.
Was it this man whom Peter the Great befriended, rather than the
John Colson of Wapping? The two mathematicians were apparently
related, and in 1698 the two Colsons may have shared the same house
in Goodman's Fields. From a letter by John Flamsteed it is also quite
clear that Bruce was living in that house.[29] Whether the 150 guineas
for tutoring Bruce ("including board and lodging") were paid to the
older or the younger "Ivan Kolsun" cannot be ascertained from the
surviving Russian documents.

John Colson the younger was a friend and disciple of Newton,
through whom Bruce could have become well acquainted with the
principles of Newtonian natural philosophy. According to the *Dictionary of National Biography,* which devotes a cursory paragraph to his
life, he was born in 1680, the son of Francis Colson of Lichfield. He
attended Christ Church, Oxford, which the Russian Tsar, together
with Bruce, visited on April 8, 1698.[30] Colson matriculated in May
1699, but he left Oxford without taking a degree, so the date of his
departure is uncertain.[31] His academic career began in 1709, when he
was appointed Master of the Free Grammar School, Rochester, en-

27. N. G. Ustrialov, *Istoriia tsarstvovaniia Petra Velikago* (St. Petersburg, 1858), III,
Appendix IX, 603.
28. Unfortunately, it has not survived. Bruce wrote to Peter the Great, June 29, 1699,
as quoted in Pekarskii, *Nauka i literatura Rossii pri Petre Velikom,* I, 292.
29. See *Correspondence,* IV, no. 594.
30. See A. I. Andreev, "Petr v Anglii v 1698 v.," in *Petr Velikii Sbornik statei,* Vol. I
(Moscow and Leningrad: Akademiia Nauk, 1947), 80.
31. There appears to be some confusion over the date at which Colson went to Oxford.
Correspondence of Sir Isaac Newton, ed. Scott, IV, 285, n.l, gives it as 1699, but this is
certainly erroneous.

dowed by Sir Joseph Williamson, former Secretary of State and president of the Royal Society. In 1713 Colson was elected a Fellow of the Royal Society and in middle age was called to Cambridge, wrote a few articles for the *Philosophical Transactions,* and, after being granted a degree, was appointed in 1739 to the Lucasian chair. He died in 1760.[32]

So much for the bare facts of Colson's life. His character is described in most economic terms by Cole the antiquary, who knew him personally: "I do not know that he was regularly of either university originally. He was a very worthy, honest man." And, says Cole, he was already "an old bachelor" when he came to Cambridge. We are also told that Colson was a "humourist and peevish, and afterwards removed to a house in Jesus lane, where a sister lived with him very uncomfortably, as their tempers did not suit." According to Newton, Colson was given his post at Sir Joseph Williamson's Mathematical School through "interest rather than merit," but he appears to have been precocious in his youth. Thus, in 1698 when Bruce met him Colson was already teaching mathematics privately. Indeed, Flamsteed's letter is addressed to "Mr. Colson Teacher of the Mathematickes at his house in Goodmans Feilds London."

How did Bruce meet Colson? Flamsteed's letter to Colson is dated October 10, 1698; as we have seen, however, the Astronomer Royal had already been visited by Bruce, together with the Tsar, at the Greenwich Observatory in February and March. It may, therefore, be more likely that Flamsteed introduced Bruce to Colson rather than the other way round, but it is also possible that their meeting came about independently at Oxford.

Presumably either Peter or Bruce thought highly of Colson, for the contract the Tsar concluded with him for training Bruce over a period of six months was by the standards of the time extremely generous. Spread over an entire year, the 48 guineas Colson received (for board and lodging, as well as instruction) represented almost as much as Flamsteed received for discharging his manifold duties and expenses at the Greenwich Observatory. And if Colson received any part of the additional 50 guineas Peter gave Bruce ("who learnt the mathematical trade from Ivan Kolsun," as noted in the second entry above) then Colson's remuneration was positively munificent. His salary will seem

32. Cf. E. G. R. Taylor, *The Mathematical Practitioners of Tudor and Stuart England* (Cambridge, Eng.: the University Press, 1954), p. 298, no. 530.

all the more striking when it is recalled that Newton offered to procure employment for Halley to teach mathematics in return for what amounted to one shilling an hour (or two pounds a month for two hours of instruction a day).

With these figures in mind, it can surely be assumed that the 50 additional guineas did not all go to Colson. Bruce no doubt took instruction elsewhere, too, and the "new Table of Refractions" which Flamsteed promised for the "Collonell" to go along with his lunar theory may perhaps be taken as an indication that Bruce's interests were much wider than the uninformative Russian record of the Tsar's expenditures reveals. To what extent natural philosophy and astronomy were also included in his "training" can only be surmised; but there is a vital clue concealed in the inventory of Bruce's library which tells us much both about the direction his mathematical studies in London may have taken and about his association with John Colson the younger.

After Newton's death Colson published his treatise on calculus, "fluxions" being the name Newton gave the celebrated method he invented: *"The Method of Fluxions and Infinite Series; with its Application to the Geometry of Curve-lines. By the Inventor Sir Isaac Newton, Kt. Late President of the Royal Society . . ."* (see fig. 14). This was not the earliest English work to explain Newton's infinitesimals. That honor goes to Charles Hayes's *Treatise on Fluxions* (1704), which Bruce also probably acquired. But John Colson's edition had the incalculable advantage of containing Newton's own text.[33] This work appeared in 1736; Bruce died in 1735. How did Newton's *Method of Fluxions* find its way into the inventory of Bruce's library? Either John Colson or another acquaintance in England must have sent this important work to Moscow before news of Bruce's death had reached London. And whoever supplied Bruce with the *Method of Fluxions* would also presumably have sent him some of the other English books in the inventory. This neglected source tells us more about Newton in Russia than any other document of the Petrine era.

33. MIAN, V, no. 569. The treatise on fluxions by Charles Hayes is not itemized in any recognizable form in the inventory of Bruce's books; there is, however, a copy of this work at the library of the USSR Academy of Sciences in Leningrad, which may originally have come from his collection.

3

Russia's First Newtonian

J. D. BRUCE AND THE COMPILATION OF HIS LIBRARY

Bruce's collection of books was acquired with a purpose. It clearly reflects the interest of a natural philosopher, mathematician, astronomer, and engineer who on his return from London stood at the head of that complex movement of intellectual and social transformation during which Russian science and language began to adopt a "modern" character. As director of the new School of Mathematics and Navigation in Moscow, Bruce laid the foundation for mathematical and scientific education, paving the way for the establishment of the St. Petersburg Academy of Sciences (in which he too was to play a part). This is why the collection of books and scientific instruments which Bruce built up over the years was, for Petrine Russia, unique.

The inventory has been published in two versions. It first appeared in 1859, with a brief and misleading introduction by the historian I. Zabelin.[1] The second version was published in 1889, in Volume V of the *Materials for the History of the Imperial Academy of Sciences.*[2] Before dealing with the discrepancies between these two lists, it is important to recapitulate the conditions under which the original inventory was compiled. After Bruce's death in May 1735, Count Salty-

1. Nikolai Tikhonravov (ed.), *Letopisi russkoi literatury i drevnosti* (Moscow, 1859), I, 28–62. Bruce's collection of books fared better than Newton's own, part of which was recovered after a disappearance of two centuries by Colonel de Villamil. See L. T. More, *Isaac Newton: A Biography* (New York: C. Scribner's Sons, 1934), p. 542.
2. MIAN, V, 152–245.

kov became temporarily responsible for the library. On June 6 of the same year (and not 1755, as Zabelin states) Saltykov was told by the cabinet of ministers "to appoint a specially qualified man" to make a list of the library's holdings. He was told in the same letter that "a special courier" would also be dispatched by the Academy to help sort out "all the curious things" for the *Kunstkammer*. The special courier failed to arrive, however, and Saltykov entrusted the task to two of his acquaintances, state councillor Bogdan Aladin and Lieutenant Captain Gur'ev, who little knew the complexity of their task.

Before completing the catalogue on July 31 they were joined by the Academy's representatives, the notary Christoph Tiedemann and Ivan Pukhort, his scribe. The final list comprises more than 1,600 items, including books, rare manuscripts, geographical curiosities, maps, mineral and botanical specimens, and astronomical and mathematical instruments. The catalogue is, in fact, the sole historical record of its kind for the first half of the eighteenth century.[3] It serves a vital purpose not only in elucidating the range of Bruce's own interests—the Navigation School, his astronomical observatory, his translation of scientific works, his practical activities as president of the College of Mining and Manufactures, his reform of the coinage, his pioneering work in optics and mathematics—but also in showing how Newtonian science first became known in Russia.

Why the catalogue has remained neglected is owing partly to Zabelin, who conveyed the impression that his incomplete list, containing less than half the items in the original list, comprised Bruce's entire collection, and also to the very real difficulties posed by the language of both versions. Even if S. E. Fel', Iu. Kh. Kopelevich, V. L. Chenakal, S. L. Sobol', B. A. Vorontsov-Vel'iaminov, and other recent scholars who have written about Bruce, were aware that Zabelin's catalogue with its 789 items was only part of a larger and far more revealing list, there were still formidable obstacles (as S. E. Fel' notes) to understanding the catalogue: "the titles are not given textually, but

---

3. Some other collections of the period—D. M. Golitsyn's and Feofan Prokopovich's, in particular—were larger but, from a scientific point of view, far less significant. See *Istoricheskii ocherk i obzor fondov rukopisnogo otdela biblioteki Akademii nauk*. Vol. I, *XVIII vek* (Moscow and Leningrad; Akademiia nauk, 1956). On the role private collections played in forming the Academy of Sciences library, see G. A. Kniazev and K. I. Shafronovskii, "Istoriia Biblioteki Peterburgskoi Akademii nauk I. Bakmeistera 1776 goda" in *Trudy Biblioteki Akademii nauk i fundamental 'noi biblioteki obshchesvennykh nauk Akademii nauk SSSR*, Vol. VI (Moscow and Leningrad: Akademia nauk, 1962), 251–264.

in an extremely abbreviated form, in addition to which the date of publication and the names of authors are cited rarely."[4]

The difficulties are further compounded by the fact that original books were in as many as twelve languages (if one unidentified work is excluded).[5] The transcription, however, is Cyrillic in spite of the fact that only thirty-four of the books in the entire collection are Russian. Hence, the transcriptions are sometimes phonetic and at other times attempts at translating the original title. Curiously enough, each of the quartet who compiled the list, brought his own personal idiosyncrasy to the ensemble. The fact that "amateurs" Aladin and Gur'ev were on the scene first accounts for the omission of dates in most of the initial six hundred volumes. Tiedemann surely knew better, and the German books in the second half are on the whole more conscientiously listed. In the case of Newton's works, the author is usually not given at all, and the date is similarly omitted. The language of the original, however, is generally supplied, and it turns out that Bruce had in his possession not only all of Newton's major scientific writings, but also many of the more important commentaries on his work by disciples and friends.

THE *Principia* AND RELATED COMMENTARIES

That Bruce had at least one copy of the *Principia* in his possession is known from sources other than the catalogue of his library. In the list of books he gave to Peter the Great in 1717, one entry refers evidently to the 1714 (Amsterdam) edition of Newton's great work.[6] The famous second edition, which Newton presented to Queen Anne, had been published by Dr. Bentley in 1713.[7] The version the Tsar acquired from Bruce was the first Amsterdam reprint of this second edition. Before identifying Bruce's second copy of the *Principia,* however, it may be appropriatae to describe some of his related commentaries.

4. S. E. Fel', "Petrovskaia Geometriia," TIIE, IV (1952), 155.
5. Total of German works: 663 (including 326 items with dates or authors unmarked); English: 302 (186 unmarked); Latin: 188 (75 unmarked); Dutch: 85 (47 unmarked); Russian: 34 (33 unmarked); Spanish: 19 (14 unmarked); Italian: 7 (7 unmarked); French: 6 (2 unmarked); Finnish: 4 (3 unmarked); Swedish: 8 (5 unmarked); Polish: 6 (6 unmarked); Czech: 1 (unmarked); 1 unidentified.
6. See E. V. Bobrova, "Obzor inostrannykh pechatnykh knig sobraniia Petra I" in *Istoricheskii ocherk i obzor fondov rukopisnogo otdela biblioteki Akademii nauk,* I, 156.
7. Gray, no. 8; David Brewster, *Memoirs of Sir Isaac Newton,* I, 318.

One of the earliest expositions of Newton's ideas was William Whiston's *Astronomical Lectures,* originally written in Latin, and brought out in English translation in 1715. This is the edition listed in Bruce's library. "Wicked Will Whiston," as Swift called him, succeeded Newton as Lucasian professor of mathematics in Cambridge in 1701, and for twenty years he was one of Newton's closest disciples and friends. His first notable work had been *A New Theory of the Earth,* which was read in manuscript by both Locke and Newton and dedicated to the latter. In it Whiston tried to account for the deluge by postulating a near collision with a comet and explained Genesis on Newtonian lines. As a result, he was strongly attacked by Anglican divines, and even tried for blasphemy, being expelled from Cambridge in 1710.

According to De Morgan, Whiston became one of the first to give lectures on natural philosophy with experiments in London.[8] He edited Newton's *Arithmetica Universalis* in 1707, and in 1716 he published a commentary on Newtonian physics, *Sir Isaac Newton's Mathematick Philosophy,* together with a discussion of Halley's work on comets. These three works are all listed in Bruce's catalogue (together with a later edition of the *Universal Arithmetick,* discussed later):

I. 1396) Лекціи астрономическія въ публичныхъ школахъ въ Камбриджѣ, на аглинскомъ языкѣ в Лондонѣ. 1715.

Astronomical Lectures, read in the Public Schools at Cambridge: by William Whiston, M.A. Mr. Lucas's Professor of the Mathematicks in that University. Whereunto is added a Collection of Astronomical Tables; Being those of Mr. Flamsteed, Corrected; Dr. Halley; Monsieur Cassini; and Mr. Street. For the Use of young Students in the Universities. And now done into English (London, 1715). (This is a translation from the first Latin edition of 1707, *Praelectiones Astronomicae Cantabrigiae in Scholis Publicis Habitae* a Gulielmo Whiston; see fig. 5).

II. 858) Новая теорія о землѣ, на аглинскомъ языкѣ, въ Лондонѣ. 1969.

A New Theory of the Earth, From its Original, to the Consummation of all Things. Wherein the Creation of the World in Six Days, the Uni-

---

8. On the relationship of Whiston (1667–1752) with Newton, see Augustus de Morgan's *Essays on the Life and Work of Newton* (London and Chicago: Open Court Publishing Company, 1914). (Also, DNB, where the *Astronomical Lectures* are erroneously referred to as *Astronomical Principles.* The English edition is relatively rare. Both Gray, no. 166, 167, and Babson, no. 126, list Latin editions only. See also H. Zeitlinger and H.C.S., *Bibliotheca Chemico-Mathematica* [London: H. Sotheran and Co., 1921], no. 5382).

versal Deluge, and the General Conflagration, As laid down in the Holy Scriptures, Are shewn to be perfectly agreeable to Reason and Philosophy. With a large Introductory Discourse concerning the Genuine Nature, Stile, and Extent of the Mosaick History of the Creation. By William Whiston, M.A. . . . [See fig. 4].

III. 1420) Госп. Невтона математическія и филозофическія настав-ленія, на аглинскомъ языкѣ, въ Лондонѣ. 1716.

Sir Isaac Newton's Mathematick Philosophy more easily Demonstrated: with Dr. Halley's Account of Comets illustrated. Being Forty Lectures Read in the Publick Schools at Cambridge. By William Whiston, M.A., Mr. Lucas's Professor of the Mathematicks in that University. For the Use of the Young Students there . . . (London, 1716).

Besides these three works by Whiston, Bruce had two more of his books, both containing astronomical data and tables, which he may have found useful in compiling the almanacs he published annually after 1709. These—a venture new to Russia—contained the more essential information on solar and lunar motion.[9]

Another full commentary on the *Principia* (dealing with the laws of motion, centripetal forces, the attractive forces of spherical bodies, bodies vibrating in pendulums, and so on) was brought out by John Clarke, Dean of Sarum (1682–1757), a distinguished mathematician and an early exponent of Newtonian science. His work, *A Demonstration of some of the Principal Sections of Sir Isaac Newton's Principles of Natural Philosophy* . . . was published in London in 1730 (see fig. 7). This is the only edition to appear in Bruce's lifetime and clearly the one noted in the catalogue as:

570) Демонстранціонъ принцибале сексіонъ Исака Нефтона филозо-фическая, на аглинскомъ языкѣ.[10]

Clarke's brother was the famous Samuel, who played so eminent a part in the Newton-Leibniz controversy. Samuel Clarke edited Rohault's *System of Natural Philosophy,* historically one of the most important texts to be associated with the Newtonian enlightenment. It should be recalled that for more than thirty years after the publication of the first edition of the *Principia* (1687), the Cartesian cosmology (with its theory of vortices) kept its ground. In England as elsewhere in Europe, Jacques Rohault's manual (translated from French

9. Cf. MIAN, V, 229, no. 41 (new serial numbering), and no. 585. Bruce's almanac is described by Pekarskii in *Nauka i literatura v Rossii pri Petre Velikom,* I. pp. 304–309.
10. Cf. Babson, no. 46; Gray, no. 62.

to Latin) continued even in Cambridge to be the text in philosophical instruction as late as 1718. It was translated from French to Latin by Samuel Clarke in 1697, but it contained notes which were a virtual refutation of the text. Thus, the Newtonian system made its official entry into the place of its birth by the back door. Clarke's stratagem completely succeeded. A tutor might assign passages from the Cartesian text, but the student would consult the marginal notes which exposed its fallacies.[11]

Rohault's work was translated into English by John Clarke in 1723, along with his brother's notes. A third English edition came out in 1735.[12] Though no date is given in the Russian entry, the translation of the title is close and suggests that it was taken from this third edition:

710) Ронаусть, Сустема о натуральной филозофіи, усмотрѣна доктором Самуйломъ Кларкерсомъ, взята отъ Ридера Исако Нейтонс, филозофа, съ прибавленіемъ [sic].

Cр. *Rohault's System of Natural Philosophy, illustrated with Dr. Samuel Clarke's Notes, taken mostly out of Sir Isaac Newton's Philosophy* [see fig. 6].

One of the best-known commentaries on the *Principia* outside England was written by W. J.'s Gravesande (1688–1742), a Dutch scientist and friend of Newton, who introduced the Newtonian philosophy into Leyden. His *Physices elementa Mathematica, experimentis confirmata: Sive Introductio ad Philosophiam Newtonianam* was first published in 1710–1721 and translated into English the same year by J. T. Desaguliers, in two volumes. A fourth edition appeared in 1731. The title indicated in the Russian catalogue could be any one of these:

532) Математикаль элиманс натураль филозофіи, чрезъ Исаак Навтона, въ двухъ томахъ, на аглинскомъ.

Cр. *Mathematical Elements of Natural Philosophy confirm'd by Experiments: or, an Introduction to Sir Isaac Newton's Philosophy*. Written in Latin by . . . W. James's Gravesande, LL.D. Professor of Mathematiks at Leyden, and F. R. S. Translated into English by the late J. T. Desaguliers F. R. S. and published by his son J. T. Desaguliers[13]

An even more important work was David Gregory's *Astronomiae Physicae & Geometricae Elementa,* published with a dedication to

11. See Brewster, *Memoirs of . . . Sir Isaac Newton,* I. p. 333 ff.; cf. Babson, nos. 103, 104: Gray lists the third English edition only (no. 143).
12. A second Latin edition came out in 1703; a third edition was published in 1710.
13. Cf. Gray, no. 83; Babson, no. 68.

Prince George of Denmark at Oxford in 1702 (see fig. 35). Here the Russian transcription is least cooperative:

706) Элементъ о звѣдосмотрѣніи, медицинскія и огенетринскія описанія.

The translation of *Astronomiae* into Russian makes use of the archaic Russian term for astronomy. There is a misunderstood attempt to render *Physicae* in Gregory's work, that is, physiologiae, by the Russian word for medical. This mistake was easy enough to make in view of Gregory's M.D. The translation of *geometricae* probably resulted from a mishearing. The scribe transposed the last word of the original title (*Elementa*) to the beginning of the Russian transcription (*Element o*), making the remaining parts of the title progressively more unreliable.

This was the first textbook composed on gravitational principles, remodeling astronomy in conformity with Newton's physical theory (Cf. *Philosophical Translations*, XXIII, p. 1312; *Acta Eruditorum*, 1703, p. 452). Newton thought highly of the book, and communicated for insertion (p. 332) his "lunar theory," which, as we have seen, Bruce would certainly have heard of during his stay with John Colson in London. The discussion in the preface of the book, in which the doctrine of gravitation was defended on the score of its antiquity, likewise emanated from Newton.[14] Gregory was the first professor who publicly lectured on Newtonian philosophy, and he owed his reputation mainly to his promptitude and zeal in adopting it. His enthusiasm for the *Principia* was boundless, and Whiston relates that he himself was led to its study by Gregory's "prodigious commendations." He left his native city of Aberdeen for London in 1691, was introduced to Newton, became one of his few intimate friends, and was elected to the Royal Society the following year (November 30, 1692).

Finally, as noted above, Bruce had a second copy of the *Principia* besides the Dutch edition of 1714 (which passed into Peter the Great's possession in 1717). Here the catalogue would seem to be of little help, for it mentions neither author nor date nor place of publication, and the title, too, is incomplete. All we have is:

315) Философія натуралисъ, на аглинскомъ языкѣ.

14. The materials on which the preface was based were found in Newton's handwriting among Gregory's posthumous papers.

What is peculiar about this transcription is that the title suggests a Latin original, but we are told that the book was in English. Since the first English translation of the *Principia,* by Andrew Motte, appeared in 1729 under the title subsequently used in all English translations (*The Mathematical Principles of Natural Philosophy*), "Philosophiia naturalis" could not be an abbreviation suggested by an English translation. "Aglinskii iazyk" probably refers to the place of origin, and the first part of the title must refer to a Latin edition.

It should be emphasized that, in spite of numerous other errors, there is not one single instance in the catalogue of a Latin title being confused with an English title. How did this particular error come about? Many of the Latin titles listed involve the use of the genitive case, this being correctly rendered by the appropriate feminine singular ending: «ae» = ай, a vowel sound which presents no difficulties to the Russian ear. The pertinent ending of "Philosoph-," however, ends in "ia," a mistake which is easily explained when it is remembered that in the original bindings of the *Principia,* the title was often abbreviated for lack of room to PHILOSOPH NATURAL (see fig. 9a). The scribe, it might well be imagined, looked no further; adding the Russian feminine ending to the noun, he maladroitly russified a word he considered to be English.

We find this item (No. 315) next to a series of other works from the same source, including Newton's *Opticks* and several volumes of the *Philosophical Transactions.* The question is, to which edition of the *Principia* does it refer? Since Bruce died in 1735, we are left with three possibilities: the rare first edition of 1687, the second edition of 1713, and the third edition of 1726, each of which was published in England. As is well known, approximately 250 copies of the first edition were published, and, by the last decade of the seventeenth century, procuring one of these copies was no easy task. Although it seems more likely, therefore, that Bruce had the second edition rather than the first, his possession of the latter should not be discounted.[15] Bruce was in England in 1698, before the second edition appeared, and we know from other sources that he was given considerable sums of money especially for the acquisition of books and scientific instruments.

15. The only known copy of the first edition of the *Principia* in the USSR was presented to the Academy of Sciences by the Royal Society in 1943. See Sir Robert Robinson's "Address of Welcome to the Delegates," *The Royal Society Newton Tercentenary Celebrations 15–16 July, 1946* (Cambridge, Eng.: the University Press, 1947), p. 2.

These expenditures suggest that Bruce was charged with buying books more or less systematically, the motive being dictated no doubt both by the Tsar's private interest and the prospective demands of the school he was to organize in Moscow the following year. Newton's renowned masterpiece would have been a natural choice for anyone at all well acquainted with the contemporary state of natural philosophy. Indeed, there is in the Library of the USSR Academy of Sciences in Leningrad today a copy of this rare first edition of the *Principia,* which I uncovered and which could be the one described in the inventory of Bruce's collection.[16]

Before turning to the work on mathematics one more item must be mentioned: *A Treatise of the System of the World.* This was intended to form the third book of the *Principia,* and Newton said it was drawn up "in the popular method, so that it might be widely read."[17] Published in 1728, the *Treatise* was in fact a translation of *De Mundi Systemate,* which appeared the same year. It contains an interesting preface, omitted in later editions, with an extract from the beginning of the *Principia* and a short account of the state of astronomy at the time Newton wrote. In one prophetic passage Newton points to the possibility of Terrestrial Tide Effects, which were discovered by Michelson in 1919; in another he indicates the existence of the planet Uranus, which was actually seen by Herschel in 1781. The Russian transcription omits the date of publication, and it could conceivably also be the second edition of 1731. There can be no doubt, however, about the identity of the work itself:

590) Треатисъ офъ те системъ ворлдъ Исакъ Невтонъ, на аглін-
скомъ.

A Treatise of the System of the World, by Isaac Newton, translated into English (London, 1728–1731; see Fig. 22).

16. The librarians of the rare book room of the library of the USSR Academy of Sciences were good enough to permit my identification of this work to be noted by a tab inside the front cover bearing my signature and noting the circumstances of its discovery: cp. $\frac{8269}{2016}$ q. This copy is in good condition, but does not contain any of Bruce's marginalia (which has facilitated the task of identifying some of the other books in his library). Nor does it carry his "Ex Libris" (which is also the case with some of the other books originally in his possession). The history of this particular copy of the *Principia* before it became a part of the USSR Academy collection is obscure.

17. J. Edleston, *Correspondence of Sir Isaac Newton and Professor Cotes, including Letters of Other Eminent Men, now first published from the originals in the Library of Trinity College, Cambridge* . . . (Cambridge and London, 1850), p. xcviii.

41

THE "FLUXIONARY" METHOD AND THE MATHEMATICAL WRITINGS

How well Bruce knew Newton's mathematical writings is more difficult to ascertain. He wrote to Euler in 1732 that he had always been "ein grosser Freund und Liebhaber der Mathematic," and two surviving letters show that Bruce had corresponded with him on problems involving tautochtronous curves.[18] What is clear from the catalogue, however, is that, in addition to copies of the *Principia* and virtually all of Newton's major work on light, Bruce also had most of Newton's published mathematical writings. Of these, that best known is the *Arithmetica Universalis,* first brought out by Whiston in 1707—it was said, against Newton's consent, which Whiston denies.[19] An English translation was made in 1720, and a second edition of it— much corrected—was published in 1728. The full title of the first English edition reads: "*Universal Arithmetick: or, A Treatise of Arithmetical Composition and Resolution. To which is added Dr. Halley's Method of finding Roots of Aequations Arithmetically. Translated from the Latin by the late Mr. Raphson, and revised and corrected by Mr. Cunn.*" This is transcribed in Bruce's catalogue as:

539) Уневерсалъ арифметики, на аглинскомъ языкѣ.

Newton's name is not indicated on the title page of the first English edition, though it is noted in the second (see fig. 10). It is more likely, therefore, that Bruce had the earlier edition.

The *Universal Arithmetick* was based on lectures Newton gave at Cambridge. "This book was originally writ," notes the editor of the first English edition, "for the private Use of the Gentlemen of Cambridge, and was deliver'd in Lectures at the public Schools, by the Author, then Lucasian Professor in that University. Thus, not being immediately intended for the Press, the Author had not prosecuted his Subject so far as might be otherwise expected; nor indeed did he ever find Leisure to bring his Work to a Conclusion . . . In this unfinished State it continu'd till the Year 1707, when Mr. Whiston, the Author's Successor in the Lucasian Chair . . . thought it a pity so noble and useful a Work should be doom'd to a College-Confinement, and obtain'd Leave to make it Publick. And in order to supply what the Author had

18. Iu. Kh. Kopelevich, "Perepiska L. Eulera i Ia. V. Briusa," in *Istoriko-Matematicheskie issledovaniia.* (Moscow), X (1957), 96–116.
19. De Morgan, *Essays on . . . Newton,* p. 13; see also Gray, no. 277; and Babson, no. 199.

left undone, subjoyn'd the General and truly Noble Method of extracting the Roots of Aequations published by Dr. Halley."

Of all the mathematical works in Bruce's library perhaps the most interesting one is briefly described as:

569) Методъ офъ флецкомъ, на аглинскомъ языкѣ.

For this almost certainly refers to John Colson's edition of Newton's posthumous treatise on calculus described above.[20]

Bruce could also find a description of Newton's "fluxionary method" in another volume listed in his library: the Leipzig *Acta Eruditorum* for 1712:

1064) Дѣла ученых людей, 1, 2, 3, 4, 5 и 6 часть въ Лейпцигѣ, 1712.

The pertinent item, *Isaaci Newtoni, Analysis per quantitatum series fluxiones ac differentias,* may be found in Book IV.

When, specifically, Bruce acquired Newton's various works cannot be definitely established, for his books and instruments were dispersed among other collections in the second half of the eighteenth century.[21] Many items were evidently acquired in the latter half of his life.

Other works, such as Whiston's *A New Theory of the Earth* (1696) (see fig. 4) were probably acquired much earlier, belonging perhaps to those books which Bruce bought when he was in London in 1698/99. Several such items may be seen in the Library of the USSR Academy of Sciences in Leningrad today, including a variety of English works on navigation, a subject obviously dear to Bruce's heart. Besides the manuals by James Atkinson (1698), Henry Gillibrand (1693), Perkins (1682), and Norwood (1667), whose work was well known to Newton, there is an item by Nathaniel Colson (1697), probably a kinsman of John Colson.[22] As might be expected, Bruce also possessed many of the

20. Colson wrote in his foreword: "I gladly embraced the opportunity that was put into my hand of publishing this posthumous Work because I found it had been composed with that in view and design. And that my own Countrymen might first enjoy the benefit of this publication, I resolved upon giving it an English translation with some additional Remarks of my own." On Colson, see Chapter 2.

21. This is also true of Bruce's manuscripts. See V. A. Petrov, "Istoriia rukopisnykh fondov Biblioteki Akademii nauk s 1730 do kontsa XVIII veka," in *Istoricheskii ocherk i obzor fondov rukopisnogo otdela biblioteki Akademii nauk*, I, 204 ff.

22. Atkinson's and Gillibrand's manuals, both with almost the same title (*[An] Epitome of Navigation*) were not as popular as *The Mariners New Kalendar* by Nathaniel Colson, which was still being republished in the second half of the eighteenth century. On Colson, see E. G. R. Taylor, *The Mathematical Practitioners of Tudor and Stuart England* (Cambridge, Eng.: the University Press, 1954) p. 264. There is as yet no larger sequel to Taylor's most useful work, and it is pleasant to record that Professor P. J. Wallis of

better-known mathematical works published in England prior to his visit, including an edition of Euclid by Newton's friend and predecessor in the Lucasian chair, Isaac Barrow (1660).[23] He also had virtually all the major mathematical writings of John Wallis (1647, 1685, 1695, and 1699), the third volume of whose collected works appeared in the year Bruce left England. (As Bruce may have known, both Flamsteed and Newton were indirectly involved in its preparation.[24])

Yet the most arresting entry in the inventory is not a published work, but a manuscript described as:

1177) Госп. генерала фельдцейхмейстера графа фонъ Брюса математическія письма, на немецкомъ языкѣ.

If these lost "Mathematical Letters" by "Count von Bruce" were to be unearthed, we would probably have a clearer picture of Newton's influence in the Petrine era. For the "Mathematical Letters" were apparently known, at least by repute, to J. G. Leutmann in Germany, who urged Bruce to publish them. Bruce declined on the ground that they were "suggested" by that "English author," presumably either Wallis or Newton.[25]

---

the University of Newcastle upon Tyne is presently engaged on a monumental "Bio-bibliography" of the mathematical and scientific writings of the period not covered by Taylor's book which should make it clear how much the lesser known contemporaries of Newton and Wallis contributed to the education, culture, and enlightenment of the time.

23. Bruce also had another English edition of Euclid, published in the following year.

24. Johannis Wallis, *Operum Mathematicorum volumen Tertium* (Oxford, Eng., 1699). On the controversy emanating from this work, see Chapter 2, above. Several of Bruce's surviving books in the library of the USSR Academy of Sciences (Leningrad) contain interesting marginalia in Bruce's own hand, but this volume is not one of them. The notes are usually in English, but occasionally in German (the language in which the "Mathematical Letters" were written, according to the inventory). The marginalia, more common in Bruce's mathematical books, are usually descriptive in character, such as the following example taken from Philip Ronayne's *A Treatise of Algebra* (London, 1727): "If any Magnitude (or Number, as the whole) be divided into such parts, that are to each other as a Number to a Number, the product of those powers of the parts, that are of the same degree, as the parts parts themselves denominate, is the greatest of all products of the like powers of the parts of the same magnitude when otherwise divided" (pp. 406–407).

25. On Leutmann, see Chapter 6, below.

4

Prince Menshikov's Exchange of Letters with Newton

Those of Newton's writings that were published after Bruce's return from London continued to find their way to Russia, but it is difficult to establish to what extent and degree they were known outside of Bruce's own immediate circle. One tantalizing clue crops up in a book by Captain John Perry, a hydraulic engineer recruited by the Great Embassy during Peter's sojourn in England. "His Majesty has always his Lords about him," writes Perry of Peter the Great, "and is himself very anxious in observing the Eclipses that happen, and in describing and discoursing of the natural causes of them to his Lords and People about him, and of the Motions of those other Heavenly Bodies within the System of the Sun, according as the great Sir Isaac Newton has indisputably demonstrated to the modern world."[1]

It would be tempting to dismiss this passage as yet one more attempt to show the Tsar, whom Perry knew and admired, in a flattering light. For what could better express the rationalist spirit of the Petrine reforms than the juxtaposition of Newton's name with Peter the Great's? Yet Perry's vignette was almost certainly drawn from life, as is suggested by the fact that in 1717, within a year of the time that his account was published, Peter "re-acquired" a copy of the *Principia*.[2] This

1. John Perry, *The State of Russia under the Present Czar* (London, 1716), p. 212.
2. See E. V. Bobrova, "Obzor inostrannykh pechatnykh knig sobraniia Petra I," in *Istoricheskii ocherk i obzor fondov rukopisnogo otdela Biblioteki Akademii Nauk*, I, 156.

work had been in his library previously; presumably he acquired it after the second edition was published in 1713 (see fig. 9). Could it have been one of the presentation copies Newton intended to send to "The Czar 6 for himself & ye principal Libraries in Muscovy?"[3] We cannot be sure of the answer, but it is probable that by then Newton's reputation among the Tsar's immediate companions must have stood very high. Of this tangible proof has survived in the form of a letter which Prince Alexander Menshikov addressed to Newton as president of the Royal Society (see fig. 17):

Monsieur,

L'Inclination particuliere, que j'ay toujours eu pour la nation angloise jointe à la juste admiration, ou Elle a mis tout le monde par sa Sagesse, par Sa valeur par tant d'autres excellentes qualités m'a fait chercher avec Empressement non Seulement les occasions de la Servir, mais m'a aussi donné l'envie de m'unir plus étroitement avec Elle.

Vous avez moyen Monsieur, de me donner cette Satisfaction, Si vous voulez bien me faire l'honneur de me recevoir dans l'Illustre Societé, dans laquelle vous occupez si dignement la premiere place, honneur que je sai fort bien qu'on ne doit pas pretendre à la legere & qui doit être attaché à beaucoup de merites, aussi je Vous assure, que j'en aurais toujours la reconnaissance düe, & qui je tacheray par toutes sortes de moyens de ne vous être par un membre inutile, c'est de quoi je Vous prie de vouloir être persuadé & que je Suis,

Monsieur,
Votre tres obeisant
Alexander Menzicoff

St Petersbourg
  ce 23 d'Aoust
    1714

Newton wrote at least three versions of a reply to Menshikov, of which the following, a fair copy, has not previously been published (see fig. 19):

Potentissimo et maxime colendo Domino, Dno Alexandro Menzicoff Romani et Russi Imperii Principi, Dno de Oranienbourgh,[4] Czarianae Majestati Primo a Conciliis, Equitum Stratego, Devietarum Provinciarum Dynasti, Ordinis Elephantis, Aquilae Altae Nigrae Equiti etc.

3. Cited by I. Bernard Cohen, *Introduction to Newton's "Principia,"* (Cambridge, Eng.: the University Press, 1971), p. 247. This phrase of Newton's is taken from a list of prospective recipients of the second edition of the *Principia* which the author was working on, and serves, therefore, as only a rough index of Newton's plan for distributing his book.

4. This refers to Menshikov's estate in the Ukraine—Oranienburg (or Rauenburg as it was sometimes called). Menshikov also resided in palatial splendor at Oranienbaum, a few miles from St. Petersburg (now Lomonosov).

Isaacus Newton salutem.

Cum Societati Regiae dudum innotuerit Imperatorem vestrum, Czariensem Majestatem, Artes et Scientias in Regnis suis quam maxime promovere, eumque ministerio vestro, non solum in rebus bellicis et civilibus administrandis, sed etiam in bonis literis et scientiis propagandis, apprime adjuvari: maximo omnes suffusi fuimus gaudio, quando Mercatores Angli nobis significarent, Excellentiam vestram pro humanitate summa, et singulari suo in scientias affectu, et erga Gentem nostram amore, in corpus Societatis nostrae cooptari, se dignari. Atque eo quidem tempore coetibus nostris finem, pro more, donec aestiva et autumnalis praeteriret impositum eramus. Sed hoc audito semel adhuc convenemus ut Excellentiam vestram suffragiis nostris eligeremus: id quod fecimus unanimo consensu. Et jam, ut primum coetus nostros prorogatos renovare licuit, Electionem Diplomate sub Sigillo nostro communi ratam fecimus. Societas autem Secretario suo in mandatis dedit, ut transmisso ad vos Diplomate, Electionem vobis notam faceret. Vale.[5]

Dabam Londini
xxv Octob. Anno
MDCCXIV

(Translation:

To the most mighty and greatly honored lord, Lrd. Alexander Menshikov, Prince of the Roman and Russian Empire, Lrd. of Oranienbourgh, first in the Councils of his Majesty the Tsar, Field Marshal, Ruler over Conquered Provinces, Knight of the Order of the Elephant and of the White and Black Eagle, etc.

Isaac Newton sends greeting.

Since it had for some time become known to the Royal Society that the Tsar, your Emperor, was promoting the arts and sciences in his dominions, and that he was especially assisted by your personal ministration, not only in the affairs of war and peace, but also in propagating literature and science, it gave us all particular pleasure to learn from the reports of English merchants that your Excellency was willing and worthy to be elected into the fellowship of our Society, by reason of your distinguished humanity, your singular disposition towards the sciences, and your love for our nation. At that time however we had, as is our custom, ceased to hold meetings until summer and autumn had passed. But we have again convened to elect your Excellency by ballot; this we have done by unanimous vote. And now, at the first meeting that it has been possible for us to hold after our adjournment, we have confirmed the election by a Diploma under our common Seal. The Society moreover has ordered its Secretary to convey the Diploma to you and advise you of the election. Farewell.

London
October 25,
1714)

5. Another rough draft in Newton's hand may be found in Appendix I.

Thus, Menshikov became the first Russian Fellow of the Royal Society. The letter suggests, of course, that Newton had been badgered by "English's merchants" some months before the Royal Society's summer recess about electing him. Apparently Newton was somewhat bothered in drawing up the draft of an appropriate response. But the proposal itself would not necessarily have sounded strange to the other Fellows. Election to the society had not yet become an unimpeachable mark of scientific ability, and, in praising Menshikov's "singular disposition" toward the sciences, Newton was saying no less—and no more—than the truth warranted.[6]

Newton may even have remembered Menshikov's visit to London in the company of Peter the Great a little over a decade a half earlier. At that time Menshikov already enjoyed that peculiar confidence of the Tsar which was to make him one of the most powerful men in Russia. Peter may have taken him along on his visits to the Tower and the Royal Society, though it was only later that Menshikov developed an interest in natural philosophy. He began his career, like Bruce, by serving in the Tsar's regiment of boy soldiers, but his origins were humble and he had no formal schooling. Menshikov never learned how to write properly as the clumsy signature appended to "his" letter to Newton indicates, but he did have considerable native intelligence and ability. He championed Peter's reforms with an immense and ruthless energy and imitated the Tsar by taking a fashionable interest in the new science. Of this he gave proof by amassing a huge library,

6. Contemporary and other eighteenth-century accounts of Menshikov's life tend to be either panegyrical evocations or melodramatic cautionary tales concerned with his tragic rise and fall, which appealed to French playwrights and German writers. Thus, the historical value of the *Historische Nachricht von dem Ehemahligen grossen Russischen Staats-Ministro, Alexandro Danielowiz, Fuerst von Menzikov, nebst dessen Abwechslenden curieusen Fatalitaeten* (by an anonymous writer, n.p., 1728), the *Leben und todt des erst hoch erhabenen aber desto tieffer wiederum gestuerzteten Fuerst Menzikof, worinnen die ganze notable Geshicht [sic] dieses grosses Gluecks-Favortiten, nebts der Beschaffenheit des Koenigreichs Siberien, in welchem er sein Leiben kuemmerlich beschlossen kurz und gut beschrieben ist* (also anonymous, Frankfurt, 1730), and Michael Ranft's later *Das merkwurdige Leben des beruehmten Fuerstens Menschikov welches mit vielen Anekdoten ans Licht stellt ein Liebhaber der Wahrheit* (Leipzig, 1774), is very limited. There is, moreover, no authoritative biography of Menshikov, though some of his archival papers are in the process of being sifted by Soviet scholars. One of the best short accounts is A. A. Golumbievskii's "Sotrudniki Petra Velikogo: Kniaz' A. D. Menshikov," in *Russkii Arkhiv*, (No. 2, 1903), pp. 371–415, 481–549; yet this, like the articles of Count Bludov ("Dnevnye zapiski kniazia Menshikova" in *Sochineniia E. P. Kovalevskago, graf Bludov i ego vremia* [St. Petersburg, 1871], pp. 219–233), and G. V. Esipov ("Zhizneopisanie kniazia A. D. Menshikova, po novootkrytym bumagam," *Russkii Arkhiv* [No. 7, 1875], pp. 233–247; [No. 9, 1875], pp. 47–74), did not prove helpful in the context examined here.

acquired—unlike the Tsar's *Kunstkammer*—mainly by plunder. He mimicked Peter's love of astronomy by adding an observatory to the palatial mansions he built for himself at Oranienbaum.[7]

At Peter's death Menshikov found himself presiding over the fate of the dynasty (see fig. 18), but in the short period that Peter's widow sat on the throne and Menshikov was virtual ruler of Russia he did not abandon his earlier aspirations and tastes. He continued to dabble in astronomy and saw to it that Peter II and his own son were introduced to scientific experiments involving air pumps and other instruments that were considered an embellishment of the age. He also liked to be known as a patron of men of learning. Christian Goldbach, one of the earliest mathematicians to join the St. Petersburg Academy, was made welcome at Menshikov's table. How much Newton knew of this it is impossible to surmise. Indeed, it is possible that the painstaking civility of his response was abetted by acquaintances engaged in the Muscovy trade. If so, it is pleasant to be able to record that Menshikov's election to the Royal Society is unlikely to have yielded any commercial dividends. At the time that he wrote to Newton, Menshikov's fate hung in the balance. Three years earlier Peter had become aware once more of Menshikov's venality. In 1714 the Tsar appointed a commision to investigate his alleged embezzlement of government revenue, which soon began to unearth massive evidence of his misconduct.

Eventually Menshikov was able to ingratiate himself once again with Peter the Great, who found his old friend too valuable to discard. The year before the commission was set up saw the publication of the second edition of the *Principia,* of which—as we have seen—both the Tsar and Bruce quickly acquired copies. It may thus have seemed appropriate for Menshikov to add to his own pretensions by addressing the illustrious president of the Royal Society in person.

7. This is briefly described in M. S. Vanderbech's "Praesens Russiae Literariae Status," in *Acta Physico-Medica Caesarea Leopoldino-Carolina naturae curiosorum,* I (1727), Appendix, 131–149—the earliest formal survey by a foreigner of the state of the arts and sciences in Russia. Menshikov's vast library was dispersed on his arrest.

«Мы теперь в большом мире; этот мир есть в нас; такие миры суть в мире».—Peter I

5

"Worlds within Worlds"

THE FIRST RUSSIAN EXPLANATION OF GRAVITY

Not long after the publication of the second edition of the *Principia,* Peter the Great ordered the translation of Christian Huygens' *Kosmotheoros* . . . which appears to be the first book in the Russian language to describe the Newtonian cosmology.[1] It was also the first work of a scientific nature to be translated into Russian. The author, one of the greatest scientist-mathematicians of the age, was a friend of Newton; each man admired the other's work. Indeed, Huygens worked out the necessary mathematics of centrifugal action some years before Newton.

1. Kniga mirozreniia ili mnenie o nebesnozennykh globusakh i ikh ukrasheniiakh (1717), described in Pekarskii's *Nauka i literatura v Rossii pri Petre Velikom,* I, 388–389, and Vol. II: *Opisanie slaviano-russkikh knig i tipografii 1698–1725 godov* (St. Petersburg. 1862), no. 349. Cf. V. S. Sopikov, *Ukazatel' k opytu russkoi bibliografii* . . . , rev. ed. (St. Petersburg: A. S. Suvorin, 1908), Part III, no. 5175. The original edition was published in Latin as *Christiani Hugenii K o s m o t h e o r o s sive de terris coelestibus earumque ornatu conjecturae* (The Hague, 1696). (Pekarskii wrongly gives the date as 1688, but a second Latin edition came out in Liège in 1704.) Quotations in the pages that follow are taken from the Russian introduction and from the first English edition of *The Celestial Worlds Discover'd: or, Conjectures concerning the Inhabitants, Plants, and Productions of the Worlds in the Planets . . . by Christianus Huygens . . .* (London, 1698). T. I. Rainov, in *Nauka v rossii, XI–XVII vekov; ocherki po istorii do nauchnykh i estestvenno-nauchnykh vozzenii na prirodu* (Moscow and Leningrad: Akademiia nauk, 1940) shows that the Copernican system became known in Russia through manuscript translations sometime before 1680, probably between 1650–1670 (pp. 437–438), but Pekarskii's view that the *Kosmotheoros* was the first *printed* work in Russia to expound the heliocentric theory still stands. (Rainov's evidence rests on a half-dozen geographies known to have been translated in this period, but only two of these have been located so far: the *Theatrum orbis terrarum* by Guiljelmus and Joh. Blaeu, and the *Selenographia* by Hevelius.)

Yet their views, notably on the conservation of energy and on space and time varied widely. The Dutch scientist, after reading the *Principia,* became strongly critical both of Newton's postulate of universal gravitation and of his belief in the existence of absolute space and motion. As Descartes's intellectual heir at the Académie des Sciences, he tried to preserve the basic Cartesian framework intact. Even his study of centripetal forces and the laws of impact was suggested by Descartes rather than Newton, and he could not reconcile his mechanistic theories with the inverse square law of gravitational force which, Huygens insisted, was "a new and very remarkable property of gravity of which it was very necessary to search out the reason."[2]

But the *Kosmotheoros* was written at the very end of Huygens' life, when he confessed to having been too greatly influenced by Cartesian fictions.[3] The Latin original, brought out a year after his death in 1695, therefore represents Huygens' final views. These views proved fascinating not only for the subtle way they trace a median course between Descartes and Newton but also because they represented the "profession de foi" of a great seventeenth-century scientist. There was a great demand for the book, and it was soon translated into English, French, Russian, and German. Daunted by the scientific reputation of its author, the translator of the Russian edition assured his readers that he tried "to follow . . . the writer's intentions in everything, and as far as was possible, without changing his turn of speech." That some importance was attached to its publication is further confirmed by the fact that Peter the Great read the manuscript of the new introduction, and may even have collaborated in writing it.[4]

In Russia Huygens' book represented an entirely novel type of literature. The translation was obviously aimed at making the discoveries of Western science clear and intelligible to the still minuscule reading public. Fontenelle with his *Pluralité des Mondes* (1686) had been the first to succeed in this new genre in Europe, and Huygens was partly inspired by his example.[5] When Dufour translated the *Kosmo-*

2. Quoted by A. E. Bell, *Christian Huygens and the Development of Science in the 17th Century* (London: E. Arnold, 1947), p. 85.

3. See Paul Mouy, *Le Développement de la Physique Cartesienne* (Paris: J. Vrin, 1934), chapter xi. Also, on Huygens' reaction to the *Principia,* see Brewster, *Memoirs of . . . Sir Isaac Newton,* I, 339.

4. See Bruce's letter to Peter, November 2, 1716, in Pekarskii, *Nauka i literatura v Rossii pri Petre Velikom,* I, 300.

5. For a general discussion of Fontenelle's historical importance, see Herbert Butterfield, *The Origins of Modern Science, 1300–1800* (London: Bell, 1957), p. 172 *et passim.* In

theoros into French in 1702, he entitled it *Nouveau Traité de la Pluralité des Mondes,* and in his introduction Huygens referred to the "ingenious French author of the Dialogues," a gesture which Fontenelle returned in later editions of his own masterpiece.

The Russian introduction opens with a number of Biblical quotations that were absent in the original. Gently it takes the reader into its confidence "about a mystery worthy of amazement." Namely, our world is not in the centre of the universe—"as it may seem to us living on it." In fact, our earth is like the other planets (globusy) and revolves around the sun. A blessing affixed to the beginning of the translation (see fig. 27) and another—'Soli Deo Gloria'—added to the end would suggest an attempt at giving this daring revelation the aura of ecclesiastical approval. Indeed, the whole purpose of the introduction, apparently, is to conciliate those liable to be offended by the startling candor of the text,[6] but the attempt was not altogether successful. The work was vehemently attacked by Old Believers, and even its unhappy publisher described the book as the invention of the devil.[7] Their intense dismay is understandable, for the revolutionary idea that prompted the book is stated on the very first page: "A Man that is of *Copernicus's* Opinion, that this Earth of ours is a Planet, carry'd round and enlighten'd by the Sun, like the rest of them, cannot but sometimes have a fancy that it's not improbable that the rest of the Planets have their Dress and Furniture, nay and their Inhabitants too as well as this Earth of ours; Especially if he considers the later Discoveries made since *Copernicus's* time of the Attendants of *Jupiter* and *Saturn,* and Champain and hilly countries of the Moon, which are an Argument

a brilliant passage Butterfield describes Fontenelle as in a real sense the founder of the "philosophe" movement, the source "of the transfer of the scientific spirit, and the application of methodical doubt" to political thought (p. 173). The French translation of Huygens' book by D . . . , was entitled *Nouveau Traité sur la Pluralité des Mondes* (with deference to Fontenelle). It came out in Paris in 1702, and Wurzelbaur's German translation was published in Leipzig in 1703. In the Russian edition the translator is not specified and is the subject of controversy. This problem is dealt with later.

6. For reasons stated below, the Russian editions of this work are extremely rare. Passages from the Russian are quoted in Pekarskii, *Nauka i literatura v Rossii pri Petre Velikom,* and in B. E. Raikov, *Ocherki po istorii geliotsentricheskogo mirovozzreniia v Rossii, iz proshlogo russkogo estestvoznaniia,* 2nd ed. (Moscow and Leningrad: Akademiia nauk, 1947), pp. 165–167 et passim. A reproduction of the title page of the first edition of the Russian translation of *Kosmotheoros* is contained in M. I. Radovskii's *Antiokh Kantemir i Peterburgskaia Akademiia Nauk* (Moscow and Leningrad: Akademiia nauk, 1959), p. 65.

7. Pekarskii, *Nauka i literatura v Rossii pri Petre Velikom,* I, 283.

of a Relation and Kin between our Earth and Them, as well as Proof of that System . . ."[8]

Earlier writers such as Cardinal Cusanus, Kepler, Giordano Bruno, "and if we may believe him, *Tycho* was of that Opinion too" had "furnished" the planets with inhabitants,[9] but their fault was to commit fairy tales about the men in the moon. Huygens, on the other hand, had "a while ago [been] thinking somewhat seriously of this matter," and had decided that this "Enquiry was not so impracticable nor the way so stopt up with Difficulties, but that there was very good room left for probable Conjectures . . ."[10]

From some people "whose Ignorance or Zeal is too great" he expected a "dreadful sentence":[11]

There's one sort who knowing nothing of Geometry or Mathematics, will laugh at it as a whimsical and ridiculous undertaking. It's mere Conjuration to them to talk of measuring the Distance or Magnitude of the Stars; And for the Motion of the Earth, they count it, if not false, at least a precarious Opinion; and no wonder then if they take what's built upon such a slippery Foundation for the Dreams of a fanciful Head and a distemper'd Brain.[12]

To these men, writes Huygens, we should answer that their "Ignorance is the cause of their Dislike, and that if they had more Sense they would have fewer Scruples" (meaning doubts). But he would not take them to task for their ignorance. "[I]f they resolve to find fault with us for spending time in such matters because they do not understand the use of them, we must appeal to properer Judges."[13]

So much for the amateurs, the dilettantes, and the untutored public.

8. Huygens, *Celestial Worlds*, pp. 1–2. The reference to Saturn is prompted by Huygens' celebrated discovery of its satellites; he also pioneered in the use of exceptionally powerful, mechanically operated telescopes (which made such discoveries possible).
9. *Ibid.*, p. 3. The reference to Tycho Brahe stems from the dispute between Kepler and Brahe's heirs over publication of the latter's posthumous works. Kepler ardently championed the Copernican system, but Brahe did not. It was claimed that Kepler interpolated his writings to suit his own views. On this, see Max Casper, *Kepler* (London: Abelard Schuman, 1959), pp. 139–140.
10. Huygens, *Celestial Worlds*, pp. 3–4.
11. *Ibid.* By "sentence" he presumably refers to Giordano Bruno's fate at the stake, to which Huygens does not explicitly allude. On pp. 155–156 he does cite Bruno's belief in an infinite universe, with which he finds himself in agreement. ("Indeed it seems to me certain, that the Universe is infinitely extended.")
12. *Ibid.*, p. 5.
13. *Ibid.*, p. 6.

The second group he had in mind was the clergy. Against them he uses the same arguments as Galileo, though here, as elsewhere, Huygens is more candid, employing that "clarity of expression" which in his style Newton so much admired.

The other sort when they hear us talk of new Lands and Animals endued with as much Reason as themselves, will be ready to fly out into religious Exclamations, that we set up our own Conjectures against the Word of God, and broach Opinion directly opposite to the Holy Writ . . . But as for Worlds in the Sky, 'tis wholly silent. Either these Men resolve not to understand, or they are very ignorant: For they have been answered so often, I am almost asham'd to repeat it: That it's evident God had no design to make a particular Enumeration in the Holy Scriptures, of all the Works of his Creation.[14]

"But perhaps," Huygens adds ironically, "they'll say, it does not become us to be so curious and inquisitive in these things which the Supreme Creator seems to have kept for his own knowledge . . ." And he answers them in even stronger language:

These gentlemen must be told that they take too much upon themselves when they pretend to appoint how far and no farther Men shall go in their Searches, and to set bounds to other Men's Industry . . . as if they knew the marks that God has plac'd on Knowledge. If our Forefathers had been at this rate scrupulous, we might have been ignorant of the Magnitude and Figure of the Earth or of such a place as America. The Moon might have shone with her own Light for all of us, and we might have stood up to the ears in Water, like the Indians at every Eclipse: and a hundred other things brought to light by the late Discoveries in Astronomy had still been unknown to us.[15]

Next comes an interesting defense of the right of the natural philosopher "to search into the Nature of Things." Conjectures—the Latin phrase is "hypotheses"—are not useless because they are not certain; provided they are tested by observation, they are "of no small help to the advancement of Wisdom and Morality."

One biographer of Newton, Louis Trenchard More, considers that men like Robert Hooke and Huygens relied on an inward sentiment of knowledge and in opposing Newton "were merely opposing theory by hypothesis." Whatever the significance of this distinction, in the *Kosmotheoros* it is clear that Huygens was still vacillating between the Cartesian view that objects of scientific calculation are products of

14. *Ibid.,* pp. 6–7.
15. *Ibid.,* p. 8–9.

thought and the Newtonian phenomenalism which regards them as external realities. Thus, if the Russian reader had his first intimation of modern scientific method in Huygens' book, it is surely ironic that "observation" should be urged in favor of a proposition so patently unverifiable, empirically, as the plurality of "worlds."

For all his apparent show of impartiality, therefore, Huygens' objectivity is defined by the nature of the "conjecture" he is trying to prove. Christianity has existed throughout the centuries on the assumption that man is the pinnacle of God's creation. If an expanded universe and the probability of intelligent beings on other worlds does not destroy the uniqueness of Christ, it certainly jeopardizes Christian theology. Huygens attempts to evade this dangerous issue by circumventing religious considerations. He draws all his arguments from secular concepts or "ideas" such as reason, justice, morality, and so on. The paradox of his position is that in order to prove what prima facie may seem absurd, he must emphasize the absolute rationality of our own solar system.

The Copernican system, Kepler's discovery of the elliptic course of planetary motion, and gravitation—"which *Mr. Isaac Newton* has more fully explained with a great deal of plains and subtilty"—all seemed to prove that Nature has an intelligible structure.[16] The corollary to this is the Cartesian notion that Nature is a cipher or cryptogram, fixed and determinable, at least in broad outline, and that in mathematics, that monument of human reason, we possess the key.

This is the underlying theme of Huygens' whole work, and his implicit assumptions are best illustrated in his attack on Athanasius Kircher, the Jesuit author of the *Iter Exstaticus.* Like Huygens, Kircher believed in a pluraltiy of worlds. This he made the thread of a book in which he visits the various planets in his dreams.[17] But instead of finding a valuable ally in so orthodox a quarter, Huygens dismisses the *Iter Exstaticum* as gullible and unscientific. Kircher's conclusions may have been identical, but he committed the unpardonable blunder of clinging to Tycho Brahe, Aristotle, and a company of

16. *Ibid.*, p. 159.
17. Athan Kircher (1602–1680) was a German mathematician and astronomer, with unusually varied interests, including physics, alchemy, geology, astrology, and more. He wrote a number of books besides the *Iter Exstaticum,* the best known being *Magnes, sive de arte magnetica* (1641) and *Mundus subterraneus* (1664). In philosophy he stood close to Raymond Lull. (See also Bruce's library, which includes the *Iter Exstaticum* as well as Kircher's *Arco Noë;* see also MIAN, V, 155, no. 77).

doctors, when it was obvious that "*Copernicus* has set them all at liberty . . . by bringing in the Motion of the Earth: which if upon no other account, every one that is not blind purposely must own to be necessary upon this."[18]

Kircher's book, then, is "idle, unreasonable stuff"—not because the conclusions were wrong, but because they were based on the wrong premises. Huygens' main purpose in the *Kosmotheoros* was not to prove the validity of the Copernican thesis, but, rather, to show that the same inherent logic with which it defined the role of *Nature* in relation to the earth was equally applicable to other planets. The real beauty of the Copernican system lay in its mathematical symmetry and aesthetic perfection. This is why "the chief Argument for the proof" of the plurality of worlds could "be taken from the disposition of the Planets."[19] In this view, the earth becomes, in Huygens' words, merely "a small speck of Dirt."[20]

The other planets *must* be like the earth since they have their light from the sun and they revolve around their axes, as Huygens' discovery of the satellites of Saturn had proved.

Now since in so many things they thus agree, what can be more probable than that in others they agree too; and that the other Planets are as beautiful and as well stock'd with Inhabitants as the Earth? Or what Shadow of Reason can there be why they should not?

If any one should be at the dissection of a Dog and there be shewn in Intrails, the Heart, Stomach, Liver, Lungs and Guts, all the Veins, Arteries, and Nerves; could such a Man reasonably doubt whether there were the same Contexture and Variety of Parts in a Bullock, Hog, or any other Beast, tho' he had never chanc'd to see the like opening of them?[21]

On the basis of other such "arguments from their similitude," as he terms them, he thinks it self-evident that the other planets must be solid just like the earth, and in every way conform to Newton's laws of gravitation, "that Virtue, which like a Loadstone attracts whatever is near the Body to its Center."[22]

The interesting point here is that Huygens unreservedly accepts Newtonian gravitation, which earlier in his life he had rejected,

18. Huygens, *Celestial Worlds*, p. 105ff: "If the (reader) cannot be satisfied with all this, he is either one whose Dullness can't comprehend it, or who has his Faith at another Man's disposal, & so for fear of Galileo's fate dare not own it."
19. *Ibid.*, p. 11.
20. *Ibid.*, p. 10.
21. *Ibid.*, pp. 17–18.
22. *Ibid.*, p. 19.

though acceptance is limited to the celestial bodies, that is, in his view it is inapplicable to individual particles of matter.[23]

Since the planets obey the same physical laws as the earth, there is no reason why they should not also contain other animate matter. "Who doubts but that God, if he had pleased might have made the Animals in A m e r i c a and other distant countries nothing like ours? (And Nature you know affects Variety)—yet we see he has not done it." Therefore it follows that the planets must have animals much like the earth, and that they have water and plants too. This gives rise to an interesting attack on the Cartesians for deducing everything from "Atoms and Motions." For when it comes to plants and animals, they find themselves "nonplus'd, and give you no likely account of their Production."

PLANETARIANS

All this is prologue to Huygens' most startling announcement: the planets are inhabited with "Planetarians who have some sort of Reason." Must it be like ours? He replies in the affirmative. It must be so whether "we consider it as applied to Justice and Morality, or exercised in the Principle and Foundations of Science . . . For the aim and Design of the Creator is everywhere the preservation and safety of his Creatures." In the rest of the book, the "arguments by similitude" are carried to staggering lengths. If fear of eclipses had first led mankind to the study of astronomy, then there is all the more reason why it should be highly developed on Jupiter and Saturn owing to the frequent eclipses of their moons! Now if they have astronomy, they must also know geometry, arithmetic, and optics "for the studies of Arts and sciences cannot be said to be contrary to Nature."[24] For the same reason, "the Arts of Engraving and Painting, which from mean beginnings have improved to that excellence that nothing that ever sprang from the Wit of Man can claim Pre-Eminence to them," must also be practised by these Planetarians.

They also know music, a proposition based on the quaintest argu-

23. In *ibid.,* p. 19, Huygens refers the reader to his essay on gravitation, where he qualifies his views further. Newton was never concerned with the cause of gravity, that is, he was not interested in the philosophic *why,* but simply in *how* a natural phenomenon behaves. This dissatisfied Huygens, who, like other critics of Newton, wanted to know why gravitation behaved the way it did.
24. *Ibid.,* pp. 67–68.

ment of all, which shows how far Huygens and his age were imbued with the noetic conception of Nature. "[T]here's a sort of Bird in A m e r i c a," he writes, "that can plainly sing in order six Musical Notes: whence it follows that the Laws of Musick are unchangeably fix'd by Nature."[25] Where did Huygens hear of this exquisite notion? Possibly he was thinking of the *American Tomineius,* or humming-bird, which was too eloquently described by Dr. Nehemiah Grew in a paper delivered before the Royal Society in 1693, at about the same time Huygens was writing his book. In the abridgement of the Society's *Philosophical Transactions* published a century later, the editors remarked: "There is nothing in this paper that can justify re-printing it, nor is it possible, from the description given, to determine the particular species meant by the writer.[26]

With disarming frankness, Huygens adds that some people might object that the Planetarians are "destitute of all refined Knowledge, just as the Americans were before they had commerce with the Europeans."[27] Yet he thinks this unlikely. It seems far more plausible that they are much like us, that they have houses, ships, and, he insists again, geometry, and "our inventions of the Tables of Sines, of logarithms, and algebra" (inventions not introduced into Russia until Peter the Great's reign). He even suggests that there may be "such plenty and affluence" on the planets, "as they neither need or desire to steal from one another; perhaps they may be so just and so good as to be at perpetual peace, and never lie in wait for, or take away the life of their Neighbour."[28]

And why not, asks Huygens confidently: "I know I should be laugh'd at for an idle Discoverer of nothing but ridiculous Whimsery, and yet there's no reason but the old one, of our being better than all the World to hinder them from being as happy in their Inventions as we ourselves are . . ." With pride and wonder Huygens enumerates the accomplishments of contemporary science, which the Planetarians may well have surpassed. There was his own invention—the pendulum clock, the invention of the stenthrophonic tube, the discovery of the circulation of the blood, and so on. "[I]n the seed of the Male are discover'd, by the help of Glasses [Huygens' microscope] Millions of

25. *Ibid.,* p. 86.
26. *The Philosophical Transactions of the Royal Society of London,* abridged by Charles Hutton, George Shaw, Richard Pearson (London, 1809), III, 540.
27. Huygens, *Celestial Worlds,* p. 66.
28. *Ibid.,* pp. 79–80.

Sprightly little Animals, which it's probable are the very Offspring of the Animals themselves: a wonderful thing and never before known!"[29] There was the discovery that air has weight, chemical experiments that have brought to light "a way of making liquors that shine in the dark"; and, above all, Borelli's intimation of the laws of gravity and their proof by Isaac Newton.[30]

Huygens had written to Newton five years after the appearance of the *Principia* to say that he thought the idea of universal gravitation "absurd."[31] But the *Kosmotheoros* represented his final verdict, and he was forced to admit that Newton's explanation of comets was incomparably better than anything imagined by Descartes. It was difficult to see how comets could cut across the vortices postulated by Descartes (see fig. 29), or explain the eccentricity of the planetary orbits and the real accelerations and retardations of the planets in their orbits, except on the lines laid down by Newton. He was also in accord with Newton over the shape of the earth and his idea of the universe.

Huygens' book ends with a sad reprimand to his former teacher. "I cannot but wonder," Huygens wrote, "that C a r t e s, the first Man that ever began to talk reasonably . . . about the causes of . . . Gravitation . . . should never meddle with, or light on it."[32] Descartes, he admitted, had no idea of the incredible dimensions of the stellar system, of the distances between the stars, as is proved by the fact that "he would have a Comet as soon as ever it comes into our Vortex to be seen by us. Which is as absurd as can be. For how could a star, which gives us such a vast Light only from the Reflection of the Beams of the Sun, as he himself owns they do; how I say could that be so plainly seen at a distance ten thousand times larger than the Diameter of the Earth's Orbit?"[33]

Huygens does not, as Voltaire would three decades later, cast the

29. *Ibid.,* p. 99.
30. The one sinister development to cloud Huygens' modern faith in scientific progress was, significantly, that "Devlish Powder made of Nitre and Brimstone" through which "it seem'd as if we had got a more secure Defence than former Ages Against all Assaults, and could easily guard our Towns, by the wonderful strength of that Invention, against all hostile Invasions; but now we find it has rather encouraged them . . . I think we had better have bin [*sic*] without the Discovery. Yet, when we were talking of our Discoveries, it was not to be pass'd over, for the Planets too may have their mischeivous as well as usefull inventions." (*Ibid.,* pp. 95–96).
31. Quoted by A. E. Bell, *Huygens and the Development of Science,* p. 89.
32. Huygens, *Celestial Worlds,* p. 158.
33. *Ibid.,* pp. 160–161.

whole Cartesian structure out of the window and speak of "the pretended philosophical principles of Descartes."[34] But did Descartes not realize that around the sun there is an "Extensum" so vast that "in Copernicus's system the m a g n u s O r b i s is counted but a front in comparison with it?" But indeed, he adds humbly, "the whole story of Comets and Planets and the Production of the World, is founded upon such poor and trifling Grounds, that I have often wonder'd how an ingenious Man could spend all that pains in making such Fancies hang together."[35] And Huygens concludes that he for his part would be well contented if he could arrive at "a knowledge of the nature of things, as they now are, never troubling my Head about their beginning, or how they were made, knowing them to be out of reach of human knowledge, or even conjecture."

34. See Voltaire's *Dictionaire Philosophique.*
35. Huygens, *Celestial Worlds,* p. 161.

6

Bruce as Translator and Instrument Maker

THE ORIGIN OF THE TRANSLATION AND ITS RECEPTION

The sense of shock Huygens' work provoked may be gauged from the complaints of Mikhail Petrovich Avramov, its unwilling publisher, who described it as "Satanic perfidy." It was to "atheistic and blasphemous books" such as Fontenelle's and Huygens', he protested, that the decline of society was due. Avramov could not even bring himself to make any theoretical criticism of the ideas he found so objectionable. He merely repeats the claims of the "atheists" in the belief that this is enough to condemn them. "Of one star alone one author [presumably Fontenelle] has written that almost 25,000 years will pass . . . before it returns to the same place where it now stands." "Such fables," concludes Avramov, are just "fancies," and he ends by suggesting that if elsewhere "such authors . . . are permitted to occupy positions of esteem and dignity," in Russia "their lips should be sealed."[1]

Newtonian gravitation, or the physical aspects of the Cartesian cosmology as Huygens described it, were to Avramov either meaningless or insignificant side by side with the notion of transplanetary life. But he also realized with instinctive insight what was the one crucial characteristic of the secular-scientific outlook:

И между тем всем о натуре воспоминают: яко бы натура всякое благодеяние и дарование жителем и всей дает твари: и тако вкрадчися,

---

1. Pekarskii, *Nauka i literatura v Rossii pri Petre Velikom*, I, 511.

хитрят везде прославить и утвердить натуру, еже есть жизнь самобытную.[2]

Avramov was, of course, an Old Believer though he had first welcomed the secular innovations of the Petrine era.[3] Then, like Stefan Iavorskii and others who had initially greeted the Tsar's coming to the throne, Avramov suffered a change of heart. He became deeply religious and devoted his energies to various constitutional projects designed to restore the patriarchate and the waning power of the clergy. He intrigued against Theophan Prokopovich, whom he held responsible for Peter the Great's secularization of the church, and was repeatedly imprisoned in the course of his long life. With pathetic obstinacy, he seized every opportunity to press for a theocratic reaction. Under Peter II, Anne, and the worldly Bühren, he was persecuted and ignored. When clerical influence rose again in the reign of Elizabeth, his voice was finally heard, and he addressed a petition to the Empress describing how the heretical ideas of the *Kosmotheoros* and the *Pluralité des Mondes* were part of the same sinister foreign stratagem that was aimed at subverting his countrymen. The same ubiquitous plot had already resulted in such evils as the suppression of ecclesiastical autonomy.

As a reformed addict of the Petrine enlightenment, he recalled ruefully how as a young man he had read Ovid and Virgil "and through reading them became quite unbalanced (*obezumilsia*)."[4] He had even read them "with pleasure and eagerness." At one time, in sheer ignorance, he had praised these authors to the Tsar. He had even begged Peter to lend him recent translations, which "were brought to His Highness by supposedly reasonable people." To compound the Devil's machinations, he had one of these published at his own expense and initiative—"a short one with pictures of horrible gods and their insensate doings."[5] These horrible gods haunted Avramov's conscious-

---

2. *Ibid.* Translation: "And all the while they keep reminding everyone of nature as if nature was responsible for all human goodness and other attributes given to all things; thus they are stealthily scheming to glorify and enhance nature, affirming the existence of life to be dependent on itself alone."

3. On Avramov and documents relating to him, see *ibid.*, pp. 498–514; also, I. A. Chistovich, *Feofan Prokopovich i ego vremia* (St. Petersburg, 1868), pp. 261, 267, 274–279, 434–437, 451–459, 683–687, et passim. Cf. I. Shishkin "Mikhail Avramov, odin iz protivnikov petrovskoi reformy," *Nevskii Sbornik*, Vol. I (St. Petersburg, 1887), 378–429 (largely based on material by Pekarskii).

4. Pekarskii, *Nauka i literatura v Rossii pri Petre Velikom*, I, 511.

5. This is probably the book listed by Pekarskii (*Nauka i literatura v Rossii pri Petre*

ness for the rest of his life. He tells elsewhere, for instance, of his feelings of disgust at seeing the ceiling of a certain mansion in Moscow decorated with the indelible images of Greek mythology. "A story about Christ, or King David, or Solomon from the Bible," he adds primly, would have been more fitting. Thus, in his mind the pagan gods and the new scientific ideas were indissolubly linked, and the passionate objections with which he turned against his earlier dalliance with Western secular culture was now aimed against Huygens and Fontenelle.

Though Avramov's petition was written more than two decades after the Russian publication of the *Kosmotheoros,* his testimony is particularly valuable because it is one of the few clues as to the identity of the book's translator. Though Huygens' work is somewhat familiar to scholars of the period by repute, both editions of the translation are rare, and no serious study of it exists. Indeed, one recent writer attributes the book to Fontenelle; another writes that it "discusses the habitability of the planets and the stars, and proves that they are not created for man, as many of them are not even accessible to the human eye" [*sic*][6] Yet another scholar ascribes the translation to Baron Huyssen, while in Sopikov's bibliography it is credited to Johann Werner Paus.[7] Neither Paus nor Huyssen were the translators of the *Kosmotheoros;* it was done instead, by Jacob Daniel Bruce, as we hope to show.

"In 1716," wrote Avramov, "General Iakov Bruce presented His Imperial Highness . . . with a newly translated book (*knizhichikha*). Affecting the secretive and cunning flattery that was habitual with him when before the Sovereign, thus concealing his godless, frenzied, and atheistic heart, Bruce praised the book by the delirious author

Velikom, II, p. 532, no. 486) as *Ovidievy Figury, V 226 Izobrazheniiakh.* No editor is indicated, but it coincides with Avramov's description and appeared in 1721. The date is significant because it shows that Avramov's "conversion" was later than he implied; see below.

6. V. Ia. Stoiunin, "Kniaz Antiokh Kantemir v Londone," *Vestnik Evropy,* II (June 1867), 106. K. K. Baumgart in a supplement to his translation of Huygens' *Tri Memuary po mekhanike* (Moscow: Akademiia nauk, 1951), *Kratkii biograficheskii ocherk,* pp. 281–288, erroneously attributes the translation to Antiokh Kantemir.

7. Eduard Winter, "Bericht von Johann Werner Paus aus dem Jahre 1732 über seine Tätigkeit auf dem Gebiete der russischen Sprache, der Literatur und der Geschichte Russlands," *Zeitschrift für Slawistik,* III (No. 5, 1958), 746; Sopikov, *Ukazatel' k opytu rossiiskoi bibliografii . . .,* Part III, no. 5176; cf. Pekarskii, *Nauka i literatura v Rossii pri Petre Velikom,* II, nos. 349, 572 (he omits a translator and does not deal with the issue).

Kristofor Huiens, and others like it, pretending that it was very clever and wholesome for the educating of all the people, and moreover very necessary to navigation, and with such habitual and godless flattery deceived (*okral*) the Sovereign."[8]

The Tsar, according to this story, accepted Bruce's translation without even looking at it. Beckoning Avramov, Peter "emphatically" (*nakrepko*) ordered him to print the maximum issue, which amounted to 1,200 copies. But at this juncture in the fate of the *Kosmotheoros*, Avramov's low opinion of its merits turned out to be more important than that of Peter the Great. Though not a man of influence at court, he happened to be director of the St. Petersburg typography, to which the printing of the book was entrusted.[9] The Tsar was away in Europe, revisiting Holland, and receiving homage at the Paris Academy, where he was praised by Fontenelle for his services to science. From Avramov's point of view, this happy coincidence was too good to miss. "So the Tsar being absent," he confessed with engaging naïveté, "I examined this book, that was contrary to God in all ways, and with my heart quaking and my soul overawed, I fell before the Mother of God, with the sobbing of bitter tears, frightened to publish and frightened not to publish." This state of indecision was soon resolved "with the help of Jesus Christ." He decided in his heart "to enlighten these madcap-atheists (*sumazbrody-bezbozhniki*), these bare-faced theomachists (*bogobortsy*)," by withholding publication, actually printing only thirty copies. Even these he was careful to hide.[10]

Bruce was not aware of Avramov's devious intentions though he wrote to Peter, who was then still in France, complaining of delays and broken promises. In the summer of 1717 Bruce was engaged in translating not only Huygens but two other books, a Dutch grammar and an alphabet for children, which had both been requested by the Tsar, and Bruce hoped that all three would be published before his return.[11] According to his letter he was under the impression that he had succeeded in persuading Avramov to bring out both the *Kosmotheoros* and the children's alphabet, but the Dutch grammar, which should have been ready before Peter set out on his voyage, "would not be

8. Chistovich, *Feofan Prokopovich*, p. 264.
9. See "Tipograpficheskoe delo," in Brockhaus-Efron, *Entsiklopedicheskii slovar'* (St. Petersburg, 1901), XXXIII, 204–220.
10. Chistovich, *Feofan Prokopovich*, pp. 264–265.
11. Pekarskii, *Nauka i literatura v Rossii pri Petre Velikom*, I, 302–303.

printed before September, though in a foreign (*nemezkaia*) country, they could do it in two weeks."[12]

Avramov's version was very different. On the Tsar's return from Europe, Avramov "took the delirious (*sumazbrodnaia*) . . . book heretofore named, and with trembling hands brought it before His Highness, telling him at length and in detail that the book (*knizhichikha*) was most contrary to God, loathsome to God, worthy only—with its frenzied, lying . . . translator Bruce,—of being burnt in a screw-clamp." The Tsar listened, so Avramov wrote, and after taking a copy of the *Kosmotheoros* away with him, pondered on the matter . . . Two weeks later, "no order was issued to publish it amongst the people." Technically, the last phrase was no doubt true, for if Peter came back from abroad under the impression that the translation had been printed as arranged, the question of ordering its "publication amongst the people" for the second time did not arise. But Avramov could not resist one final flourish: the Tsar, he wrote, decreed that the *Kosmotheoros* should be returned to the "frenzied translator Bruce" who was to send all copies back to Holland, including, one would infer, the thirty copies of the sabotaged edition![13]

How much of Avramov's account is true? What Avramov neglected to mention was that a second printing of the *Kosmotheoros* was ordered in 1724, by which time he had been safely transferred from the typography to the College of Mines. This fact alone would imply that the book had the Tsar's continued support.

Avramov's petition was, of course, written many years after the events described. It was more an *apologia pro vita sua* than an attempt to describe what really happened, but the inclusion of his encounter with the Tsar was not due to any natural proclivity for exaggeration.[14] It was deliberate and served an important purpose: to have admitted his self-appointed role of public censor without implicating Peter was tantamount to a confession of treason.

The truth was probably more simple: having decided to sabotage the first edition, Avramov printed thirty copies as insurance against the contingency that his deception should be revealed. The ruse appears to have worked because in December 1724 he was recalled to his

---

12. *Ibid.*, I, p. 303.
13. Chistovich, *Feofan Prokopovich*, p. 265.
14. The precise date of Avramov's petition is not known. It was probably written shortly after the accession of Elizabeth in 1740.

previous post in the typography to supervise the printing of another ambitious translation, the *De Officio Hominis et Civis*, by Samuel Puffendorff. In all likelihood, Peter did not suspect Avramov's real views.

Indeed, it is doubtful whether Avramov was always clear about them himself. By temperament passionate and even erratic, his "conversion" at the time the *Kosmotheoros* was being published could hardly have been as sudden, final, or as irrevocable as he claimed. This is proved by the fact that his compilation on Ovid was brought out not before Huygens' book, but four years later, in 1721.[15]

Finally, there is the evidence from Bruce's correspondence with Peter on November 2, 1716, in which he wrote that Huygens' "philosophico-mathematical book is in readiness (*v gotovnosti*). Your Highness deigned to send me a little letter about it prior to your departure, after pleasing to read its introduction in my home and ordering that I should translate it myself."[16]

The "little letter" Bruce refers to has not been published, but if Peter read the introduction, there is certainly no reason for believing Avramov when he claimed that the Tsar had been tricked into printing the book. Nor incidentally, would Bruce have used navigation as an argument in its favor, for nothing in Huygens' work is even remotely connected with it. Yet none of this disproves Avramov's essential point: that the idea for translating Huygens came in the first place from Bruce.

It is clear from other letters between Bruce and the Tsar that, in spite of the Northern War and pressing economic problems, Peter retained an extraordinary interest in the translation of foreign works into Russian, often specifying not only the authors but also how and what part of their writings should be translated. In Puffendorff's case, for instance, he ordered the first part dealing with the "duties of man and the citizen [*grazhdanin*] . . . translated, the second, dealing with the Christian faith" ignored, because he saw "no need for it."[17]

He was equally explicit when he wrote to Bruce about the Dutch

---

15. Cf. Chistovich, *Feofan Prokopovich*, pp. 261–262, and Pekarski, *Nauka i literatura v Rossii pri Petre Velikom*, II, nos. 486, 532–533.

16. Pekarskii, *Nauka i literatura v Rossii pri Petre Velikom*, I, 300.

17. "o proshu, daby ne po konets ruk perevedena byla, no daby vniatna i khoroshim shtilem . . ." See the Tsar's interesting instructions on the art of translation: "Akty o vyshikh gosudarstvennykh ustanovleniiakh," in N. A. Voskresenskii's *Zakonodatel'nye akty Petra I* (Moscow and Lodnon: Akademiia nauk, 1945), I, 35, no. 7; 114, no. 145; 148, no. 203 ("O priemakh perevoda . . . knig," etc.).

grammar and the children's alphabet.[18] In the latter case, Bruce replied that he had found two "suitable-looking" books and, after translating one of them, would forward the galley proofs "so that if there is anything your Highness does not like, please [*soizvol'te*] mark it down, and it can still be altered."[19] Similarly with the Dutch grammar, Peter received parts of the uncompleted manuscript for correction and answered through his cabinet secretary Makarov that "it is well translated, and [His Highness] requests your Excellency to continue, so that by your efforts it is translated entirely."

But Bruce, harrassed by other duties, sought a way out, and a rather amusing correspondence ensued in which he offered various excuses for delay. Peter suggested that a certain Larionov could help him out, "and if he is too busy . . . take another Russian, who teaches Russian grammar." Chagrined, Bruce replied that his own Russian was quite adequate; in any case it could always be corrected by the proofreaders at the typography. What he needed was not someone who knew Russian grammar, but Dutch. Menshikov had lent him the services of a translator from the Admiralty called Hamilton, who at least knew Latin grammar and a little Dutch, but "on account of some intrigues" Hamilton had been enticed away from him.[20] Now that he had reached declensions (*deklinazii*) he was at a loss how to continue.[21] But the Tsar was equally persistent, and Bruce promised to have it finished at about the same time as the *Kosmotheoros,* having written to the author of the Dutch grammar in Amsterdam for a dictionary with which to "decipher" the remaining three hundred words.[22]

18. Probably intended for the reformed monastic schools, see *Ibid.*, pp. 53–54, no. 40, item 13 ("O uchene rebiat gramote vozle manisterei [*sic*] . . .").

19. Pekarskii, *Nauka i literatura v Rossii pri Petre Velikom,* I, 302.

20. Probably Hugo Johann Hamilton, one of the ubiquitous Scots in Russia at this time. A Major-General of Swedish cavalry, he was taken prisoner at the Dnieper in July 1709 and conveyed to Moscow. He had fought at Narva, Clissow, Frauenstadt, and Poltava. After his release, he became a field-marshal in the Russian army and died full of honors in 1748. See A. Francis Steuart, *Scottish Influence in Russian History* (Glasgow: J. Maclehose and Sons, 1913), p. 83.

21. The actual grammar, 716 pages long, is described by Pekarskii (*Nauka i literatura v Rossii pri Petre Velikom,* II, p. 395, no. 358, *Vilima Sevela iskusstvo niderlandskogo iazyka*). It was apparently a very thorough textbook, containing chapters on etymology, syntax, pronunciation, correct usage, ellipsis, pleonasms, synechdoches, metonyms, metaphors, etc.

22. Pekarskii, *Nauka i literatura v Rossii pri Petre Velikom,* I, 301.

NEWTON'S OPTICAL DISCOVERIES

The *Kosmotheoros* was Bruce's most important translation though he was more widely known through didactic works of public enlightenment. Thus, in 1707 he was responsible for the earliest Russian popularization of the Copernican theory, which was followed in 1709 by his celebrated almanac (*Briusov's Kalendar'*). This appeared annually thereafter,[23] while Bruce tried to combine his activities as a publicist and translator with that of what today might be called "scientific adviser." He was repeatedly sent abroad to recruit foreign specialists, engineers, officers, and artisans, and he induced the Tsar to correspond with Leibniz. Bruce himself conducted a spirited correspondence with Peter the Great over a long period of years, much of it concerning the Tsar's favorite pastime, astronomy.[24] For much of his life this was also Bruce's chief scientific pursuit,[25] though he also developed a strong interest in optics. He was the first in Russia to realize the significance of Newton's optical discoveries.

Newton's treatise on optics, his second major work, did not appear until 1704, but Bruce may already have heard of the new theory of light and color when he was studying in London. For Newton had formulated his theories of colors as early as 1666. Three years later, in 1669, he made this discovery the subject of his public lectures at Cambridge.

In 1671 he began to communicate it to the world, and it was at this time that he intended to publish his optical lectures, where these matters were handled more fully, together with a treatise of "Series and Fluxions." But the controversy which surrounded the publication of the *Principia* and his antipathy toward Robert Hooke, who died in 1703, prevented him from doing so. Thus, the *Opticks* appeared long after the discoveries on which the book was based had been made. Similarly, the lectures which had been the basis for it came out only in 1728.

Virtually all of Newton's optical work is listed in the catalogue of Bruce's library. Thus, the *"Optical Lectures Read in the Publick*

23. Cf. B. E. Raikov, *Ocherki po istorii geliotsentricheskogo mirovozzreniia v Rossii*, pp. 155–160.
24. Bruce also acquired rare books and instruments for the royal *Kunstkammer*. On one memorable occasion in Königsberg, he recovered a rare version of *Nestor's Chronicle*.
25. Cf. V. I. Chenakal, *Ocherki po istorii russkoi astronomii, nabliudatel'naia astronomiia v Rossii XVII i nachala XVIII v.* (Moscow and Leningrad: Akademiia nauk, 1951), chapters v, vi.

Schools of the University of Cambridge, Anno Domini, 1669. By the late Sir Isaac Newton, Then Lucasian Professor of the Mathematicks . . . (London, 1728)," appears in the Russian catalogue as:

580) Оптикаль лектюль, на аглинскомъ языкѣ[26] [see fig. 21].

Bruce also had two copies of the *Opticks*. In neither case is the author indicated, nor the place or date of publication:

577) Оптиксъ оръ, атреатисъ, на аглинскомъ языкѣ.

and,

312) Оптика, на аглинскомъ языкѣ.

In this case, however, the anonymity need not necessarily be ascribed to the negligence of the catalogue's authors. The first edition of this work, which appeared as *Opticks: or, A Treatise of the Reflexions, Refractions, Inflexions and Colours of Light* . . . (see fig. 20) had no name on the title page, though Newton attached the initials I. N. at the end of the advertisement.[27]

Both this work and the *Optical Lectures,* therefore, fall outside the period of Bruce's stay in England, though Bruce could have read of Newton's optical discoveries before either work appeared in the *Philosophical Transactions* published by the Royal Society, which are entered in the catalogue under six separate entries, comprising at least twenty-six volumes:

I. 306) Трансакцыи филозофическія, на аглинскомъ языкѣ, в двухъ томахъ.

II. 310) Трансакцыонъ филофической, на аглинскомъ языкѣ в трехъ томахъ.

III. p. 226, 11) Философическія трансакціи на ноябрь мѣсяцъ, 1693, въ полдестъ, на аглинском языкѣ.

26. The *Optical Lectures* were first published in Latin in 1729, a year after the English translation of the original manuscript. In the preface to the English translation it is stated that "the present treatise is a faithful Translation of a very correct Copy, taken from the Latin Original as it was read in 1669 . . ." (cf. Gray, no. 190). But in actual fact, the anonymous English translation omits the second half of the original Latin text entirely. Realizing this, S. I. Vavilov, late President of the USSR Academy of Sciences, translated the entire Latin version into Russian (*Lektsii po optike,* 2nd ed. [Moscow and Leningrad: Akademiia nauk, 1946; 1st ed., 1927]).

27. A second edition of the *Opticks* appeared in 1717, a reprint in 1718, a third edition in 1721, a fourth edition in 1730.

IV. p. 226, 14) Философическія трансакціи не генварь и февраль месяцы 1730, на аглинскомъ языкѣ.

V. 501) Фило зо фикаль трансакцыонъ, на аглинскомъ языкѣ.

VI. 475) Философикаль трансакцыонъ, на аглинскомъ в двадцати шести томахъ.

Now the first printed communication by Newton is contained in the *Philosophical Transactions* for February 19th, 1672, in which he tells of his "New Theory about *Light* and *Colors:* Where *Light* is declared to be not Similar or Homogeneal, but consisting of difform rays, some of which are more refrangible than others . . ."[28] Volume VII of the *Transactions* for the same year also contains Newton's famous communication concerning his "New Kind of Telescope," which at the time created far more interest than the revolutionary theory of light. Succeeding volumes of the *Transactions* contain a series of Newton's communications, which indicate accurately enough the course of the scientific battle and Newton's more important contributions:

Farther suggestions about his reflecting telescope: Together with his table of apertures and charges for the several lengths of that instrument	VII, 4032
	I, 200
Answer to some objections made by an ingenious French philosopher, to the new reflecting telescope	VIII, 4032
	I, 200
Some considerations upon part of the letter of M. de Vercé concerning the catadioptrical telescope pretended to be improved and refined by M. Cassegrain	VII, 4056
	I, 204
Experiments proposed in relation to Mr. Newton's theory of light: With observations	VII, 4059
	I, 135
Answer to Mr. Pardie's letter on Newton's theory of light	VII, 5014
	I, 142
Answer to some considerations on Newton's doctrine of light and colours	VII, 5084
	I, 202
Answer, further explaining his theory of light & colours	VIII, 6087
	I, 158

28. *Philosophical Transactions: Giving some Account of the Present Undertakings, Studies, and Labours of the Ingenious in many considerable parts of the World,* VI (No. 80, 1671), 3075.

Hopes of perfecting telescopes by reflections rather than re-
fractions .... *Ibid.*
On the number of colours .... *Ibid.*
Answer to Mr. Linus's letter, animadverting on the theory of
light and colours .... IX, 218
                                                         I, 161
Considerations on Mr. Linus's reply .... X, 500
Another letter on the same argument .... X, 503
A particular answer to M. Lucas's letter .... XI, 556
                                                         I, 163[29]

These issues of the *Transactions* contain Newton's more important writings on light, many of which are not published elsewhere. Others also contain Newton's contributions on other subjects: on chronology, for example, or his invention "of an instrument for observing the Moon's distance from the fixed stars at sea."[30]

How many of these volumes were in Bruce's possession? Of the entries listed in his catalogue, No. VI is no doubt the most significant, for it comprises twenty-six volumes. There is no way of telling, however, from the transcription itself whether these are consecutive. If they are, and if the set is complete, volume XXVI would take the unabridged *Transactions* to 1703, which is to say that Bruce had all of Newton's major contributions to the Royal Society in his library. This can be no more than an assumption because by the turn of the century, as the *Transactions* became a profitable publishing undertaking, a variety of abridgements came into being, to any one of which the Russian entries might refer. Some were published in popular three-volume condensations (and similar formats). Thus, the second entry in Bruce's catalogue—(310) V Trekh TOMAKH—might well be John Lowthorp's successful edition of 1716, which carried Newton's imprimatur as president of the Royal Society.

THE NEWTONIAN TELESCOPE

The most important immediate outcome of Newton's work on optics and light was undoubtedly his telescope, and it is here that we can best trace Newton's direct influence on Bruce. For the history of astronomy, the invention of the Newtonian reflector opened a new

29. Editions cited according to Paul Henry Maty's *General Index to the Philosophical Transactions from the first to the end of the Seventieth Volume* (London, 1787).
30. *Philosophical Transactions*, XIII, 155; VIII, 129.

era, banishing the large, cumbersome, and expensive telescopes of Hevelius' day. The idea behind it was essentially simple. Instead of using an objective lens, which invariably produced chromatic aberration, Newton allowed the image to be thrown on a convex speculum situated at the base of the tube. It was reflected to a second mirror (later, Newton used a prism) and then to the viewer's eyepiece, placed at right angles to the horizontal axis of the tube (at the end opposite the speculum; see fig. 40).

Ingenious as the idea was, it was basically an extension of the principles involved in Newton's theory of light, and the success of the telescope seemed to confirm its general truth. In Russia this advanced type of telescope was not, or so it is generally assumed, used till after the opening of the Academy of Sciences in 1726, when the French astronomer Joseph Nicolas de l'Isle organized the Academy's observatory along "modern" lines.[31] But Bruce was already familiar with the Newtonian reflector; this is established beyond doubt by an item in the catalogue:

39, p. 189)   Зрительная трубка Неутоновой инвенціи. Цена 71 рубль 70 копеек. Тамесу не плачено.

In addition to this telescope of "Newtonian invention," the inventory cites an impressive number of other telescopes, instruments, oculars, objective lenses, varying in focal length from twenty-seven to a few inches. Unfortunately, few technical details are supplied. Where a description is given (as in the case cited above), it is due, one suspects, to the bill attached to the instrument. Hence, next to the Newtonian telescope another item is noted, and here, too, the creditor is "Tames":

40, p. 189) Гидоставикальныя вѣски. Цена 9 рублей 15 копеекъ. Тамесу не плачено.

This "Tames" was evidently the source for the Newtonian telescope. His identity is partly revealed by the poet Antiokh Cantemir, who was Russia's emissary in London between 1732 and 1738. Cantemir acquired the services of a certain "Jean Thomas" as his

31. The earliest reference to a Newtonian telescope, according to P. P. Pekarskii (*Istoriia Imperatorskoi Akademii Nauk v Peterburge*, 2 vols. [St. Petersburg, 1873], I, 130), is 1735, when Empress Anne acquired a seven-foot model from Joseph de l'Isle. Pekarskii is supported by B. A. Vorontzov-Vel'iaminov (*Ocherki istorii astronomii v Rossii* [Moscow: Gosudarstvennoe izdatel'stvo tekhniko-teoreticheskoi literatury, 1956]); and also by Raikov in the *Ocherki po istorii geliotsentricheskogo mirovozzreniia v Rossii*, p. 194.

instructor in mathematics. In a letter to Lord Harrington on February 18, 1735, he describes him as a "cy-devant capitaine d'artillerie au service de sa majesté britannique."[32] Three years later, in a note to the historian V. N. Tatishchev, "Toms" crops up again in Cantemir's correspondence.[33] Finally "Thomas," alias "Toms," alias "Ivan Tames," appears as a signatory to a petition addressed to Peter the Great in 1720, in which Tatishchev—a protégé and intimate friend of J. D. Bruce—was also mentioned.[34] Presumably, therefore, Captain John Thomas was the latter's contact in London, through whom he acquired both the Newtonian telescope and the balances. Neither instrument was ever paid for, probably because Bruce received the items not long before his death. All available evidence suggests that he began to preoccupy himself seriously with optics only after 1726, when he resigned his position as director of the College of Mines and Manufactures, retired from politics and his official duties, and assumed the life of a scientist and astronomer on his large estate near Moscow.

BRUCE'S NEWTONIAN TELESCOPE

This is confirmed by the recent recovery of a series of six letters, five of them to the German scientist Johann-Georg Leutmann, which Bruce wrote between January 1726 and February 1731.[35] The letters to Leutmann are particularly revealing, for in light of Newton's works on optics, as listed in the catalogue, the scope of Bruce's interests and professional competency can be established.

Bruce, it appears, was working on the construction of a telescope. "As for the making of catadioptrical tubes," he writes to Leutmann on August 17, 1727, "it should be recalled that they are not as easy to

32. L. N. Maikov, "Materialy dlia biografi kn. A. D. Kantemira." in *Sbornik otdeleniia russkago iazyka i slovesnosti Imperatorskoi Akademii Nauk* (St. Petersburg), LXXIII (1903), 27.

33. Cantemir to V. N. Tatischev, March 1738, *ibid.*, p. 104.
О требуемых от вас часах я по си пору ни от кого ничего не слыхал, и понеже не ведаю, к кому господин Томсъ о том сюды писал, не могу никакой услуги вам в сем показать; к тому жь, будут ли часы к вам отправленны, теперь в Самаре вас застать не будут.

34. See Chapter 13, n. 14, below.

35. V. L. Chenakal, ed., "Pis'ma Iakova Vilimovicha Briusa k Iogannu-Georgu Leitmanu," in *Nauchnoe nasledstvo*, Vol. II, ed. S. I. Vavilov (Moscow: Akademiia nauk, 1951). pp. 1083–1101.

make as some imagined, and it is hardly surprising that the English dislike making them. Nevertheless, this summer I ordered two telescopes of that kind to be made, one of metal, the other of glass, altogether only six inches in size, and the greatest difficulty consisted, in the first place, in finding the composition of the metal, whose microscopic pores (*poros*) might be polished as easily as glass."[36] What kind of telescope Bruce was referring to is indicated in the next letter of July 27, 1729, where he thanks Leutmann for sending him an example of a "Speculum Newtoniani."[37]

The correspondence is evidently incomplete. In the two years that elapsed between the two published letters, Bruce presumably continued his experiments with alloys, which he hoped could be used in making the metallic mirror for his telescope. What he had in mind and the problems he faced may be realized by remembering Newton's difficulties when he built his own celebrated prototype.

Newton made his first telescope some time during 1668. In response to an urgent request from the Royal Society, he constructed a second one (which is still in its possession to-day and one of its greatest treasures). He made the alloy from which the speculum was formed and carried through the entire process of construction, polishing the mirror himself and also making the mounting. The whole telescope, a little less than six and a quarter inches in length, gave a magnification of approximately thirty-eight times. It aroused great excitement in London and scientific circles on the Continent.[38] Newton promised to send the Society an account of the discovery which led him to make the telescope—the discovery being, of course, the decomposition of white light into colours by the prism. This paper appeared in the *Philosophical Translations* for February 19, 1671/72; in the *Opticks* Newton returned to the technical problems the immense success of his telescope had raised.

Using instead of Object-glass a concave Metal," wrote Newton, he

36. *Ibid.*, p. 1089. The letters are printed in Russian translation, but were originally written in German.
37. *Ibid.*, p. 1090. On Johann Georg Leutmann (born in 1667 at Wittenberg, died in 1736 at St. Petersburg), see *Russkii biograficheskii slovar'* (St. Petersburg, 1914), x, 175–176; also MIAN, Vols. 1–III, VI, VIII (St. Petersburg: Akademiia nauk, 1885–1900), passim. Six of his books are listed in Bruce's library, including *Anmerkungen zum Glasschleifen* (Wittenberg, 1729), no. 867; *Instrumenta meteorognosiae inservienta* . . . (Wittenberg, 1725), no. 867; *Geometriia repetita* (sic) *oder kurzgefasste geometr. Grundlehren* (Wittenberg, 1725), nos. 1069, 877.
38. Newton's own drawings of his telescope (and related correspondence) are reproduced in the *Correspondence*, I, Plates I and II, pp. 74–75, 76–77.

found "Objects appeared much darker in it than in the Glass, and that partly because more Light was lost by Reflexion in the Metal, than by Refraction in the Glass . . ."[39] The problem, then, was to find a suitable material for the speculum which might combine high reflectivity, good thermal conductivity, and ease of manufacture. The main drawback of glass was that, as a poor conductor of heat, it tended to lose its true shape owing to uneven expansion and contraction. Its poor thermal conductivity causes a temperature gradient during changes of heat and cold, particularly when those changes are sudden. Thus, during a fall of temperature, the edge loses heat more rapidly than the center of the mirror or lens, owing to the additional radiating surface at its periphery. Consequently, such a speculum, if concave, will flatten, and its focal length will increase. Metallic mirrors, however, stand up better to climatic extremes because, by responding almost immediately to temperature changes, they simply enlarge or contract, as the temperature rises or falls, without any appreciable change in the figure. Metal, nevertheless, had other drawbacks. In Newton's own words, "because Metal is more difficult to polish than Glass, and is afterwards very apt to be spoiled by tarnishing, and reflects not so much Light as Glass quick-silver'd over does": for the future, therefore, Newton suggested that "instead of the Metal, a Glass ground concave on the foreside, and as much convex on the backside, and quick-silver'd over on the convex side" be used.[40] Further,

The Glass must be every where of the same thickness exactly. Otherwise it will make Objects look colour'd and indistinct. By such a Glass I tried about five or Six Years ago to make a reflecting Telescope of four feet in length to magnify about 150 times, and I satisfied myself that there wants nothing but a good Artist to bring the Design to perfection. For the Glass being wrought by one of our London Artists after such a manner as they grind Glasses for Telescopes, though it seemed as well wrought as the Object-glasses use to be, yet when it was quick-silver'd, the Reflexion discovered innumerable Inequalities all over the Glass. And by Reason of these Inequalities, Objects appeared indistinct in this Instrument. For the Errors of reflected rays caused by any Inequality of the Glass, are about six times greater than the errors of refracted rays caused by the like Inequalities. Yet by this Experiment I satisfied myself that the Reflexion on the concave side of the Glass, which I feared would disturb the Vision, did no sensible prejudice to it, and by consequence that nothing is wanting to perfect

39. Newton's *Opticks,* Book I, pp. 91–92.
40. *Ibid.,* p. 94.

these Telescopes, but good Workmen who can grind and polish Glasses truly spherical.[41]

Owing to the extremities of heat and cold, it was particularly important to solve the problem of finding a good material for the speculum in a place such as St. Petersburg or Moscow; and it was this that clearly preoccupied Bruce. In the two telescopes he built, he tried both the solutions Newton had suggested, inserting a metallic speculum in one telescope and a glass speculum in the other. He even tells Leutmann in one of the surviving letters that *"the English author both gave rise and directed the course of these experiments* [with the lenses (emphasis mine)]."[42] Though the name is not indicated, it is clear that Newton is "the English author" Bruce had in mind. It is equally clear that he was familiar with the details of Newton's achievement, for in another letter to Leutmann, he writes: "Had I known that your Excellence was engaged in using Newton's composition for the metal, I would have written to you about it."[43]

Bruce was making independent experiments on the very same lines, and he tells Leutmann that "the white composition has more pores than all the other alloys, and polishing the metal, when the same method is used as with the glass, requires a good deal of work. Nevertheless," he adds, "one must say that the metallic (mirrors) give better results than the Glass ones . . ."[44]

Bruce may have known of Newton's own experiments with alloys from the *Philosophical Transactions*. A particularly graphic passage may be found in a letter Newton sent to Oldenburg, where he gives full particulars of the composition he used:

The way wch I used it is this. I first melted the Copper alone, then put in ye Arsenick, which being melted I stirred them a little together, bewaring in the meane time that I drew not in breath neare the pernicious fumes. After that I put in the Tin, & again, so soon as that was melted, wch was very suddenly, I stirred them well together, & immediately powered them of.[45]

Bruce also used arsenic, though the specific composition is not revealed:

41. *Ibid.*, pp. 94–95.
42. Bruce to Leutmann, January 8, 1726, in Chenakal, ed., "Pis'ma Iakova Vilimovicha Briusa k Iogannu-Georgu Leitmanu," p. 1088.
43. *Ibid.*, p. 1090.
44. *Ibid.*, pp. 1089–1090.
45. Newton to Oldenburg, January 29, 1671/72, from Cambridge, in the *Correspondence*, I, 84.

Here I am sending two small specula, of which the lesser is made with arsenic, and the other without. Please send them to Professor Leutmann and tell him that I managed just last year to make a *Tubus Catadioptricus.* Both these mirrors, which I made at home, are of different type. One is made of metal and arsenic; the other is made of glass. The glass speculum is much better because the polishing of the metal is harder than that of the glass.[46]

Bruce adds that he also used a metallic composition for the smaller mirror required by the Newtonian telescope—the mirror which reflects the image of the speculum to the eyepiece on the side of the tube. This, however, "was not as effective as a prism of mountain crystal," presumably because "mountain crystal" (probably quartz) had a lower coefficient of expansion.[47]

Neither the date nor the recipient of the letter are known. But that Bruce actually completed his Newtonian telescope is suggested by the recent recovery of a metallic speculum at the Hermitage Museum in Leningrad.[48] The reverse side of the speculum has an inscription (see figs. 23, 24) which, translated, reads:

> Made by the own striving of Count
> Iakov Vilimovich Bruce:
> in the year 1733, in the month of August.

Bruce died less than two years later, on April 19, 1735, the first constructor of a Newtonian telescope in Russia.

46. Chenakal, ed., "Pis'ma Iakova Vilimovicha Briusa k Iogannu-Georgu Leitmanu," pp. 1094–1096.
47. *Ibid.*, p. 1096.
48. Recovered by V. L. Chenakal and reproduced in the *Astronomicheskii Zhurnal,* XXVIII (1951).

7

Newton and the Naval Academy

NEWTON AND ANDREW FERQUHARSON

A few years after Bruce constructed his Newtonian telescope, a work published by the Naval Academy in St. Petersburg credited Newton on its title page with books he had in fact never written. This curious work, whose author was given as "Andrei Farvarson," turns out to be one of the more important surviving clues in tracing Newton's influence through the pedagogic establishments set up in the Petrine era. As we have seen, Bruce founded the Navigation School in Moscow on his return from London, and it became in turn the nucleus for a network of mathematical schools around the country, eventually giving birth to the Naval Academy.

This Academy was presided over by a friend of Bruce, Andrew Ferquharson, who was known in his adopted country as "Andrei Danilovich Farvarson." He had taken Bruce's place when the latter relinquished his charge of the Navigational School in 1703 and later was transferred from Moscow to the new Naval Academy in St. Petersburg, which he guided for more than two decades. He was recruited into Russian service during the Tsar's stay in England, together with John Dean, the master shipwright with whom Peter worked at Deptford,[1] and John Perry, the engineer. Perry, who lived and worked in Russia for many years, described Ferquharson as the first person in

1. John Dean wrote an account of his experiences in Russia, now in the possession of Sir John Ingram, to whom I am grateful for letting me see it.

Muscovy to teach mathematics. He is similarly described in Russian sources. Thus, a recommendation by the Admiralty Board in 1737 requesting his promotion to the rank of brigadier speaks of Ferquharson's selflessness and of his merits in introducing "the first teaching (*obuchenie*) of mathematics" into Russia.[2] It goes on to say that "almost all of the Russian subjects in Her Imperial Majesty's fleet, from the highest to the lowest, [had been] taught seafaring sciences through him."[3] Yet it is not wholly clear how Peter came to hear of Ferquharson and to invite him to Moscow. It is generally assumed that he was recommended to the Tsar by Sir David Mitchell,[4] a fellow Aberdonian, for whom Peter developed a great affection; it is perhaps more likely that Bruce was involved too, since it was in his School of Mathematics and Navigation that Ferquharson first worked.

Two young assistants, Stephen Gwyn and Richard Grice, both of whom had just completed their course at the Royal Mathematical School of Christ's Hospital, were also recruited to help Ferquharson. This school, founded in London by Charles II in 1673, was probably the first nonclassical school in the world in that it had a scientific bias and prepared its pupils for the technical professions. Indeed, most of the mathematical masters in the Royal Navy in the eighteenth century were its alumni, and it has been convincingly suggested that the school in Moscow was modeled on it.[5] Certainly the English school was well known in the naval circles in which Peter liked to move with his friend Vice Admiral Peregrine Osborne. Marquis of Carmarthen. It also had a connection with the Greenwich Observatory and John Flamsteed, who was responsible (as we have seen) for training two pupils from Christ's Hospital every year. The task of inspecting the school and recruiting Gwyn and Grice, then fifteen and seventeen years old, respectively, was entrusted to Dr. Pëtr Postnikov.

Postnikov's activities in England are obscure. It is almost certain, as Andreev suggests, that he visited Oxford, but Postnikov could not

2. Cf. Al. Sokolov, "Andrei Danilovich Farvarson," *Morskoi Sbornik*, No. 15 (December 1856), 173.

3. *Ibid.*, p. 173.

4. Admiral Mitchell was one of the Tsar's constant attendants in England, but the connection with Ferquharson rests on mere conjecture.

5. Similar schools, for instance, the Mathematical School in Halle (1708) and the School of St. Jean Baptiste de la Salle in Rouen (1705), appeared in other countries after the Moscow School had been established, but Professor E. Medynsky's claim of priority on the latter's behalf has been disputed by Nicholas Hans. See his "The Moscow School of Mathematics and Navigation (1701)," *The Slavonic (and East European) Review*, XXIX (1950), 532.

have recruited Gwyn and Grice at "Christ Church College"—"gospital' tserkvi Khristovoy" clearly refers to the school ("Christ's Hospital").[6] They arrived in Moscow together with Ferquharson in August 1699, but, according to F. F. Veselago in his *Ocherk istorii morskogo kadetskogo korpusa* (1852), they were not paid their salaries and were more or less forgotten until they complained officially of their neglect to Peter. The Tsar was then engaged in preparations for a major war with Sweden and Turkey, but he issued an *ukaz* on January 14, 1701, establishing the Moscow School, the primary purpose of which was to provide instructors in mathematics and navigation and to train qualified naval personnel; subsequently, architects, artillery officers, teachers, engineers, civil servants, clerks, artisans, and others were also included among its graduates.

This was made possible by dividing the institution into preparatory departments and the naval school proper. The former were called the Schools of Russian and Ciphering, their best pupils being transferred to the naval department where they studied English, algebra, geometry, trigonometry, geography, astronomy, navigation, and surveying. In 1715 the Naval Department was moved from the Sukharev Tower in Moscow to St. Petersburg.[7] The Russian government employed Ferquharson in many technical duties, though it is as a teacher that he left his mark. His writings, which represent some of the earliest contributions to mathematical knowledge in Russia, were used as manuals of instruction by several generations of students at the St. Petersburg Naval Academy.[8] The extent of his influence was due in

6. They may have brought with them some of the elementary text books on navigation, arithmetic, geometry, etc., to be found in the USSR Academy of Sciences library in Leningrad. The following manual by Perkins, for example, was also used in Christ's Hospital: *The Seamans Tutor, Explaining Geometry, Cosmography, and Trigonometry: with Requisite Tables of Longitude & Latitude of Seaports, Traverse Tables, Tables of Easting and Westing* . . . (London, 1682). This particular copy contains the signature of William Hill, an English mathematician whom Postnikov may have met in his quest for instructors to teach mathematics in Russia. The Tsar furnished him with funds for this purpose.

7. Richard Grice was dead by this time. He had been killed by robbers in a Moscow street one night in January 1709. Gwyn, who died in 1720, accompanied Ferquharson to St. Petersburg.

8. Most of Ferquharson's works were used in MSS form. Two of his books were brought out by Magnitskii, author of the first Russian manual on arithmetic: a book on logarithms (the earliest of its kind) appeared in 1716 (as "Logarifmy"); the other one in 1722. Its contents are adequately described by the convoluted title:
Таблицы горизонтальных северныя и южныя широты восхождения солнца, с изъяснением: чрез которые зело удобно, кроме труднаго арифметического

part to his longevity and merits as a teacher and in part to his extraordinary versatility as a pedagogue. After his death in 1739, the Admiralty Board complained of the difficulty of finding an adequate replacement, for he taught at least ten subjects, including arithmetic, geometry (plane and solid), trigonometry, astronomy, and geography.

Newton's doctrines found some of their earliest exponents, such as David Gregory, in Scotland, and it would be valuable to ascertain something of Ferquharson's intellectual background before he left for Moscow. Unfortunately little is known. At the University of Aberdeen, Ferquharson was Milne Bursar at Marishal College in 1691–1695 and, after graduation, was Liddell Mathematical Tutor at the College for two or three years before he left for Russia. He probably took over Bruce's duties as director of the Moscow School when the latter resumed his military career in 1703, and he remained there until 1716, when he was transferred to the Naval Academy in St. Petersburg. It appears that Peter was on intimate terms with Ferquharson, and according to Veselago he also belonged to the secret club of which the Tsar, Bruce, Menshikov, and others were members—the "Society of Neptune."[9] Unlike many foreigners who came to Russia in the Petrine era, Ferquharson mastered the Russian language sufficiently well to teach in it. This much may be gathered from the fact that, when the time came to seek a successor, the prospect of finding one in England was considered slim because no one there, it was assumed, could instruct in Russian as he had done.[10]

Of Ferquharson's contacts with Bruce, close though they probably were, too little is known to be of appreciable help in establishing the tenor of his own cosmological and physical views. Though he edited or "corrected" several works of a scientific nature, few of them ap-

---

исчисления неправильное или непорядочное указание компасовъ ... чрез сихъ-же зело удобно найти и скоро зело возможно тем, которые в восточную Индию морешествуют ...
This was published by Vasilii Kipriianov in Moscow "pod protektsieiu . . . ego siiatel'stva vysokorozhdennago grafa Iakova Vilimovicha Briusa."

9. The society may have had Masonic connections; indeed, Russian Masons believed that Masonry was introduced into Russia by Peter the Great. See A. N. Pypin, *Russkoe masonstvo* . . . (Petrograd: Ogoi, 1916), p. 88. According to *Latomia XIII, Handbuch III,* p. 106, the Tsar was accepted into Masonic society by Newton's friend, Sir Christopher Wren, in 1698, and on his return he established a lodge with Lefort as Venerable Master and Patrick Gordon as Warden. There is no confirmation of this in Dr. Moritz Posselt's *Der General und Admiral Lefort—sein Leben und seine Zeit,* 2 vols. (Frankfurt am Main, 1866), II. Posselt does refer to Lefort's interest in natural philosophy and alchemy.
10. Sokolov, "Andrei Danilovich Farvarson," p. 174.

peared in print. Ferquharson's claim to be considered a Newtonian rests on a curious translation by Ivan Satarov, published under the following title:

Эвклидовы элементы изъ двенатцати нефтоновыхъ книг выбранныя, ь осмь книгъ чрезъ профессора маөематіки Андрея Фархарсона сокращенныя . . . the translation of which is: *Elements of Euclid, selected from Newton's twelve books, and condensed to eight books by Professor of mathematics Andrew Ferquharson* . . . [see fig. 30].

There are several intriguing features to this work. To begin with, there is the title itself. Newton was never responsible for any edition of Euclid. That Ferquharson should have "condensed" his "twelve books" into "eight" (as the title states) is, therefore, false. Equally misleading is the claim that Ferquharson was responsible for the selection. For Satarov's translation was in fact taken from a popular abridgement belonging to Andrew Tacquet, a Jesuit and distinguished mathematician, whose edition of Euclid and Archimedes was first published in 1654, six years before the author's death.[12]

Reprinted several times thereafter, Tacquet's work was still in use at the end of the eighteenth century. The Russian edition of 1739 does not even mention the Jesuit's name, though its dependence on Tacquet's text can be established easily enough. In 1745 Satarov brought out his translation of certain theorems by Archimedes, which were originally published by Tacquet as a supplement to the *Elements* (included in the same volume). In the Russian edition of Archimedes, the debt to Tacquet is freely acknowledged; moreover, the pagination (pp. 286–475) continues from that of the *Elementy* (as in the Latin edition of the Jesuit's work). Why then was Tacquet's name omitted in the translation of the first part of his book in 1739? And why was Newton credited with the original for the Russian edition, of Euclid?

Dr. A. P. Iushkevich suggested in 1948 that Ferquharson may well have been responsible. He may, "out of vanity," have removed Tacquet's name. Iushkevich further surmises that, after Ferquharson died in 1739, the identity of the real author could be revealed without

11. *Ibid.*, p. 173.
12. The first edition of Cl. A. Tacquet's work is rare. I have used the corrected second edition of 1665: *Elementa Geometriae Planae ac solidae. Quibus accedunt selecta Ex Archimede Theoremata. Auctore Andrea Tacquet Societatis Iesu Sacerdote & Matheseos Professore . . . Antverpiae, apud Iacobum Meursium.* This, and several later editions. may be found in the British Museum. There is no indication in the Russian translation which edition Satarov used.

hindrance.[13] The truth is that Ferquharson cannot even be credited with "condensing" Euclid into "eight books"; Tacquet had already done this in his Latin edition. And yet, as Iushkevich himself admits, the *Elementy* are not merely a translation from the Latin. There are several major changes in the Russian for which Ferquharson was possibly responsible, though even this did not seem entirely certain to Iushkevich because of further ambiguities in the title of the Russian edition of Archimedes.

The title not only credits Tacquet for the selection of the theorems published, but it also mentions "Georg Petr Domkio" as the one responsible for "condensing" them (as Ferquharson had been credited for doing in the edition of Euclid):

Архимедовы теоремы Андреем Таккветом езуитом выбранные и Георгием Петром Домкио сокращенные с латинского на российский язык хирургиусом Иваном Сатаровым преложенные. Напечатаны при Санкт-петербурге в Морской Академической Типографии Первым Тиснением 1745 лета.

which is, *The theorems of Archimedes selected by Andrew Tacquet the Jesuit and condensed by Georg Petr Domkio, put into the Russian tongue from the Latin by Ivan Satarov, the surgeon. Published at St. Petersburg, in the typography of the Naval Academy,* [with this] *first edition in the year 1745.* But who was Domkio? His name (as Domki*no*) also appears once in the *Elementy,* at the end of the introduction, which provides a historical sketch of the development of mathematics down to Descartes and Gassendi. Domkio's actual identity long remained wrapped in mystery, however.

Bobynin, in his excellent bibliography of Russian works on physics and mathematics, merely describes "Domkino" as the author of "an interesting introduction" to the *Elementy.*[14] More recently, Dr. Iushkevich confessed his ignorance of this author's works, further lamenting that he had not been able to find the Latin original of the edition of Tacquet which "Domkino" had "condensed." It seemed to him that this author, rather than Ferquharson, may well have been ultimately responsible for the variations of the Russian version of the *Elementy* from the one initially brought out by Tachquet.[15] There-

13. Cf. A. P. Iushkevich, "O pervom russkom izdanii trudov Evklida i Arkhimeda," TIIE, II (1948), 570.
14. Cf. V. V. Bobynin, *Russkaia fiziko-matematicheskaia bibliografiia,* I, Part II, *Knigi i stat'i vyshedshie v gody 1726–1745,* p. 67.
15. Iushkevich, "O pervom russkom izdanii trudov Evklida i Arkhimeda," p. 570.

fore, Ferquharson's honor, already blemished by the charge that he omitted Tacquet's name in order to enhance his own credit, was further exposed to the suggestion that he was not even responsible for the changes in the Russian translation. Conceivably these could be traced to "Domkino's" "condensation" of Tacquet's abridgement of Euclid.

And in the absence of any primary evidence, apart from the title, concerning Ferquharson's association with the *Elementy*—the manuscript has not apparently survived—it might have been tempting to let the argument rest where Iushkevich left it. The absence of any positive proof was one weakness of the argument; it was also its strength. "Georg Petr Domkino" could not be traced. As it later transpired, this was hardly surprising. No such author existed.

"DOMKINO" ALIAS GEORGE PETER DOMCKE

In an article appearing after that by Iushkevich, I. Ia. Depman identified "Domkino"; he was none other than George Peter Domcke, author of a two-volume work expounding the Newtonian system. This book is in Latin, and on the title page Domcke's name is given in its ablative form as Domckio. The Russian translator presumably took this to be the nominative and transliterated accordingly (adding an "n" by way of orthographic error in the one place where his name appears in the *Elementy*). But was Georg Petr Domkio, alias Georgius Petrus Domckius, actually responsible for a "condensed" edition of Tacquet's abridgement of Euclid, as Iushkevich had implied? Depman did not take his inquiries beyond establishing "Domkino's" identity. Yet this is sufficient to vindicate Ferquharson's reputation from some of Iushkevich's aspersions. For Domcke did not "condense" any edition of Tacquet's Euclid, which implies that Ferquharson, and no one else, was responsible for the changes in the Russian translation.[16]

This being so, is it fair to assume that Ferquharson was also responsible for the title, with its omission of Tacquet's name and its inclusion of his own and Newton's? Was it merely "out of vanity" that Ferquhar-

---

16. Cf. C. G. Thomas, later Thomas-Stanford (Sr C.) Bart., *Early Editions of Euclid's Elements* (London: the Bibliographical Society, 1926). This contains a useful bibliography. The British Museum, which houses a rich collection of Euclid editions (some not listed by Thomas), has two works by Domcke, one of them a translation, but he was not, it appears, associated with any edition of Euclid.

son chose to have Newton's name included in the title of a work to which the author of the *Principia* bore no direct relation?

To make this claim is to imply that Ferquharson wished to impress his colleagues by associating his own with the weight of Newton's reputation. Yet if it stood so high in Moscow and St. Petersburg, was there not a better way of sharing in his reflected glory than by ascribing to Newton "twelve books" of a work he never wrote? There is a more credible explanation.

The misleading title is not the only "fraudulent" claim in the *Elementy.* We have already mentioned the book's "interesting introduction," to which both Iushkevich, Depman, and other scholars refer: what they have all overlooked is that, though it is attributed to "Domkino," the original of the Russian text is actually to be found in Tacquet's *Elementa Geometriae.* This work contains an identical introduction, of which Tacquet and not Domcke is the real author.[17] The Russian introduction is merely a translation of the "Historica Narratio—de ortu & progressae Matheseos," preceding Tacquet's exposition of Euclid. The question arises, therefore: Who was responsible for substituting "Domkino"? If we assume that Ferquharson wanted to steal Tacquet's laurels and claim them for his own, it is odd surely that "Domkino" should have been brought in to share them along with Newton. The answer must lie elsewhere.

Georg Peter Domcke is chiefly remembered for his exposition of Newtonianism, which he published in 1730, when such works were still relatively uncommon. It is hardly likely that he could have come to Ferquharson's attention any other way, for Domcke published almost nothing else. He was, however, a friend of William Whiston, as can be determined from Whiston's introductory note to the *Philosophiae Mathematicae Newtonianae illustratae* (see fig. 31). There Whiston praises Domcke's opus for its value as an introduction to Newtonian philosophy: "Perlegi hujusce libri quae mihi visa sunt ad ferendum de eodem judicium necessaria: Et Authorem P H I L O S O P H I A M N E W T O N I A N A M tum clare intellexisse, tum perspicue explicasse; quin & Elementa Mathematica, eidem Philoso-

17. In Tacquet's *Elementa Geometriae Planae ac solidae,* the *Historica Narratio* follows some preliminary remarks addressed to the reader, and precedes the imprimatur by Guillielmus Bolognino. The original account (unnumbered) takes up eleven pages and is somewhat condensed in the Russian translation (pp. 1–7). It leans heavily on Petrus Ramus, who was Milton's favourite pedagogue and represents the first such essay in Russian to acquaint readers with the historical development of mathematics.

phia necessario praemittenda, breviter satis & feliciter exposuisse censeo[.]" Such attempts to expound the Newtonian system had appeared as early as 1703, when John Keill at Oxford brought out his manual on it, and other comprehensive works were devoted to it at a later date by Whiston himself and by David Gregory, all of whom were duly praised by Domcke in his own preface. What made Domcke's own treatment of Newton's "mathematical philosophy" so different from other commentaries, however, was its scope. His second volume opens with Newton's celebrated "regulae philosophandi," taken from the third book of the *Principia*. The rest of the volume carefully summarizes its contents, and, after explaining Newton's laws of motion, of gravity and centripetal force, closes with a précis of the *System of the World*.

But Domcke's first volume is not directly concerned with the *Principia* at all, but rather with the mathematics on which this work is based. His purpose, as Domcke says in the introduction to the first volume, was to lay bare the foundations, to make it easier for readers to follow Newton's own argument. This part of his book therefore covers some of the same ground as the *Arithmetica Universalis*. It begins by defining elementary mathematical procedures and then proceeds step by step to an analysis of more elaborate ones, for example, geometry, plane and solid, the use of logarithms, algebra, and Newton's method of "fluxions" (I, pp. 81–97). In other words, the first volume of the *Philosophiae Mathematicae Newtonianae illustratae tomi duo* does exactly what the subtitle claims: it deals with the mathematical elements (Elementa Matheseωs) "ad comprehendam demonstrationem hujus Philosophiae scitu necessaria . . . ," that is, with those things necessary for an understanding of Newtonian philosophy. Hence Domcke relied on Euclid, to whose *Elements* he pays obeisance both in his preface (p. vii) and in his "annotationes observatu per necessariae" (pp. xiv–xvi). And it is here that the key to the ambiguities of the title of the Russian edition is to be found.

As we have seen, the Russian title of the *Elements* refers to "Newtonian books" which Ferquharson had supposedly condensed. This is no longer so mystifying if we assume that Ferquharson did indeed attempt to provide his Russian pupils with an introduction to the mathematics of the *Principia* of the kind Domcke had so painlessly furnished in the first volume of his textbook. No other commentary of that type devoted as much attention to the principles of Euclidean geometry, an understanding of which was needed to make the proofs employed by Newton in the *Principia,* where even the results arrived at by "fluxions" are couched in geometrical form, intelligible. Even Domcke's treatment of the subject may not have been basic enough for Ferquharson's requirements, which is why it may have occurred to him to expand on it by adding material from Tacquet's abridgement of Euclid's *Elements.* This, after all, was the edition Whiston himself had chosen to bring out for similar reasons.[18] It is doubtful whether Ferquharson ever intended to translate Tacquet's entire edition of Euclid, with its inclusion of the second part containing Archimedes' theorems.

This much can be surmised from the title under which it appeared in Russian in 1745, which, prima facie, is just as puzzling as the title of the first part: "The Theories of the Archimedes selected by Andreas Tacquet and condensed by George Peter Domkio . . ." Here, Ferquharson is not given any credit at all, while Satarov is conspicuously presented as the translator. Was this intended as a slight to Ferquharson's memory? The internal evidence does not support such an interpretation. The theorems from Archimedes, as added to Euclid's *Elements* by Tacquet, comprise only a small portion of his book (pp. 277–350 in the second edition), and Satarov may well have translated them independently after Ferquharson's death. This hypothesis is supported both by the date of publication and by the fact that, unlike the *Elementy,* the *Arkhimedovy teoremy* of 1745 makes no major departures from Tacquet's text (pp. 286–475 in the Russian version).

Then why, it may be asked, did Satarov bother to include "Domkio" on the title page of the *Arkhimedovy teoremy,* when Domcke had clearly as little to do with it as with the translation of Euclid? This is probably to be explained by the circumstances under which the first part of the *Elementy* was composed. If Ferquharson had originally planned a work similar to Domcke's, it is easy to see why Domcke's shadow haunts both the title and the introduction of the *Elementy.* It is conceivable that Satarov was working from a manuscript which, at its inception, may have owed something to the *Philosophiae Mathematicae Newtonianae illustratae.* Ferquharson may well have intended to supply his pupils with an elementary text

18. Cf. Cl. A. Tacquet, *Elementa Euclidea Geometriæ planæ ac solidæ et selecta ex Archimede theoremata,* 1st Cambridge ed. (Cambridge, Eng., 1722), pp. 2–3, where Newton is duly lauded.

on mathematics and Newtonian physics similar to that by Domcke. Part of the Russian title, therefore, quite correctly reflects the provenance of his effort, whereby Ferquharson actually intended to "condense" some of Domcke's "Newtonian books." Indeed, this is indirectly confirmed by the symbols used in the *Elementy*, which correspond to Domcke's practice rather than to Tacquet's.[19] Why, then, did Ferquharson decide to abandon what may have been his original intention? It is by no means certain that he did.

According to archival records, Ferquharson left no fewer than eight "corrected" books at his death, that is, works not written by himself, but abridged, edited, or rephrased by him. These were meant for publication, and at one point the order to have them printed was actually given.[20] It was never executed, however, and the manuscripts unfortunately appear to have been lost. That is why there is no way of ascertaining to what extent Ferquharson was concerned with popularizing the Newtonian system. The title of the *Elementy* remains our most tantalizing clue of his activities in that regard. Although he wrote several other works, including one on trigonometry, one entitled "Definitions," others on navigation, algebra, and geometry, and one on an invention of his own to facilitate observations at sea,[21] only a minuscule proportion of these writings were ever published. This does not mean that Ferquharson's efforts on Newton's behalf necessarily went unnoticed. His manuscripts and manuals of instruction were used by teachers at the Naval Academy in St. Petersburg. In their judgment, Ferquharson's works were "written for those people, who were well versed in the mathematical sciences."[22]

Was Ferquharson, like Bruce, an ardent Newtonian? The surviving evidence is too scant to be sure. Of his ties with other disciples of Newton in England and Scotland nothing is known, though it would be surprising if he came to Russia without some awareness of the *Principia's* significance. The *Opticks,* it is true, was published after Ferquharson had already begun teaching in Moscow, and we have no way of knowing to what extent he remained in touch with the latest

19. For example, Tacquet preferred to use the letters AE rather than the equal sign(=). Similarly, geometrical proportions in the Russian book are indicated, as in Domcke's, in a way that was then relatively novel.
20. Sokolov, "Andrei Danilovich Farvarson," p. 173.
21. *Ibid.*, pp. 174 ff.
22. I.e., "sochineny radi takikh liudei, kotorye v matematicheshikh naukakh dovol'no uprazhnialis'."

scientific literature. No inventory is known to have been made of Ferquharson's library, which was dispersed at his death. It was thought to be a valuable collection by contemporaries. There were some six hundred items: 84 in Latin; the rest in English, French, German, and Dutch; 1 in Russian. About 78 of the works belonged to the Academy of Sciences; about 277 of them to the Naval Academy. Most of the remainder were left to Ferquharson's heir, one Alexander (by some accounts, his nephew; by others, his half-brother), and returned to England.[23] The importance of Ferquharson's role in introducing modern mathematics into Russia, though widely acknowledged, remains obscure. The same is true of the early years of the Naval Academy; apparently no records survive to indicate how natural philosophy was taught in these crucial but neglected years when Russians were first exposed to ideas associated with Newton and the rise of modern science.

's GRAVESANDE's *Introductio ad Philosophiam Newtonianam*

The lack of records is not as surprising as it may seem. There were as yet no manuals on science in Russian. Nor could there have been at a time when the vocabulary to match the abstract terminology used by Newton and his contemporaries was still in the process of painful gestation. Even so straightforward a work as the *Kosmotheoros*, in which Huygens wholly eschewed technical language, presented immense problems to the translator. Ferquharson found it necessary to devise his own texts, and his ability to do so must have seemed a valuable asset to students and colleagues.

Peter the Great was keenly aware of this problem, and after 1718, when Bruce became president of the new "Collegium of Mining and Manufacture," he commissioned J. D. Schumacher to undertake what would earlier have fallen on Bruce's shoulders. Schumacher entered the Tsar's service after studying at the University of Strassburg, which he left after being accused of atheism; in 1721 Peter sent him to England, France, Holland, and Germany to acquire instruments, technical and biological specimens of various descriptions—and books.[24] Unfortunately Schumacher did not record his travels in any great detail.

23. Cf. *Zhurnal Admiralteiskoi Kollegii* (December 1743), no. 3774.
24. On Schumacher's journey, see Pekarskii, *Nauka i literatura v Rossii pri Petre Velikom*, I, 533–558.

We do know that his quest took him to Leyden where he set out to meet Jakob 's Gravesande, a friend of Newton who also happened to be one of his earliest champions on the Continent.

In the previous year, 's Gravesande had published a pioneering textbook on natural philosophy, which was to carry his influence far beyond the Netherlands: the *Physices elementa mathematica experimentis confirmata: sive Introductio ad Philosophiam Newtonianam* (Leyden, 1720). This work was reprinted several times, and translations of it appeared in most European languages, including French and English. It exercised a considerable impression on contemporary thought, which was probably due in part to the ingenuity of its experiments,[25] but even more no doubt to 's Gravesande's priority in giving a detached and able explanation of Newton's doctrines. Certainly Schumacher went out of his way to praise the book; and when the St. Petersburg Academy of Sciences opened shortly after Schumacher's return, it adopted one of s' Gravesande's books as its official text on physics.

This fact is all the more remarkable because it appears that none of the principal founding members of the St. Petersburg Academy were themselves favorably disposed toward Newton's doctrines. The *Introductio ad Philosophiam Newtonianam* had possibly already been used in the Naval Academy; Bruce had certainly lost no time in acquiring a copy. The Naval Academy, however, was soon eclipsed as a focus of scientific activity in Russia by the St. Petersburg Academy of Sciences, where Newton's doctrines became from its very opening the subject of acrimonious debate.

25. Georg-Wilhelm Richmann, for example, took his design for a static electrical machine from this book. See G. V. Rikhman, *Trudy po fizike,* ed. by A. T. Grigorian (Moscow: Akademiia nauk, 1956), p. 604. On Richmann, see below pp. 146–151, 153–155.

PART II

Newton and the St. Petersburg Academy of Sciences

26. Title page from *Kniga mirozreniia ili mnenie o nebesnozemnykh globusakh* (St. Petersburg, 1717; Moscow, 1724).

27. First page from *Kniga mirozreniia ili mnenie o nebesnozemnykh globusakh*, with blessing added.

КНІГА
МІРОЗРѢНІЯ,
ИЛИ
МНѢНІЕ,
о
небесноземныхъ глобусахъ,
и ихъ украшеніяхъ.

___

Напечатася
въ Санктъ пітербургскои
Тvпографіи, 1717 году,
Октября 27 дня.

А въ Московскои противъ тогожъ перво
1724 году Марта въ 31 день.

ВО ИМЯ ІИСУСОВО АМІНЬ.
Господіна хрістіана гюенса.

МІРОЗРѢНІЕ
ИЛИ
МНѢНІЕ,
о
НЕБЕСНОЗЕМНЫХЪ ГЛОБУСАХЪ,
и украшеніи ихъ,
пісаное къ господіну константіну
ГЮЕНСУ.
Его Господіну брату.

___

КНІГА ПЕРВАЯ.

Нітся мнѣ не возможно
есть, дражайшіи Господіне
брате, чтобъ тому, иже по
мнѣнію Копернікову сію
землю, на неіже обіщаемъ, между
А планета...

262 О Небесноземныхъ

тягость къ солнцу, и отъ чего оная
проісходітъ, тамож показали.
И удівляюсь я толь болши что
Картезіюсъ оную обшелъ; ибо онъ
отъ тяжелости, сюже гіблеса ко
землѣ пригоняемы, прежде всѣхъ
началъ лучше пісаті, нежели до него
обыкновенно было. Плутархусъ по-
вѣствуетъ въ вышереченнои своеи
книгѣ, о лицѣ круга луны, De facie
in orbe lunæ, былъ еще древле нѣкто,
иже мыслілъ, что луна того для
въ своемъ крузѣ пребываетъ, понеже
въ кругъ движущая сіла земли, ко от-
хожденію луны равною сілою тяжелости, сюже бы къ землѣ могло
пріблізішся, отымается. И равно
томужъ и въ наши времена не
точію о лунѣ, но и о протчіхъ
планетахъ Алфонзусъ бореліусъ
учілъ. А имянно что главнѣйшыя
планеты,

Глобусахъ. 263

планеты, тяжелости къ солнцу
имѣютъ, луны же убо ко землѣ
ко Юпітеру и Сатурну: около ко-
торыхъ ходятъ многімъ тщателнѣе
и остроумнѣе. Таковоежъ Господінъ
Ісакъ Нютонъ вновь изъяснілъ, како
отъ сіхъ прічінъ эклітпіческія круги
планетъ, свое проісхожденіе имѣютъ,
въ ихже одномъ фокусъ, [точка
зажіганія] солнце мѣсто свое
имѣетъ, якоже Кеплерусъ вымы-
слілъ. Подобаетъ же убо, по мнѣнію моему, [которое о натурѣ
тяжелости вящше имѣю, и отъ
чего планеты вѣсомъ своімъ къ
солнцу клоняпся] чтобъ віхогному
кругу небесныя матеріи около
оного не вовсе на сдінои странѣ,
но тако вершілісь, да бы разлíч-
ными и зѣло прыткіми дьіженіями,
на всѣ страны по разлíчнымъ его
П 4 частямъ

28. The first explanation, in Russian, of Newtonian gravitation: "Takovoezh Gospodin Isak Niuton vnov' iz'iasnil, kako ot sikh prichin eklipticheskiia krugi planet, svoe proiskhozhdenie imeiut . . ."

ЭѴКЛІДОВЫ

ЭЛЕМЕНТЫ

Изъ двенатцати нефтоновыхъ книгъ
выбранныя,

И

въ осмь книгъ

чрезъ профессора маѳематіки
АНДРЕЯ ФАРХВАРСОНА
с о к р а щ е н н ы я;

СЪ ЛАТІНСКАГО на россійскіи языкъ
кіругусомъ іваномъ сатаровымъ
ПРЕЛОЖЕННЫЯ

НАПЕЧАТАНЫ при САНКТЪПЕТЕРБУРГѢ
въ Морскои Академіческои Тvпографіи
Первымъ Тvсненіемъ 1739 Лѣпа.

30. Title page from *Evklidovy elementy iz dvenatzati neftonovykh knig vybrannyia* (Moscow, 1739)—the first Russian title page to bear

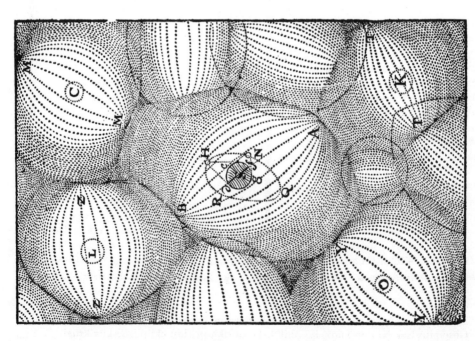

29. The Cartesian theory of "vortices"—from Descartes's *Principia Philosophiæ*.

Philoſophiæ Mathematicæ

NEWTONIANÆ

ILLUSTRATÆ.

TOMI DUO.

Quorum prior tradit Elementa Matheſeos
ad comprehendendam demonſtrationem
hujus Philoſophiæ ſcitu neceſſaria:

Poſterior continet 1) Definitiones & Leges motus generaliores; 2) Leges virium centripe-
tarum & Theoriam attractionis ſeu gravita-
tionis corporum in ſe mutuo; 3) Mundi
Syſtema.

A GEORGIO PETRO DOMCKIO.

LONDINI:

Sumptibus THO. MEIGHAN, Bibliopolæ in
vico *Drury-Lane*, & JER. BATLEY, ſub
Signo Columbæ, in vico *Pater-Noſter-Row.*

MDCCXXX.

31. Title page from George Peter Domcke's *Philosophiæ mathematicæ
Newtonianæ illustratæ* (London, 1730).

Inclitae Societatis Anglicanae Per Illustri Praesidi
et Nobilissimis Doctissimisque Sodalibus
S. D.

Academia Petropolitana.

5

Quando institutum Vestrae Societatis et praeclara quae in lucem protulit inventa tam Gallis quam Germanis ita placuerunt ut Anglorum exemplo excitati similia et ipsi collegia atque Academias fundarent, non dubitamus quin eruditi omnes summis laudibus dignum censeant Petri Magni Imperatoris consilium munificentia Augustae confirmatum de condenda Petropoli Academia quae non solum litteras in Imperio Russico parum antea cultas doceret, sed etiam medicinam et mathesin novis augeret inventis novisque methodis illustraret. Ad Astronomiae vero incrementum inprimis profuturas speramus observationes nostras quas tanto gratiores vobis fore confidimus quo rariores adhuc in Septentrione fuerunt. interea dum aliæ Academiæ specimina in iustum volumen colligantur; binas differtationes e Academicorum nostrorum ad has litteras adiungere visum fuit; nobis vero honorificentius nihil accidere poterit, quam si nostri conatus vestræ Eruditorum Societati quæ et tempore prima est et auctoritate numero atque meritis Sodalium nulli secunda, in dies magis probentur. D. Petropoli d. XI. Oct. 1726.

Christianus Goldbach
Acad. Secretar.

32. The St. Petersburg Academy's inaugural letter to Newton.

G. J. 's GRAVESANDE,

PHILOSOPHIÆ
NEWTONIANÆ
INSTITUTIONES,

In Usus

ACADEMICOS.

LUGDUNI BATAVORUM
Apud PETRUM VANDER Aa,
Bibliopolam & Typographum Academia atque Civitatis,

MDCCXXIII.
Cum Speciali Privilegio Præpotent. Ordd. Hollandiæ &
West-Frisia

33. Title page from the first Newtonian textbook to be used in the St. Petersburg Academy.

FRANCISCVS ALGAROTTVS

34a. Antiokh Cantemir (1709?–1744).

34b. Francesco Algarotti (1712–1764).

DI MY LORD HERVEY.

WHEN the gay Sun no more his Rays shall boaſt,
And human Eyes their Faculty have loſt;
Then shall theſe Colours and theſe Opticks die,
Thy Wit and Learning in oblivion lie;
England no more record her *Newton's* Fame,
And *Algarotti* be an unknown name.

34c. Lord Hervey's lines to Algarotti,
inserted in Algarotti's *Neutonianismo per
le Dame* (Naples, 1739).

ASTRONOMIÆ

PHYSICÆ & GEOMETRICÆ

ELEMENTA.

Auctore DAVIDE GREGORIO M. D.
Aftronomiæ Profeffore Saviliano OXONIÆ,
& Regalis Societatis Sodali.

OXONIÆ,

E THEATRO SHELDONIANO, An. Dom. MDCCII.

35. Title page from David Gregory's *Astronomiæ Physicæ & Geometricæ Elementa* (Oxford, Eng., 1702).

Johannis Wallis S. T. D.

Geometriæ Professoris SAVILIANI, in Celeberrima
Academia OXONIENSI,

OPERUM MATHEMATICORUM

Volumen Tertium.

QUO CONTINENTUR

CLAUDII PTOLEMÆI ⎫
PORPHYRII ⎬ Harmonica :
MANUELIS BRYENNII ⎭

ARCHIMEDIS ⎰Arenarius, &
⎱Dimensio Circuli;

Cum EUTOCII Commentario:

ARISTARCHI SAMII, de Magnitudinibus & Distantiis
Solis & Lunæ, Liber :

PAPPI ALEXANDRINI, Libri Secundi Collectaneorum,
hactenus desiderati, Fragmentum :

Græce & Latine Edita, cum Notis.

ACCEDUNT
EPISTOLÆ nonnullæ, rem Mathematicam spectantes;

E T
OPUSCULA quædam MISCELLANEA.

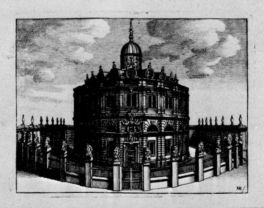

O X O N I Æ,
E THEATRO SHELDONIANO, An. Dom. MDCXCIX.

36. Title page from John Wallis' *Operum Mathematicorum
Volumen Tertium* (Oxford, Eng., 1699).

IL
NEWTONIANISMO
PER LE DAME,
OVVERO
DIALOGHI
SOPRA
LA LUCE, I COLORI,
E L'ATTRAZIONE.

———————— quæ legat ipſa Lycoris.
Virg. Egl. X.

Novella Edizione emendata ed accreſciuta.

IN NAPOLI
MDCCXXXIX.

A ſpeſe di GIAMBATISTA PASQUALI
Libraro e Stampatore di Venezia.

37. Title page from Francesco Algarotti's *Neutonianismo* (as the author spelled it in his first edition), dedicated to Empress Anne.

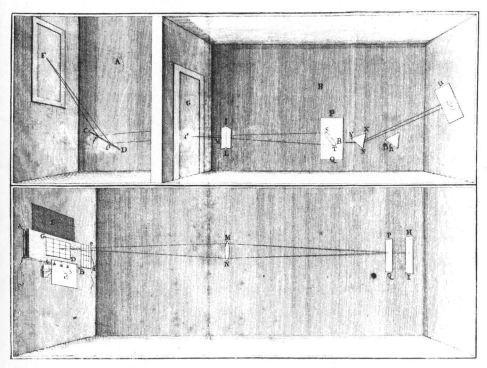

38. Sketch illustrating Newton's "experimentum crucis" from Francesco Algarotti's *Neutonianismo,* with Cantemir's commentary (from his *Fourth Ode*): "If a sunbeam is let into a darkened room through a small slit, so that it falls on a trihedral piece of glass, normally called a *prism* . . . that beam will break up into several other beams of different colors, of which one will be violet, another purple, a third blue, the fourth green, the fifth yellow, the sixth orange, the seventh red. This phenomenon was first discovered by the renowned English philosopher Newton."

"We hope that our observations will be useful above all for the development of astronomy, which should be all the more interesting to you in that up till now these have rarely been prosecuted in the North . . ."—Goldbach to Newton, 1726

8

The Foundation of the Academy
and the Death of Newton

It is commonly assumed that Leibniz had much to do with the original project of founding the St. Petersburg Academy.[1] It is more likely, however, that the idea for such an institution occurred to the Tsar long before his meeting with the great German philosopher at Torgau in 1711. In this context it is worth recalling Peter's visit to the Royal Society in 1698, as well as his contact, through Bruce, with John Colson, whose lost missive to the Tsar was probably concerned with the problem of introducing mathematical learning into his country.

The extent to which Peter was interested in this matter is further confirmed by his supposed invitation to Edmond Halley to come to Russia. Halley declined, but it appears that the Tsar asked him "many questions concerning the fleet which he intended to build, the sciences and arts which he wish'd to introduce into his dominions . . ."[2] It is in any case undeniable that the Tsar began gathering material for his

1. Liselotte Richter (following Moritz Posselt's earlier study, *Der General und Admiral Lefort*) gives credit for the Academy entirely to Leibniz; see her *Leibniz und sein Russlandbild* (Berlin: Akademie-Verlag, 1946), chapter iii. Leibniz' projected plan for the Academy may be found in N. A. Voskresenskii, *Zakonodatel'nye akty Petra I*, I, 270, no. 330. (This is the anonymous Russian version of Leibniz' original proposal.) On Leibniz' influence in Russia, see V. I. Chuchmarev, "G. V. Leibnits i russkaia kul'tura 18 stoletiia," *Vestnik istorii mirovoi kul'tury* (No. 4, 1957), pp. 120–132.

2. See "The Eulogy of Dr. Halley," in *The Gentleman's Magazine and Historical Chronicle*, XVII (November 1747), 506. There appear to be no extant contemporary records confirming this story, either in Russian or English: but, as noted above (see p. 22). Flamsteed implied that Halley had some sort of association with Bruce.

Kunstkammer—subsequently to serve as the nucleus of the Academy collection—long before his contact with Leibniz. He was already purchasing books on a grand scale during his first European journey through intermediaries such as Bruce and, later, others such as Boris Lebiadnikov. Later still, as we have seen, Schumacher was given this task; he was to become the Academy's long-lived secretary.

Bruce's library also passed into the possession of the St. Petersburg Academy. Subsequently much of it was lost or destroyed, but for a long time the Newtonian collection it contained remained the most comprehensive of its kind not only in Russia but in all of Eastern Europe; it served to acquaint generations of readers with Newton's writings. Through Bruce's "Ex libris," it has been possible to trace several of these in the present library of the Academy of Sciences in Leningrad[3] (see fig. 12). The historian V. N. Tatishchev (1686–1750), who always spoke of Bruce in the most respectful terms, was one of those who used it (hence, the references to Newton in Tatishchev's own writings). So did Lomonosov and Richmann, who, when they studied the *Principia,* almost certainly read one of Bruce's copies.

Bruce played an important part in setting up the St. Petersburg Academy, and it was apparently at his insistence that six of the ten founding members were mathematicians.[4] Indeed, his correspondence with Leibniz, Leonard Euler, Christian Wolff, Johann Leutmann, and some of the Academy's earliest luminaries is testimony to the significance that Bruce ascribed to mathematics. The proposed plan for the institution was signed by Peter the Great in February 1724. In the following year some of the eminent mathematicians and men of science invited from abroad began arriving in the Russian capital. Peter did not live long enough to see it officially opened, and the inauguration, which took place on August 1, 1726, was attended by his widow and successor, Catherine I.[5] Appropriately, the first communication of the new Academy (see fig. 32) was addressed to none

3. Several of Bruce's books found their way into Peter the Great's collection and are not included in the inventory made up on the former's death; Bruce's "Ex libris" is, therefore, often the only clue as to their real provenance.

4. Cf. Eduard Winter, *Halle als Ausgangspunkt der Deutschen Russlandkunde im 18 Jahrhundert* (Berlin: Akademie-Verlag, 1953), p. 186. See also Daniel Bernoulli's letter to Leonard Euler in 1734: "Say what you will, but the esteem in which the Academy is held largely depends on the mathematical sciences." (Quoted by P. P. Pekarskii, *Istoriia Imperatorskoi Akademii Nauk v Peterburge,* 2 vols. [St. Petersburg, 1870–1873]. I, 110.)

5. Meetings of the Academy had, however, already taken place in 1725; see Pekarskii, *Istoriia Imperatorskoi Akademii Nauk v Peterburge,* I, xxxviii ff.

other than Isaac Newton, who had then been president of the Royal Society for twenty-three years:

To the renowned President of the famous English society and its worthy and most honorable fellows, the St. Petersburg Academy sends its greetings. Inasmuch as the foundation of your Academy and the celebrated discoveries it has achieved have been admired by the French, no less than by the Germans, (so much so that, roused by the English example, they too have founded similar societies and academies), we doubt not that all men of learning will deem the decision of the Emperor Peter the Great (confirmed by the generosity of the Empress) to found the St. Petersburg Academy worthy of the highest praise: that it might not only disseminate learning throughout the Russe Empire (where it was but thinly spread before) but also enrich medicine no less than mathematics with new discoveries and methods. We hope that our observations will be useful above all for the development of astronomy, which should be all the more interesting to you in that up till now these have rarely been prosecuted in the North. While other transactions of the Academy are contained within the volume that is being sent presently, we thought it appropriate to enclose two contributions by our academians with this letter. For us nothing can be more gratifying than the growing approbation of Your Learned Society, first by virtue both of its foundation and the authority it enjoys, and yielding to no other Society in both the number of its fellows and their services and merit.

11th October, 1726[6]

The letter was signed by Christian Goldbach, who owed his own contributions to mathematics partly to Newton.[7] He was born in 1690, and spent his youth traveling in Germany, France, Italy, Holland, and England.[8] In 1711 he met Leibniz, who then corresponded with him. He heard of Bilfinger's and Hermann's departure for the Russian capital in 1725, which prompted him to leave for St. Petersburg too,

6. Royal Society Archives, Early Letters, G. 2.5.

7. "Es ergibt sich," write the editors of Goldbach's recently published correspondence with Euler, "dass *Goldbach* und *Euler* gleichzeitig und unabhängig voneinander von den Entdeckungen *Newtons* ausgehend alle ausreichenden Bedingungen für die Integrierbarkeit bei rationalen Exponenten feststellten. Bedingungen denen Euler später die heutige Form gab." See the introduction to A. P. Iushkevich and Eduard J. Winter, eds., *Leonhard Euler und Christian Goldbach: Briefwechsel 1729–1764* (Berlin: Abhandlungen der Deutschen Akademie der Wissenschaften, 1965), p. 10. Nonetheless, Goldbach does not appear to have been as well informed about Newton's mathematical work as J. D. Bruce. Thus, when he wrote to Leonard Euler in 1744 concerning a new book: "davon mir nur der folgende Titel bekannt ist: La méthode des fluxions par Mr *Newton* . . .," he had clearly not yet heard of John Colson's well-known book on the "fluxionary" method. See Goldbach to Euler, July 5, 1744, *ibid.*, p. 202. (Colson's *Method of Fluxions* came out in a French translation by Buffon, in 1740).

8. Cf. I. K. Luppol, "Pol' Gol'bakh—russkii akademik (k 150-letiiu so dnia smerti)," *Vestnik Akademii nauk SSSR* (Nos. 4–5, 1939), pp. 163–167.

though unlike them he had no invitation to the Academy. He was thus the first of the foreign enthusiasts to be drawn to Russia by nothing more substantial than the expectation of its promise and achievement. He offered his services to Lavrentii Blumentrost, the Academy's president, and by way of recommendation recalled his contacts with Bruce some seven years earlier, when one of his friends communicated to Bruce "some ideas of mine concerning a rather strange experiment, related by Mersenne's father in one of his letters to Descartes."[9] Goldbach spoke with "respect and wonder" of Bruce, and it is curious to note that his friendship was cultivated by Menshikov. His St. Petersburg career began when he was engaged as the Academy's historian, but Goldbach abandoned this in favor of becoming its conference secretary and it was in this capacity that he wrote his letter to Newton.

The letter was entered in the records of the Russian Academy for a meeting held on September 6, 1726, and it arrived in London just in time to be read at the last ordinary meeting of the Royal Society presided over by Newton.[10] He died a few weeks later, an event which almost certainly did not pass unnoticed in St. Petersburg, as is suggested by an item in Bruce's library.[11]

9. See Pekarskii, *Istoriia Imperatorskoi Akademii Nauk v Peterburge,* I, 156–157.

10. On March 2, 1727. Cf. C. H. Weld, *A History of the Royal Society, with memoirs of the presidents compiled from authentic documents,* 2 vols. (London, 1848), I, 437.

11. MIAN:

1238) Похвальные стихи на госп. Ісака Невтона, на аглинскомъ языкѣ, в Лондонѣ. 1728.

This might well be J. T. Desaguliers' allegorical poem, *The Newtonian System of the World, the best model of Government,* Richard Glover's *A Poem to Sir Isaac Newton,* or any one of a number of other poetic encomiums which appeared together in 1728. On J. T. Desaguliers's poem, see Gray, no. 67. There is an interesting discussion of Glover and others who "poetised" Newton in Marjorie Nicolson's masterly *Newton Demands the Muse, Newton's "Opticks" and the Eighteenth Century Poets* (Princeton, N.J.: Princeton University Press, 1946).

9

The Principal Objections to the Cartesian Theory

At the time that the Russian Academy opened, Cartesian natural philosophy on the Continent still remained largely unaffected by the critique of it contained in the *Principia*. Thus, in the *Éloge* that Fontenelle delivered before the French Academy of Sciences in 1727, he went out of his way to praise Newton and even suggested that in the future natural philosophers would be obliged to follow the method he propounded. He reserved most of his eulogy for Newton's discoveries in light and color, however. A few bold spirits like 's Gravesande had already discarded the vortices theory, but Fontenelle, like his colleagues in the Paris Academy, could not bring himself to recognize the doctrine of universal gravitation. To do so would have meant jettisoning the Cartesian cosmology. This was so closely tied to Descartes's views on motion and space that abandoning it was tantamount to rejecting Cartesianism as a whole. It is important to understand Newton's main argument against the prevailing orthodoxy, which, once breached, permitted the Newtonians to mount a concerted attack along the whole range of contemporary natural philosophy. Much of that prolonged and often bitter debate over gravitation, over "vires vivae," over the shape of the earth, over electricity and the ether, and over light and color, was to echo in St. Petersburg too.

In the *Principia* Newton not only employed methods of scientific investigation quite different from those of Descartes, but he directly attacked the Cartesian system by showing its basic hypothesis to be untenable. Thus, in Proposition LII of Book II, he wrote: "*If a solid sphere, in an uniform and infinite fluid, revolves about an axis given in position with an uniform motion, and the fluid be forced round by only this impulse of the sphere; and every part of the fluid perseveres uniformly in its motion: I say, that the periodic times of the parts of the fluid are as the squares of their distances from the centre of the sphere.*"[1] And in the Scholium he drew the appropriate implications: "I have endeavoured in this Proposition to investigate the properties of vortices, that I might find whether the celestial phænomena can be explained by them; for the phænomenon is this, that the periodic times of the planets revolving about Jupiter, are in the sesquiplicate ratio [i.e., 3/2th power] of their distances from Jupiter's centre; and the same rule obtains also among the planets that revolve about the Sun. And these rules obtain also with the greatest accuracy, as far as has yet been discovered by astronomical observation. Therefore if those Planets are carried round in vortices revolving about Jupiter and the sun, the vortices must revolve according to that law. But here we found the periodic times of the parts of the vortex to be in the duplicate ratio of the distances from the centre of motion; and this ratio cannot be diminished and reduced to the sesquiplicate, unless either the matter of the vortex be more fluid the farther it is from the centre, or the resistance arising from the want of lubricity in the parts of the fluid should, as the velocity with which the parts of the fluid are separated goes on increasing, be augmented with it in a greater ratio than that in which the velocity increases. But neither of these suppositions seems reasonable. The more gross and less fluid parts will tend to the circumference, unless they are heavy towards the centre. And tho', for the sake of demonstration, I proposed at the beginning of this Section, an Hypothesis that the resistance is proportional to the velocity, nevertheless, 'tis in truth probable that the resistance is in a less ratio than that of the velocity. Which granted, the periodic times of the parts of the vortex will be in a greater than the duplicate ratio of the distances from its centre. If, as some think, the vortices move more swiftly near the centre, then slower to a certain limit, then again swifter near the circumference, certainly neither the sesquiplicate, nor any other certain and determinate ratio can obtain them. Let philosophers then see how that phænomenon of the sesquiplicate ratio can be accounted for by vortices."[2]

1. *Principia,* Book II, Proposition LII, Theorem XL.
2. *Ibid.,* Book II, Proposition LIII. Theorem XL, Scholium.

Newton's other objection made an analogous contradiction appear between the supposed motion of vortices and Kepler's law that radii drawn to the sun describe areas proportional to the times. Proposition LIII was formulated in the following way: *"Bodies, carried about in a vortex, and returning in the same orb, are of the same density with the vortex, and are moved according to the same law with the parts of the vortex, as the velocity and direction of motion."* The Scholium at the end of this proposition therefore stated: "Hence it is manifest that the planets are not carried round in corporeal vortices. For according to the *Copernican* hypothesis, the Planets going around the Sun revolve in ellipses, having the sun in their common focus; and by radii drawn to the sun describe areas proportional to the times. But now the parts of a vortex can never revolve with such a motion." To prove it, Newton drew three orbits around the sun: the outermost concentric to it, the two innermost ones with corresponding aphelions and perihelions. Thus, a body revolving in the external orbit concentric to the sun, in a radius drawn to the sun and describing areas proportional to the times, will move with a uniform motion. But, according to the laws of astronomy, the body revolving in the orbit between the outermost concentric orbit and the innermost orbit should move more slowly at its aphelion and more quickly at its perihelion. Yet this, according to the laws of mechanics, does not conform with the way vortices would move.

In fact, the space between the aphelion of the innermost orbit and the corresponding point of the outermost concentric orbit being more narrow than the space left between the perihelion of the same innermost orbit and the corresponding point of the outermost orbit, the matter of the vortex in and around the intermediary orbit ought to move more swiftly in the narrowest space, the body revolving therefore with more speed at its aphelion than its perihelion. "For the narrower the space is, thro' which the same quantity of matter passes in the same time of one revolution, the greater will be the velocity with which it passes through it." But these conclusions, Newton adds, are mutually exclusive, "So that the hypothesis of vortices is utterly irreconcileable with astronomical phænomena, and rather serves to perplex than to explain the heavenly motions."[3] The General Scholium of Book III of the *Principia* then completed the series of objections. Having shown the incompatibility of each of Kepler's two

3. *Ibid.,* Book II, Propostion LIII. Theorem XLI, Scholium.

laws with the mechanics of vortices, Newton then says: "The hypothesis of Vortices is press'd with many difficulties. That every Planet by a radius drawn to the Sun may describe areas proportional to the times of description, the periodic times of the several parts of the Vortices should observe the duplicate proportion of their distances from the Sun. But that the periodic times of the Planets may obtain the sesquiplicate proportion of their distances from the Sun, the periodic times of the parts of the Vortex ought to be in the sesquiplicate proportion of their distances."[4]

Newton then drew a fourth objection to the idea of a multiplicity of vortices in the heavens: "That the smaller vortices may maintain their lesser revolutions about *Saturn, Jupiter,* and other Planets, and swim quietly and undisturb'd in the greater Vortex of the Sun, the periodic times of the parts of the Sun's Vortex should be equal. But the rotation of the Sun and Planets about their axes, which ought to correspond with the motions of their Vortices, recede far from all these proportions."[5] In supposing the motion of the vortices to be caused by the rotation of the sun, a new difficulty arose: "Therefore, in order to continue a vortex in the same state of motion, some active principle is required, from which the globe may receive continually the same quantity of motion which it is always communicating to the matter of the vortex. Without such a principle it will undoubtedly come to pass that the globe and the inward parts of the vortex, being always propagating their motion to the outward parts, and not receiving any new motion, will gradually move slower and slower, and at last be carried round no longer."[6] Finally, there was one other objection provided by comets: "The motions of the Comets are exceeding regular, are govern'd by the same laws with the motions of the Planets, and can by no means be accounted for by the hypothesis of Vortices. For comets are carry'd with very eccentric motions through all parts of the heavens indifferently, with a freedom that is incompatible with the motion of a Vortex."[7]

These, then, were Newton's principal objections to the Cartesian theory of vortices. Cartesians became acutely conscious of the formidable nature of the challenge they presented to their whole system only

4. *Ibid.,* Book III. General Scholium.
5. *Ibid.*
6. *Ibid.,* Book II, Proposition LII. Theorem XL, Cor. IV.
7. *Ibid.,* Book III. General Scholium.

in the last years of Newton's life. As Condorcet noted, even a decade after his death Newton's criticisms were still regarded as unacceptable in the Paris Academy of Sciences.[8] Nor is it surprising that the academicians in St. Petersburg, drawn largely from German, Swiss, and French universities where natural philosophy was still generally taught in the spirit of Descartes or Leibniz, should have shown themselves equally wary. Yet Newton's doctrines must already have seemed even in their eyes to confront the prevailing orthodoxy with certain difficulties, for the very first problem to be discussed at the earliest meeting of the Russian Academy (before the official inauguration) was taken from the *Principia*.

8. Cf. Pierre Brunet, *L'Introduction des théories de Newton en France au XVIIIe siècle. Avant 1738* (Paris: A. Blanchard, 1931), p. 341.

10

Newtonian Physics at the New Academy

Newtonian physics gained a foothold at the new Academy from the time it opened through the adoption of a textbook which was one of the earliest on the Continent to discard the Cartesian orthodoxy.

How this happened is not entirely clear. Shortly after the inauguration of the Academy a resolution was passed (January 14, 1726) stipulating that one of its professors of natural philosophy (Georg-Bernhardt Bilfinger) should start giving lectures on the subject.[1] "The commentaries and conclusions," it was resolved, were to be drawn from "'S Gravesande's 'institutiones' on Newtonian philosophy."[2] The edition Bilfinger was requested to use had been published only three years earlier as : *Philosophiae Newtonianae Institutiones in Usus Academicos* (Leyden, 1723; see fig. 33).

This book was largely based on the same author's *Physices elementa mathematica,* published, as we have seen, three years earlier, when it was still common in universities to use textbooks on natural philosophy derived from Descartes. This was true even at Cambridge, where,

1. Cf. MIAN, I, 169–170.
2. *Ibid.,* p. 171. I.e.,

Того ради конца, профессоры сея Академии, сего 1726 года в будущий 24 день месяца генваря, чтениями учение свое публичное начнут во дни: понедельник, среду, четверток и субботу, и впредь таким определением и учреждением поступать будут, о котором всем любителям добрых наук, а наипаче рачителям к учению, сим для известия объявляется... Георг Бернгард Бильфингер, физики экспериментальной и феоритической профессор, искусства физикальные восприимет с изъяснениями их и конклюзиями, следствуя в том Гравесанду в институциях философии Невтониа[н]ской, для пользы академической в 1723 году изданных...

until 1718, the text used was by an eminent French Cartesian—the *Physics* of Jacques Rohault. Indeed, his version of Descartes was still being taught there after 's Gravesande's *Philosophiae Newtonianae Institutiones* appeared, even though Dr. Samuel Clarke, an ardent Newtonian, had published a new translation of Rohault. It contained a running commentary of notes which, while it avoided the language and appearance of controversy, actually constituted a complete refutation of the text.[3]

'S Gravesande's tactic was more straightforward. He had been a Cartesian, but, after studying the *Principia,* he changed his views. He did modify some of Newton's opinions. For example, he accepted his emission theory of light, but went on to propose that every substance possesses a natural store of light corpuscles, which are expelled when a substance is heated to incandescence. Conversely, corpuscles may become united to a material body, as happens, for instance, when the human body is exposed to the rays of a fire. From this he further inferred that light corpuscles were a chemical element, a supposition that became quite customary after the time of Newton (though he himself had never made it).[4] It is unlikely that Bilfinger, to whose care 's Gravesande's textbook was entrusted, shared any of the author's Newtonian sympathies.

Why the book was chosen is suggested by a report of J. D. Schumacher, the Academy's permanent secretary, which he addressed to Peter the Great in 1722. Schumacher had been sent abroad by the Tsar to enrich his collection of books and scientific instruments. He met the Dutch mathematician in Leyden and wrote to tell Peter how highly he regarded 's Gravesande's "Elementa fisika mekhaniko Nevtoniana" [*sic*].[5] His admiration may well account for the selection of this book as the source of Professor Bilfinger's lectures. Hence the paradox that

3. Bilfinger, like other Academicians was supposed to give lectures four times a week to the youth of the university or gymnasium attached to the Academy. They were also instructed to give public lectures three times a year on the model of the Paris, London, and Berlin academies.
4. Newton suspected that light corpuscles and ponderable matter might be transmuted into each other. See *Opticks,* p. 349.
5. For example,

Азъ выше упомянул, яко помянутый Сгравесандъ Элемента физика механико Невтониана, экспериментисъ иллюстрата, в полдести издалъ, где все машины къ экспериментамъ механическимъ, статическимъ, гидростатическимъ, физическимъ, и оптическимъ срисованы и напечатаны обретаются, и разные монархи оные машины Мусенброку делать повелели, и понеже азъ зналъ, яко вашему

Newton appears to have gained, at the new Academy of St. Petersburg, a modicum of recognition denied him at the Académie des Sciences in Paris and other better-established institutions. But the paradox is more apparent than real. Russia's early exposure to Newton's doctrines through Bruce and Andrew Ferquharson was a result of Peter the Great's decision to visit England in 1698; toward the end of his reign, however, he came into increasing contact with scholars and men of science in Germany, where Newton was held in low esteem. For this Leibniz was largely responsible. It was to him that the Tsar turned when the project of founding the Academy of Sciences came close to realization, a fact which affected Newton's subsequent reputation in Russia.

Leibniz died in 1716, but the honor of advising Peter fell to his best-known follower, Christian Wolff (1679–1754). Opinions about Wolff differ. His version of Leibniz, whose views he systematized through prolific writings of his own, is generally held to be unsatisfactory, though in the absence of published works by Leibniz himself it was long the accepted one. Partly through these efforts Wolf enjoyed an estimable reputation during the eighteenth century. Nor was his own opinion of himself modest. In a letter to Blumentrost, who had been interested in acquiring his services for Peter the Great, Wolff compared himself to Aristotle and Hasan.[6] It is only fair to add that Kant regarded him as the "greatest of the dogmatic philosophers." Even Emily de Breteuil, the Marquise du Châtelet, who translated the *Principia* into French, was at one time under his influence, and, for a short while, she persuaded Voltaire to share her enthusiasm. She later dismissed Wolff as "un grand bavard en métaphysique."[7]

He gained renown throughout Europe through the scandalous cir-circumstances which forced him to flee from Halle in 1723 at the command of Frederick William of Prussia. Subsequently he taught at Marburg, where he was to supervise Lomonosov's instruction. The os-

императорскому величеству такия машины благондравны суть, того ради следовалъ образцу их ...
Pekarskii, *Nauka i literatura v Rossii pri Petre Velikom, I*, 544. Bruce, too, apparently prized this work; see p. 38 above.

6. November 12, 1724, in *Briefe von Christian Wolff aus den Jahren 1719–1753; ein Beitrag zur Geschichte der Kaiserlichen Academie der Wissenschaften zu St. Petersburg* (St. Petersburg and Leipzig, 1860), p. 27. On Blumentrost, see Pekarskii, *Istoriia Imperatorskoi Akademii Nauk v Peterburge*, I, 1–15.

7. Marquise du Châtelet to Maupertuis, Cirey, September 29, 1738, in G. E. D. du Châtelet, *Lettres de la Mse Du Châtelet réunies pour la première fois . . . par Eugène Asse* (Paris, 1878), p. 236.

tensible cause for his dimissal from Halle was a book called *Vernuenft-tige Gedanken von den Wuerckungen der Natur,* which the Pietists claimed was conducive to atheism, but the brutal fashion in which he was ousted from a chair he had held for sixteen years also lent him the halo of political martyrdom. Not only did he stand against clerical intolerance, but he also opposed the arbitrary power of the crown. Though this generous view of Wolff's acts is not entirely vindicated by the facts, his recall by Frederick the Great in 1740 was undoubtedly an important event.[8] With his triumphant restoration, philosophy achieved a long-lasting pre-eminence at German universities, and its freedom from theologico-confessional interference. Largely by his endeavors, German was recognized as a philosophical language for the first time. Yet in several respects the authority his name acquired was most unfortunate. Great as his services were to German education, Wolff's dogmatic rationalism hindered the advance of science. Above all, he failed to appreciate the growing importance of mathematics in natural philosophy, and his objections to Newtonian physics were based on the same grounds as those of Leibniz, whom he followed in rejecting the theory of universal gravitation as a restoration of "occult qualities."

From the opening of the Russian Academy, Wolff cast his shadow over its proceedings. Its professor of natural philosophy, Christian Martini (1699–17?), was appointed at his recommendation. In grati-tude, Martini had a clause inserted in his contract with the Russian Academy asserting that he would "expound physics according to Wolff's principles."[9] At its opening meeting he announced confidently that he had discovered "viam ad perpetuum mobile" and was subse-quently deemed incompetent to continue in his post, which he ex-changed for that of Logic, Metaphysics, and Ethics. The man who succeeded Martini, Georg-Bernhardt Bilfinger, was also a pupil of Wolff and a far more able one. His defense of Leibniz and the Car-tesians against Daniel Bernoulli, another founding member of the Russian Academy, brought St. Petersburg into the controversy being conducted elsewhere on the Continent over the validity of Newton's doctrines.

8. Wolff's expulsion was preceded by his request that university authorities expel a student who had been reckless enough to criticize his metaphysics. See J. C. Bluntschli, *Geschichte des Allgemein Staatsrechts und der Politik seit dem sechzehnten Jahrhundert bis zur Gegenwart,* Vol. I (Munich, 1864), 213.

9. Cf. Pekarskii, *Istoriia Imperatorskoi Akademii Nauk v Peterburge,* I, 73–77.

"... There is nothing more simple than Cartesian vortices: and if they are not deemed entirely serviceable, I should like them to be changed as little as possible."—Georg-Bernhardt Bilfinger

11

The Opening Controversy over Gravitation

The exchanges between Bernoulli and Bilfinger became so heated that minutes of Academy meetings for 1729, the year in which their disputes appear to have come to an angry climax, were thought unworthy of retention in its official records, leaving a gap in the proceedings from October 29, 1729, to September 11, 1730.[1] Bilfinger's and Bernoulli's differences eventually prompted Blumentrost to set up a committee to adjudicate their charges against each other. The disagreement over fundamental problems of natural philosophy mixed with personal antagonism was recorded with punctilious solemnity by G. V. Müller, who replaced Christian Goldbach as the Academy's historian.[2] Not all the details concerning their encounters have survived. Nor is it certain what, to use Bernoulli's phrase, the initial "pomum discordiae" between them was.[3] Yet the nature of the theoretical differences between them is readily apparent both from what Bilfinger and Daniel Bernoulli wrote before coming to St. Petersburg and from their later contributions to the *Commentarii*.

1. Cf. *Protokoly zasedanii konferentsii Imperatorskoi Akademii Nauk s 1725 po 1803 goda* ed. K. S. Veselovsky, Vol. I (St. Petersburg, 1897), 22.
2. E.g., "Herr *Bülffinger* und herr *Bernoulli* fielen hiernächst noch in unterschiede scharffe wort-wechsel, und warff sonderlich herr Bulffinger dem herrn *Bernoulli* verschieden fehler vor, die er in mathematischen sachen solle begangen haben. Beschuldigte ihn auch, dass er, herr *Bernoulli, öffentlich* gesagt habe, wenn dieses und jenes sich im experiment nicht so verhalten würde, wie es ihm der calculus mitgebracht, so wolle er sich als einen schelm von der academie jagen lassen ... Das experiment sey nachgehends gemachet worden, welches seinen calculum wiederleget habe ..." (MIAN, I, 503).
3. *Ibid.*, p. 574.

Daniel Bernoulli, the second son of John Bernoulli (1667–1748), was the bearer of one of the most illustrious names in the annals of science. The family, which came originally from Antwerp, found asylum in Basel. There, in the course of a century, eight of its members brilliantly cultivated various branches of mathematics. John, his brother Jacob (1654–1705), and Daniel (1700–1782), are the best known. Indeed, in the successful application of the calculus, Jacob Bernoulli has been regarded as worthy of a place by the side of Newton and Leibniz. He and his brother John were the first two foreign associates of the French Academy, though their relationship was marred by an unseemly public quarrel that brought John into a dispute with various English and Scottish mathematicians. He was a good friend of Leibniz,[4] and he disliked and depreciated Newton. His harsh and jealous bearing toward his son Daniel was partly inspired by the latter's refusal to support the Cartesian system.[5]

Daniel Bernoulli's emancipation from his father's domination was gradual, and when he came to St. Petersburg it was not as a mathematician but as a professor of medicine. His elder brother, Nicholas, whom his father considered the most able of his sons, had given him lessons in geometry, but Daniel studied to become a physician. His departure for St. Petersburg was occasioned by a peculiar circumstance— his invitation had not made it clear whether the president of the Russian Academy intended it for Nicholas or for Daniel. The confusion was resolved by inviting both. Nicholas assumed the chair of mathematics; Daniel, the chair of medicine. His interest in physics must already have been keen because a year before coming to St. Petersburg he had published a short mathematical treatise on dynamics, in which he defended the "incomparable Newton" against Riccati.[6] It was to this subject that he devoted much of his work in the Russian Academy. Ultimately, with the aid of the new analysis, he was able to solve difficult mechanical problems for which the geometrical methods adhered to in the *Principia* offered no prospect of solution. He is

4. Cf. G. W. Leibniz, *Virorum celeberr. Got. Guil. Leibnitii et Iohannis Bernullis, Commercium philosophicum et mathematicum*, 2 vols. (Lausanne and Geneva, 1745).

5. "Entgegen den Anschauungen seines Vaters," writes Professor Fritz Burckhardt, Daniel accepted the Newtonian theory of gravitation "und wenn dies wohl ein Grund war, dass die Annäherung des Vaters an diesen Sohn niemals eine vollständige wurde, so hatte es auch den eminenten Erfolg dass Daniel Bernoulii der Newton'schen Lehre ganz wesentlich auf dem Kontinente zum Siege verhalf." See his *Die Basler Mathematiker Daniel Bernoulli und Leonhard Euler* (Basel, 1884), pp. 16–17.

6. Cf. Daniel Bernoulli, *Exercitationes quaedam mathematicae* (Venice, 1724), p. 69.

noted as one of the founders of "mathematical physics"; by introducing into mechanics the principle of the conservation of "vis viva" (of which Huygens had already shown some inkling in his researches on the compound pendulum) he advanced one of the most important laws of science. Curiously enough, however, only the tragic death of his brother in St. Petersburg some eight months after their arrival persuaded Daniel to take up the chair thus left vacant.

In this respect Daniel Bernoulli's position in the Russian Academy was not unlike Bilfinger's, who was also invited to assume a chair that he soon relinquished. Otherwise, their backgrounds had little in common. Bilfinger (1693–1750) had been drawn initially to theology, in which he retained a lifelong interest, but he took up mathematics and natural philosophy under the influence of Christian Wolff.[7] He was not an original thinker, and if he was in fact the first to refer to the so-called Wolff-Leibnizian succession, which, through Baumgarten, Meier, and other pupils of Wolff, so long dominated philosophy at German universities (as well as at Moscow University in the second half of the eighteenth century), his own place is squarely in the same tradition. He introduced Wolff's teachings to Tübingen, where Bilfinger became professor of philosophy in 1721. There he met the same hostility that Wolff encountered at the University of Halle, though his fate at Tübingen was less dramatic. Bilfinger suffered all the odium of expounding Wolff's teaching, but he enjoyed little of the glory. His orthodoxy was held suspect, and his advancement retarded. When at

7. An account of Bilfinger's life (also Bülffinger and Billfinger) may be found in Pekarskii, *Istoriia Imperatorskoi Akademii Nauk v Peterburge*, I, 81–95. See also Windelband, Erdmann, Zeller, Ueberweg, and other authorities on German philosophy, none of whom give a comprehensive or reliable account of Bilfinger's philosophical views, as is pointed out by Richard Wahl ("Professor Bilfinger's Monadologie und prästabilirte Harmonie in ihrem Verhältniss zu Leibniz und Wolff," *Zeitschrift für Philosophie and philosophische Kritik* [Halle], new ser., LXXXV [Nos. 1 and 2, 1884]), who says that "Mehrere von den ältern philosophischen Geschichtswerken nennen nicht einmal seinen Namen; andere . . . zeichnen ihn mit so allgemeinen und leblosen Farben, dass ein bestimmtes Bild von ihm und von seiner Thätigkeit nicht zu gewinnen ist (pp. 68–69)." Wahl is critical of the accounts accorded Bilfinger in some of the standard histories of German philosophy, and has kind words only for Erdmann's pages on Bilfinger in his short monograph on Kant's teacher, Martin Knutzen (*ibid.*, p. 69). Wahl himself, however, does not deal with Bilfinger's period in St. Petersburg; and Pekarskii, who makes no attempt to discuss his scientific or philosophical views, also passes by his contributions to the *Commentarii*. Some of the latter were translated into German and published separately; some were also reissued in the original in Bilfinger's . . . *Varia in fasciculos collecta* . . . (Stuttgart, 1743). All of Bilfinger's published writings, including his polemic with Iavorskii, can be found in the British Museum.

Wolff's prompting he was invited to St. Petersburg, Bilfinger eagerly accepted.

Some months after his arrival in the Russian capital, Bilfinger addressed an epistle to Clarke giving a laudatory account of the inauguration of the Russian Academy and courteously setting forth some of their differences. This interesting letter, neglected by historians of the Academy, may be found in an appendix to the second edition of the *Commentatio hypothetica*.[8] It first appeared in 1723, and it provides some important clues to the nature of Bilfinger's philosophical differences with Daniel Bernoulli two years later.

From the *Commentatio* it transpires clearly enough how closely Bilfinger followed Leibniz' views, including his monadology and pre-established harmony. In fact, the book, though it had the distinction of being placed on the Index at Rome, is no more than a defense of Leibniz' metaphysical views. Leibniz was dead, and Bilfinger appears to have conceived the idea of doing for him what Samuel Clarke had attempted for Newton by defending him from his critics. This is why the *Commentatio* may almost be read as a supplement to Clarke's correspondence with Leibniz, for it contains sections refuting the objections of several of his more distinguished opponents, including Newton.[9] Although Bilfinger modified some of his Leibnizian positions in his next work, the *Dilucidationes*,[10] which appeared in 1725, his attitude to Newton's doctrines remained fundamentally unchanged.

This much may be gathered from the most important opus to be written by him in Russia—a memoir on the "cause" of gravitation which won the prize of the French Academy in 1728.[11] The intellectual incentive of this work is revealed by Bilfinger's statement "that there is nothing more simple than Cartesian vortices: and if they are not

8. I.e., Bilfinger, *De harmonia animi et corporis humani maxime praestabilita, commentatio hypothetica. Accedunt solutiones difficultatum a Foucherio, Baylio, Lamio, Tourneminio, Newtono, Clarkio . . .* (Tübingen, 1723). I used the second edition which was published in 1734 and has an altered title, "ex mente illustris Leibnitii," being inserted before "commentatio hypothetica." The letter to Clarke (pp. 309–318) is dated September 29/30, 1726 (St. Petersburg).

9. For the attacks on Newton and Clarke, see especially *ibid.*, section nos. 212 and 225; for his defence of Leibniz against Clarke's strictures, nos. 209 and 210.

10. I.e., Bilfinger, *Dilucidationes philosophicae de Deo' anima humana, mundo, et generalibus rerum affectionibus* (Tübingen, 1725).

11. I.e., Bilfinger, *De causa gravitatis physica generalis disquisitio experimentalis quae praemium a Regia Scientiarum Academia promulgatum, retulit: anno 1728* (Paris, 1728).

deemed entirely serviceable, I should like them to be changed as little as possible."[12] The whole purpose of the memoir—which is why the French Academy found it so much to its taste—was to circumvent Newton's demonstration that the action of vortices was incompatible with observed planetary motion. Bilfinger argued that Newton's objections to the Cartesian theory could be said to be irrelevant if it were assumed that vortices penetrate each other's paths of action. He tried to show that this was in fact the case by introducing a hypothesis concerning the ether of which the vortices were supposedly composed. This very conveniently assumed that the ether was made up of matter so fine that the cross-movement of one vortex through another in no way arrested its force or velocity.

The *De causa gravitatis . . . disquisitio* closes with the disarming admission that "Perhaps our defense of vortices will confirm some in their belief in this explanation, and others in their criticism of it."[13] In the spate of Cartesian rectifications that followed the publication of Bilfinger's memoir, his was not long remembered, though it appeared at a particularly timely moment.

The Cartesians in the Académie des Sciences were just then beginning to realise the formidable character of the Newtonian case against them. In the very same year that it awarded Bilfinger a prize for his memoir, the French Academy also came out with an *Éloge* of Newton, in which Fontenelle, its "secrétaire perpetuelle," qualified his eloquent tribute to the author of the *Principia* by indicating some of the immense difficulties its theories presented to accepted beliefs. "Il ne faut pas s'étonner que les philosophes reviennent souvent à cette matière. Rien n'est plus intéressant pour eux que de savoir si l'ingénieux système des tourbillons de Descartes, et qui se présente si agréablement à l'esprit, tombera accablé sous les difficultés qu'on lui oppose; et si l'on sera réduit à en prendre un autre qui a des difficultés aussi grandes et plus frappantes, quoi qu'il a des faces avantageuses."[14]

But to Daniel Bernoulli the "ingenious" Cartesian system no longer seemed worth saving. In the *Commentarii* he repeatedly dismisses the "hypothesis Cartesiana" as "insufficiens," while Bilfinger attempts to

12. *Ibid.*, p. 6, no. xi: "Fatendum est, nihil esse vorticibus Cartesianis simplicius: igitur omnia putem tentanda prius quam deservantur; atque si omnino servari non possint, velim, ut non nisi minima, quae fieri potest, mutatio fiat."
13. *Ibid.*, p. 36, no. lxvii.
14. *Histoire* [of the Académie Royale des Sciences] (Paris, 1728), p. 134.

prove the contrary.[15] Both of them conducted experiments with liquids and vessels of various descriptions in support of their principles, often, at least in Bernoulli's case, with the aid of theorems taken from the second book of the *Principia*. Their disagreement was not only over vortices and the inverse-square law of gravitation, which Daniel Bernoulli accepted without reservation, but over their approach to natural philosophy. To Bilfinger, who had little training in mathematics, the physical causes of phenomena seemed more important than their mathematical resolution. This is one of the main reasons for his prolonged dispute with Daniel Bernoulli. The two were formally reconciled before they left St. Petersburg, but we know from Bernoulli's later correspondence that he did not change his initial assessment of Bilfinger as a man "qui se mêle à tout, et ne comprend rien a fonds."[16] Frederick the Great, it is only fair to add, regarded Bilfinger as one of the most outstanding men he had known, though he was not, it is true, concerned with Bilfinger's qualities as a natural philosopher. His negative attitude to Newton, as Daniel Bernoulli implied in one of his letters, was most unfortunate for the Russian Academy. It was also responsible for involving Bilfinger in yet another controversy with one of Newton's friends and disciples, James Jurin.

15. Cf. Bernoulli, "Dissertatio de actione fluidorum in corpora solida et motu solidorum in fluido," *Commentarii* (St. Petersburg), II (1729), 318.
16. MIAN, I, 576.

". . . ist der gute Newton zu bedauern, als welcher nicht nur inter
Celeberrimos Goldbachios, Bulffingeros, etc. keinen Platz findet,
sonder sogar mit vieler Verachtung tractirt wird."—Daniel Bernoulli

12

Bilfinger and James Jurin

The dispute between Bilfinger and James Jurin resulted from the
words with which Bilfinger chose to describe the theory of Newtonian
gravitation. It was, he insisted in one of his papers in the *Commentarii,*
"a vulgar hypothesis." Since the *Commentarii* were regularly dispatched to the learned societies of Europe, the offending article caught
the eye of James Jurin, secretary of the Royal Society from 1721 until
Newton's death six years later. He resolved that Bilfinger's remarks
should not pass unchallenged, and he promptly sent to St. Petersburg
a riposte, which was published in the next issue of the *Commentarii*
together with Bilfinger's comments inserted by way of answer in the
footnotes.[1]

Jurin (1688–1750) was regarded as one of the most learned men of
his day. As a scholar of Trinity College he heard lectures on experimental philosophy, but after graduating from Cambridge he went on
to Leyden to study medicine. On his return to England he acquired a
large medical practice and gained considerable notoriety through his
support of inoculation against smallpox, which won for him the admiration of Voltaire and, eventually, the approbation of his profession,
which was appropriately marked when Jurin was elected to the presidency of the College of Physicians. He became a Fellow of the Royal

1. I.e., James Jurin, "Disquisitio physicae de tubulis capillaribus a Jacobo Jurino ad
Academiam transmissae ut ejusdem Commentariis insererentur, una cum notis a
Georgio Bernhardo Bülffingero, ad quem id negotium pertinuit, adjectis," *Commentarii*
(St. Petersburg), III (1728, pub. 1732), 281–282. Jurin included this paper in his
Dissertationes physico-mathematicae . . . (London, 1732), where it appeared as "Disser-
tatio III" with minor emendations, sharpening the polemical tone of the original.

Society as early as 1717 or 1718 and imbibed Newtonian philosophy from Newton himself, becoming an ardent supporter of his teaching on motion and of his system of fluxions. Jurin's admiration for Newton was related to his professional interests in that his experiments on the specific gravity of human blood and, still more, his papers on the power of the heart were original attempts at converting physiology into an exact science. John Freind at Oxford had earlier tried to apply Newtonian principles to medicine,[2] and Jurin's efforts were aimed in the same direction. Unlikely as it seems, this was the cause of his polemical encounter with Bilfinger.

Bilfinger was not interested in physiology as such, but shortly after arriving in St. Petersburg he conducted some experiments involving running water that also interested Jurin because of his concern with the behavior of human blood. Jurin had written a number of papers on the subject, which had been criticized by P. A. Michelotti in 1723.[3] Since Jurin regarded the discussion as far from closed, Bilfinger's contribution in the *Commentarii* provided an opportunity to re-state his case. The starting point for both Jurin's and Bilfinger's experiments was a phenomenon several natural philosophers of the period had found difficult to explain, that is, the suspension of water in capillary tubes. Bilfinger explained it in terms of the action of vortices; Jurin argued that it was "owing to the attraction of a small annular surface on the inside of the Water."

To prove this assertion, Jurin took "a Glass Funnell by severall Inches Diameter, having its small End drawn out into a very fine Tube, which Funnell being inverted and fill'd with Water, the whole Quantity of Water therein contain'd was sustain'd above by the Level of Attraction of that narrow *Annulus* of the Glass with which the upper surface of the Water was in contact."[4] Bilfinger believed this explanation to be totally inadequate, and, from giving specific arguments against it, he was drawn into a more general attack

---

2. On Freind, see Chapter 15, below.

3. Cf. Jurin, "De motu aquarum fluentium," *Philosophical Transactions*, XXX (No. 355, 1719–20), 748–766, and *ibid.*, XXXII (No. 373, 1722–23), 179–190. See also *Jacobo Jurini Regiae scientiarum Londinensi Societati A secretio Dissertationis de Motu Aquarum fluentium contra nonullas Petri Antoni Michelotti Animadversiones Defensio Accedit ejusdem Michelotti Ad Illustris atque Excellentiss. Virum Antonium De Comitibus Patritium Venetum Eruditissimum Epistola in qua illi ipsi Jurinianae Defensioni respondetur* (Venice, 1724). (Michelotti's reply is on pp. 15–38.)

4. Jurin, "The Cause of the Ascent and Suspension of Water in Capillary Tubes," *The Philosophical Transactions (from the Year 1700 to the Year 1720)*, abridged by Henry Jones, 2nd ed., Vol. IV, Part I (London, 1731), 428.

on the Newtonian theory of gravitation, thus bringing upon himself all of Jurin's formidable prowess as a Latinist and his indignation as a Newtonian.[5] Far from being a hypothesis, Jurin affirmed, the theory of gravitation as described by Newton was a "vera et indubitata . . . Theoria."[6] "That Great Man [Magnus ille Vir]" had explained that theory in the *Principia;* to avoid such cavilling in the future, "Clarissimus Bulfinger" would do well to look at Definition VIII, to be found in the opening pages of that work:

I likewise call attractions and impulses, in the same sense, accelerative, and motive; and use the words attraction, impulse, or propensity of any sort towards a centre, promiscuously, and indifferently, one for another; considering these forces not physically, but mathematically; wherefore the reader is not to imagine that by those words I anywhere take upon me to define the kind, or the manner of any action, the causes or the physical reason thereof, or that I attribute forces, in a true and physical sense, to certain centres (which are only mathematical points); when at any time I happen to speak of centres as attracting, or as endued with attractive power."[7]

In other words, Newton felt that his mathematical expression of gravity was not a final explanation, but a descriptive empirical law, and it was precisely this, his initial reluctance to provide a purely physical explanation for action at a distance, that was the source for Leibniz' objection that Newton's formula was a return to the discredited scholastic concept of "occult qualities." The same charge in much the same terms is repeated by Bilfinger's friend, Jakob Hermann, in the opening paper of the *Commentarii.*[8] Both he and Bilfinger were to leave St. Petersburg when their five-year term expired in 1730, after a "brevi et nervosa oratione latina" lamenting Schumacher's in-

5. "Primo itaque loco displicere videmus Viro Cl. quod attractionis voce usi simus ad exponendam actionem vitri in aquam suspensam, cum is dolere sibi profitetur, *misceri* et *officere* nostrae, quam vocat, *Hypothesi vulgarem de attractionibus litem.* Qua etiam causa excusationem quandam subiicit ad explicationem nostram *hoc vinculo soluendam.* Candide sane et perhumaniter! Eam vero nos perlibenter accepimus, quoniam tanti Viri sententia excusatione omnius indigere videmur. Ceterum si attractionis loco aliam vocem, utpote congruitatem, cohaesionem, propensionem, sive etiam impulsum aquae ad vitrum substitui placeat, nullam nos litem movebimus. Res modo, verba non motabimur," (Footnote addition: "Egregie veroiste. Nihil est, quod in hac doctrina aut desiderens aut mutatum velim.) See Jurin, "Disquisitio physicae de tubulis capillaribus a Jacobo Jurino," p. 282.

6. *Ibid.,* p. 291.

7. *Principia,* Book I, Definition VIII, as quoted by Jurin in "Disquisitio physicae de tubulis capillaribus a Jacobo Jurino," pp. 282–283.

8. I.e., "gravitas ipsa est *qualitas occulta*," cf. Jacob Hermann's "De mensura virium corporum," *Commentarii* (St. Petersburg), I (1726, pub. 1728), 37.

trigues.[9] Daniel Bernoulli followed them in 1733, noting sadly "ist der gute Newton zu bedauern, als welcher nicht nur inter Celeberrimos Goldbachios, Bulffingeros, etc. keinen Platz findet, sondern sogar mit vieler Verachtung tractirt wird."[10] With the departure of its foremost mathematicians and natural philosophers, the St. Petersburg Academy suffered a period of disarray and decline, and it was left to one of its very first students—Prince Antiokh Cantemir—to maintain the tradition that Bruce had so auspiciously begun.

9. Pekarskii, *Istoriia Imperatorskoi Akademii Nauk v Peterburge,* Cf. I, 70.

10. Daniel Bernoulli to Leonard Euler, June 20, 1742, in P. H. Fuss, ed., *Correspondance Mathématique et Physique de quelques célèbres géomètres du XVIIIème Siècle précédée d'une notice sur les travaux de Léonard Euler,* Vol. XI (St. Petersburg, 1843), p. 480.

13

"Il Propagatore del Newtonianismo"

Antiokh Cantemir is usually considered to be one of the founders of
modern Russian literature; his position as intermediary between the
philosophes and the Russian Enlightenment has often been discussed.
He was also, however, a pioneer of Newtonianism in Russia—the first
to "poetize" Newton in Russian as well as the first to attempt to popu-
larize his system in that language. If this has so far passed unnoticed,
Cantemir himself is partly responsible. Throughout the eighteenth
century he was best known to Russian readers for his translation of
Fontenelle's *Pluralité des Mondes*—a work which is sometimes regarded as the most successful popularization of the "new science" ever
written. The French original first came out as early as 1686, a year before the *Principia,* and Fontenelle, who later became a pillar of the
Cartesian establishment at the Académie des Sciences, saw no reason
to alter his entertaining account of Descartes's system.[1]

Voltaire and others made fun of his imperviousness to change, but
the long-lived secretary of the Paris Academy, though recognizing
Newton's greatness, remained opposed to the theory of universal gravitation to the very end of his days (he died in 1757). The misfortune for
Cantemir's poshumous reputation has been that he was saddled with
Fontenelle's antiquated views. Thus, Plekhanov thought of Cantemir
as a Cartesian who failed to keep abreast of his contemporaries in England and France,[2] and later studies have done nothing to dispel this

1. On Fontenelle, see J. R. Carré, *La Philosophie de Fontenelle ou le Sourire de la
Raison* (Paris: F. Alcan, 1932). Curiously enough, however, Fontenelle's celebrated *Éloge*
(1727) of Newton was the first serious account of Newton's life and achievements.
2. See G. V. Plekhanov, "A. D. Kantemir," *Istoriia russkoi obshchestvennoi mysli,* in

impression.[3] It can in fact be shown, however, that the *Pluralité des Mondes* reflected Cantemir's own views only while he was a very young man, when he was still a student at the St. Petersburg Academy. After that, circumstances conspired to remove him to England. There he met one of Bruce's friends, and in the course of his stay in London—where he lived for more than six years—Cantemir abandoned Cartesianism. Like Bruce himself some three decades earlier, he became an ardent champion of Newton's doctrines, for which he then tried to gain support within the St. Petersburg Academy.

Cantemir was born in Constantinople, probably in 1709, the year Peter the Great defeated Charles XII at the battle of Poltava. When the Turks retaliated by declaring war on Russia at the end of 1710, the ruling hospodar of Moldavia was replaced by Antiokh's father, Demetrius, a man of vast erudition and considerable accomplishment as scholar, historian, philosopher, and musicologist. His political instincts were less sound. Convinced that the star of the Ottomans was on the wane and that Russia was the rising power, Demetrius negotiated secretly with Peter. Late in 1711 he signed two treaties with him: the first providing for alliance and aid in the war; the second, for Moldavia to receive an autonomous status under the military protection of Russia. But Demetrius' ambitions were shattered by the catastrophe Peter's army suffered at the Pruth, where he was surrounded and compelled to sue for peace. Only with the greatest hazard did the Tsar manage to conceal Demetrius Kantemir from the Turks. Disguised as a woman, he and his whole family were smuggled past Turkish lines. In this way Antiokh came to be raised in Russia.

He owed his title to his father, whom Peter the Great professed to regard as the legitimate ruler of Wallachia, granting him a large an-

---

Sochineniia, ed. D. B. Riazanov, Vol. XXI (Moscow and Leningrad: Gosudarstvennoe Izdatel'stvo, 1925), 86.

3. This point is made with particular force by Helmut Grasshoff in his recent monograph on Cantemir: *Antioch Dmitrievič Kantemir und Westeuropa. Ein russischer Schriftsteller des 18 Jahrhunderts und seine Beziehungen zur westeuropaischen Literatur und Kunst* (Berlin: Akademie-Verlag, 1968), as well as in his earlier article on the poet's *Letters on Nature and Man (Pis'ma o prirode i cheloveke)* which was written in 1743 but not published until the second half of the nineteenth century. To the embarrassment of scholars who had based their analyses of Cantemir's thought exclusively on this work, Grasshoff was able to show that it could hardly be called the poet's own, that it was in fact a paraphrase of Fénélon's *De l'existence de Dieu (cf. "Kantemir und Fénélon,"* in *Zeitschrift für Slawistik,* II [1958], 382). Since Archbishop Fénélon was a Cartesian of sorts, however, Grasshoff concluded his study by substantially agreeing with those who identified Cantemir's philosophical views with those of Fontenelle.

nual pension and extensive crown lands near Kharkov. But the most curious of the privileges granted him shows how concerned Demetrius Kantemir was with the education of his children: he was promised, in view of the absence of universities in Russia, that his children could be educated abroad. When Antiokh petitioned the Tsar to make use of this right, the petition was unaccountably left unanswered. Yet his education was not neglected. Antiokh had both Greek and Russian tutors, to whom he owed some of his facility in several languages. He mastered both Latin and Greek, and, apart from Russian and modern Greek—the language he most often heard at home—he also became fluent in Italian and French. The subjects that came to dominate his interests at an early age were mathematics and philosophy. These interests remained with him for the rest of his short life and prompted his association with the Academy of Sciences, whose gymnasium he began to attend as soon as his family moved to St. Petersburg.[4]

This gymnasium consisted of five classes, and it was Peter the Great's original intention that it should function like a university. Scholars invited to the Academy would provide the faculty, and the school would in turn eventually produce some of the scholars from which members of the Academy might be drawn. What went wrong with this excellent scheme can easily be imagined. While it proved possible to provide the new Academy with scholars and scientists who enjoyed a European reputation, to provide them with students capable of benefitting from their instruction turned out to be far more difficult. There was no adequate system of lower education to support the prestigious edifice at the top.[5] And yet those who did manage to profit from the teaching they received in the gymnasium acquired a formidable education, Cantemir being an impressive example.

His instructor in mathematics was Friedrich Christian Mayer, and he appears to have formed a special affection for Friedrich Gross, who had come with Bilfinger from Tübingen. As professor of ethics and moral philosophy, Gross published a book shortly after his arrival in Russia, and he dedicated it to Antiokh Cantemir.[6] Little is known of

4. Cf. M. I. Radovskii, *Antiokh Kantemir i Peterburgskaia Akademiia nauk* (Moscow and Leningrad: Akademiia nauk, 1959), p. 17.

5. Cf. V. G. Ikonnikov, "Russkie universitety v sviazi s khodom obshchestvennogo obrazovaniia," in *Vestnik Evropy* (No. 9, 1876), pp. 161–206.

6. E.g., Friedrich Gross, *Institutiones Philosophiae rationalis seu Logicae; conscriptae in gratiam celsissimi Principis Antiochi Cantemiri* (St. Petersburg, 1726). Cantemir was close both to Gross and Mayer, but the latter fell ill and died on November 24, 1729. It was said that he taught mathematics to "the youngest son of the Wallachian hospodar

Cantemir's other contacts within the Academy at that time, though it is clear that he keenly followed some of its debates. For example, he so much sympathized with the attempt of Joseph de l'Isle and Daniel Bernoulli to propagate the heliocentric theory in 1728 that he astonished the narrow world of Moscow and St. Petersburg literati by coming out in the following year with his *First Satire*, in which he devoted some passages to ridiculing the obscurantists who still opposed the Copernican system.[7] He next turned to his translation of Fontenelle's dialogues, which could also be interpreted as a sly attempt to ridicule the Scriptures, and had it not been for the support of Theophan Prokopovich, Cantemir would no doubt have been exposed to the mercy of the Holy Synod.

As it is, his preoccupation with natural philosophy and literature was interrupted by the constitutional crisis of 1730, when the Russian nobility attempted to impose conditions on Anne of Courland, the claimant to the throne. Cantemir played an important part in the ensuing débâcle, and though he supported the new monarch against the powerful magnates who tried to restrict her authority, he was regarded, in spite of his youth, as a potential political threat. Anne therefore "rewarded" him by sending him to England in 1731 as her representative at the Court of St. James. This is why he missed witnessing some of the opening controversies over Newton's theories in the St. Petersburg Academy, but he kept in close touch with it throughout his sojourn abroad. There can be little doubt that at that point in his intellectual development he still sympathized with the Cartesians, but his views were to change radically while he was in England.

Cantemir's official link with the Academy was to keep it informed of scientific events in the English capital. From time to time he would place orders with English instrument makers, make drawings of the latest inventions for dispatch to St. Petersburg, and in other ways fulfill commissions on its behalf. Thus, when Sir Hans Sloane, who succeeded Newton as President of the Royal Society, was made an honorary member of the Russian Academy, it was Cantemir who personally

---

Demetrius Kantemir, the prince Antiokh, with such success, that in the course of something like two years, he became expert (*znatok*) in both geometry and algebra." Cf. Pekarskii, *Istoriia Imperatorskoi Akademii Nauk v Peterburge*, I, 211–212. Thus, the foundations were laid for Cantemir's later mathematical studies in London.

7. Cf. V. J. Boss, "Kantemir and Rolli-Milton's *Il Paradiso Perduto*," Slavic Review, XXI (September 1962), 454–544, and Kantemir, "Satira I," in *Sobranie stikhotvorenii*, ed. P. N. Berkov (Leningrad: Sovetskii pisatel', 1956), p. 74 (also, Cantemir's note [p. 64]).

presented him with the diploma.[8] Unfortunately, few of Cantemir's own letters from this period have survived. There is, therefore, no way of ascertaining with any precision when he first became seriously interested in Newton's work.

The inventory of his library, however, lists all of Newton's major writings: two editions of the *Principia,* a French edition of the *Opticks,* the *Arithmetica Universalis,* treatises on the "fluxionary" method, and various commentaries on Newton's work on light, as well as copious material on his controversy with Leibniz.[9] Moreover, there are in the inventory several other works which in one way or another owed their inception to Isaac Newton, such as "Wicked Will Whiston's" *Praelectiones Astronomicae,* and David Gregory's *Astronomiae Physicae & Geometricae Elementa* (see figs. 5, 35), the first textbook composed on gravitational principles (which contains a preface for which Newton is commonly held responsible).[10] Even more revealing is the fact that Cantemir appears to have gone out of his way to acquire Newton's nonscientific writings such as the geography by Bernhard Varenius. This book is often called Newton's first printed work because his name appears in large letters on the title page. Even more famous is the *Chronology* (which some of Newton's admirers would rather forget).[11]

8. Cf. M. I. Radovskii, "Angliiskii naturalist XVIII v. Gans Sloan i ego nauchnye sviazi s Peterburgskoi Akademiei Nauk," in TIIE, XXIV (No. 5, 1958), 310 ff.

9. The inventory of Cantemir's library may be found in V. N. Aleksandrenko's "Biblioteka Kantemira" in *K biografii Kn. A. D. Kantemira* (Warsaw, 1896):

i. no. 624, *Philosophiae naturalis principia mathematica auctore Neutono Amsterdam 1723 in 4* (Latin reprint of 2nd ed.);

ii. no. 705, *Newtoni Principia Philosophiae Mathematica Geneva 1739, in quarto;*

iii. no. 346, *Optique de Neuthon, traduite par la Coste* (first French ed. of the *Opticks* [*Traité d'Optique*] published in 1720; second ed., which was "beaucoup plus correct que la première," published in 1722; and a number of editions after that, any of which could have been acquired by Cantemir). He also had Smith's valuable treatise on the subject which, while not written by Newton himself, was largely based on his work: "Sistème [sic] d'Optique par Robert Smith a Cambridge, 1738 in 4 livr. anglais."

iv. Of Newton's more important mathematical writings, the inventory includes, apart from the 1707 ed. of the *Arithmetica Universalis,* the first French edition of John Colson's *Method of Fluxions and Infinite Series* (e.g., no. 323, *Méthode des fluxions par Neuthan*) and a French translation of John Clarke's revision of Humphrey Ditton's *An Institution of Fluxions . . . with some of the uses and applications of that method according to Sir Is. Newton* ("Le méthode des fluxions par Humphri Ditton corrigé par Jean Clarke in 1726 in 8").

10. Both these books are also listed in the inventory of Bruce's library. See Chapter 2.

11. Newton's chronological work was concerned with fixing the dates for the chief events of early Greek, Egyptian, Assyrian, Babylonian, Median, and Persian history and was embodied in *The Chronology of the Ancient Kingdoms Amended.* It was first published in France, without the permission of Newton who had persistently ignored

It was probably Newton's mathematical work, however, that first drew Cantemir's attention. This much may be surmised from a letter he wrote to Lord Harrington in 1735 in which he announced that he had at last found "un maitre des mathématiques," a certain "Jean Thomas" whom he described as a "cy-devant capitaine d'artillerie au service de sa Majesté brittanique."[12] Almost certainly, this is the same "Ivan Tames" who makes his appearance in the inventory of Bruce's library.[13] As one of the few Englishmen (or Welshmen?) of the era to have a professional knowledge of Russian, Captain Thomas would have been of invaluable help to Cantemir at a time when his knowledge of English was still limited.[14] This is confirmed by a curious fact: all the Newtonian items in his library, including the commentaries and the books originally written by Newton in the vernacular, are in languages other than English.

Voltaire, who only left England two years before Cantemir's arrival, appears to have had the same difficulty. For, of the English commentaries to the *Principia* that began to appear soon after Isaac Newton's death, none were widely read. The fact that as a language English had not yet become fashionable across the Channel was a further serious

---

letters addressed to him on the matter. A small part of his Biblical studies were also embodied in a book which appeared after Newton's death as *Observations upon the Prophecies of Daniel and the Apocalypse of St. John.* Among the papers he left are thousands of pages concerned with theological matters, as well as alchemical subjects. Newton's work as a historian of antiquity has recently been investigated by Professor Frank Manuel.

12. L. N. Maikov, "*Materialy dlia biografii kn. A. D. Kantemira,*" in *Sbornik otdeleniia russkogo iazyka i slovesnosti Imperatorskoi Akademii Nauk* (St. Petersburg), LXXIII (1903), 27.

13. See above, Chapter 6, pp. 72–73.

14. Thames was also a friend of Tatishchev; cf. Cantemir to V. N. Tatishchev, March 1738, in Maikov, "Materialy dlia biografii kn. A. D. Kantemira," p. 104.

О требуемых от вас часах я по си пору ни от кого ничего не слыхал, и понеже не ведаю, к кому господин Томс о том сюды писал, не могу никакой услуги вам в сем показать; к тому жь, будут ли часы к вам отправленны, теперь в Самаре вас застать не будут.

His occupation in Russia is revealed by the following ungrammatical petition to Peter the Great in N. A. Voskresenskii, *Zakonodatel'nye akty Petra I-ogo,* Vol. I, ed. Syromiatnikov, p. 85, no. 92:

Доношение иноземца Ивана Тамеса Петру I о представлении царю перевода книги английском Адмиралтействе с извинением в замедлении выполнении порученного дела, от 13 июля 1720 года.

Державнейший монарх...

С приносителем сего письма, з господином Татищевым, послал я к В. В. книгу о надлежащем отправлении Адмиралтейства, которою я по указу В. В. с аглинского на российской язык перевел...

С Москве, июля 13 д. 1720.

В 4 д. августа.   Иван Тамес.

handicap. Voltaire sought to remedy this situation through his celebrated efforts on Newton's behalf, but Cantemir disliked the sage of Ferney—and it has usually been supposed that this was due to alleged philosophical differences with him.[15]

Hence his reaction to the news that Voltaire was about to write "un roman de la philosophie de Newton . . . un roman qui sera à capacité de tout le monde" was to dismiss the idea as preposterous.[16] "La philosophie de Newton traitée par M. Voltaire, qui assurement ne sait point d'algèbre rendue à la capacité de tout le monde! Ne riez-vous pas, madame, de cette proposition? Si le pauvre homme a fait voler les armées sans marquer leur chemin et le ressort qui le faisoit agir, jugez par là ce qu'il fera, lorsqu'il voudra expliquer les mouvements des corps célestes! J'ai peur que nous ne soyons pas en sûreté sur la terre. Voilà par exemple un cas, qui vous doit donner une ample matière pour vous récréer de la mélancolie."[17]

But, as we can see from the letter Cantemir wrote to the same Marquise the following year, his dissatisfaction was in no way directed against the object of Voltaire's endeavor. He even solicited her conversion to Newtonianism, as long, presumably, as Voltaire was not her guide. In other words, he was ridiculing the *Elémens* before it was even completed. His comment on the *Lettres Philosophiques,* which so scandalized authorities in France, was equally disdainful. It was, he wrote, nothing but a collection of "discours [que Voltaire] a entendu aux cafés de Londres."[18] The author, Cantemir wrote in another letter to the same lady, "m'a paru dans la plupart de ses ouvrages un homme qui se pique d'écrire sur les matières qu'il n'entend pas." His advice to Voltaire was to stick to satires and epigrams—advice Cantemir himself was following with some success—"mais qu'il n'y entre point de la philosophie." But at the same time what could Cantemir then have read of Voltaire's "ouvrage" to which he refers with such condescension? His *Essai sur la Poésie Epique* (which Voltaire wrote while he was in England), the *Henriade,* some plays, perhaps, and some of his

15. Cf. F. Ia. Priima, "Antiokh Kantemir i ego frantsuzskie literaturnye sviazi," *Trudy otdela novoi russkoi literatury, Institut russkoi literatury,* I (Moscow and Leningrad: Akademiia nauk, 1957), 7–45.
16. Cantemir to the Marquise de Monconseil, November 29, 1736, in Maikov, "Materialy dlia biografii kn. A. D. Kantemira," pp. 63–64.
17. *Ibid.,* p. 64.
18. *Ibid.*

poetic productions, but the only work with philosophical pretensions would have been the *Lettres,* concerning which Cantemir's comments are not unjustifiable. Voltaire himself admitted that the "Letters" lacked depth. The book's great virtue was that it sparked the discussion over the theories of "M. Loke" and "Sir Newton," which at that point amply fulfilled Voltaire's expectations. He then retreated from the storm the book had aroused into the seclusion of Madame du Châtelet's house at Cirey to pursue more serious studies of Newton, of which the *Elémens* was one brilliant result.

The *Elémens de la Philosophie de Neuton,* which appeared at the beginning of 1738, was dedicated appropriately to Madame du Châtelet, whose distinguished French translation of the *Principia* was posthumously brought out by Clairaut.[19] Voltaire's work, which dealt both with Newton's discoveries in light and with the theory of universal gravitation, should of course have found a friendly response in Cantemir. It did not do so, paradoxically enough, because the Russian poet was immersed in a study of the same subject. Indeed, his mockery of Voltaire was made with the air of a man who considered himself just as knowledgeable in the matters on which the author of the *Elémens* had now set himself up as an authority. Hence his gibe at Voltaire's incompetence, as he supposed, to explain Newton because "[il] ne sait point d'algèbre." It so happens that Cantemir was himself engaged in writing a manual on algebra. Yet it would be unfair to dismiss his attitude as being wholly compounded of the envy Voltaire so often inspired in fellow "littérateurs." The fact is, Voltaire's work was not at first well received in English society. Thus, we find Marie Clare, Viscountess Bolingbroke, dismissing it in much the same tone as Cantemir in a letter she presumably wrote after scanning the *Elémens,* though the date of her letter suggests she could hardly have spent much time reading it ("There is a book of Voltaire in which he claims to have brought Newton's philosophy within the reach of all; I do not send it to you because this object is not attained in it at all and I do not believe you would claim that you understand it any more than I do, or than he himself does. All he did was to put together and trans-

19. Voltaire's elegant eulogy to this remarkable woman with whom he lived for so many years—"vaste et puissante génie, Minerve de la France, immortelle Émilie, disciple de Newton et de la verité"—is repeated in the handsome introduction he wrote to the first volume of her translation.

late into French a few written notes that Pemberton and others gave him").[20]

There was, in other words, a certain resentment in London that a Frenchman who was not even a natural philosopher, but a mere poet and dramatist should take it upon himself to explain Newton to the world. Resentment later turned into admiration, but the initial reaction explains the peculiar character of Cantemir's own remarks about the *Elémens*. That is to say, far from disagreeing with Voltaire's philosophical position, Cantemir fully endorsed it and only regretted —prematurely, as it turned out—that the man was unequal to the task. He turned instead to Francesco Algarotti, a friend and rival of Voltaire, who anticipated the *Elémens* with the *Neutonianismo per le Dame*, a work which was not only the first but also the most successful popularization of Newton to appear on the Continent.[21]

Algarotti's intellectual development bears a remarkable resemblance to Cantemir's own. He was born in Venice in 1712, and, like the Russian poet, began to write verse at an early age. In 1728, just before Cantemir began to translate the *Pluralité des Mondes*, Algarotti too read the work and became passionately interested in science. Yet his faith in Fontenelle was soon shaken through circumstances which, as in Cantemir's case, fortuitously brought him in contact with Newton's thought at a time when it was still only poorly appreciated by his countrymen. With Algarotti the intellectual incentive was provided by the Anglomania that swept Italy even earlier than France, so that, when he emerged from childhood, ideas and tastes of English provenance began to rival the dominance long held by Paris.[22] The attraction was mutual and reciprocal. The "Grand Tour" had become an essential part of an English nobleman's education, and it was a meeting with one of these English travelers, Martin Folkes, who had been vice-president of the Royal Society during Newton's presidency, that made Algarotti decide to visit England.

20. The letter (to Isabella, Countess of Denbigh) is dated June 18, 1738, and is quoted in translation from the Denbigh Papers by M. R. Hopkins in *Married to Mercury: A Sketch of Lord Bolingbroke and His Wives* (London: Constable, 1936), p. 251.

21. Francesco Algarotti, *Il Neutonianismo per le Dame, ovvero Dialoghi sopra la Luce i Colori* (Naples, 1737); cf. Babson, no. 145; Gray, no. 194, lists 1738, 1739, and 1746 eds. and also cites editions of 1752, 1757, and 1836. The English ed., translated by Elisabeth Carter (a friend of Dr. Johnson), appeared as *Sir Isaac Newton's Philosophy Explain'd for the Use of Ladies In Six Dialogues on Light and Colours,* from the Italian of Sig. Algarotti (London, 1738; 1742).

22. Cf. Arturo Graf, *L'anglomania e l'influsso inglese i Italia nel secolo XVIII* (Turin: E. Loescher, 1911).

He did so while Cantemir was himself engrossed in his study of Newton's work with Captain Thomas; in this way the Russian poet came to hear from the author's own lips about the *Neutonianismo*.[23] According to Voltaire, Algarotti benefited no end from the helpful criticisms of Emily de Breteuil, whom he judged to be "assez savant dans sa langue [Italian] pour lui donner de très bon avis."[24] In reality, however, Algarotti owed far more to Fontenelle, for though he attacked the Cartesian system, he plagiarized the format of the *Pluralité des Mondes* so that his work became a kind of parody of that by Fontenelle. The latter's feelings may well be imagined, for not only did Algarotti mercilessly expose Descartes's "errors" in the light of Newton's theories but he had the further effrontery to dedicate the *Neutonianismo* to Fontenelle.[25] Nor was Voltaire's expected endorsement one of unrestrained enthusiasm; he felt that Algarotti's dialogues had stolen the thunder from his own book on the same subject,[26] (which was still resting with his publisher in Holland).

Elsewhere, however, the *Neutonianismo* met immediate success and was soon translated into many languages. Cantemir must also have thought very highly of it, for he began translating the manuscript before it even appeared in print.[27] "Il traite fort bien sa matière," he wrote to a friend in Paris, "et il n'est pas moins clair que badin, de sorte que si vous voulez devenir Newtonienne à peu de frais, vous le serez à la seconde lecture de ce livre . . ."[28] Unfortunately, the Russian

23. Cantemir to the Marquise de Monconseil, July 10, 1738, in Maikov, "Materialy dlia biografii kn. A. D. Kantemira," p. 110.

24. Voltaire, *Oeuvres*, 72 vols., ed. Adrien Beuchot (Paris, 1834), I, 8. Voltaire was clearly peeved at Algarotti for another reason: he had originally intended to dedicate the work to his paramour, the Marquise du Châtelet.

25. "Il étoit bien juste que les Dames," he wrote in his dedication, "qui par votre secours se sont apperçues du grand changement que Descartes avoit introduit dans le *Monde Pensant,* s'apperçussent aussi du changement nouveau dont Newton est l'auteur . . . Vous avez embelli le Système des *Cartesiens,* j'ai tâché de *domter le Newtonianisme* & de lui prêter des attraits." Cf. *Le Newtonianisme pour les Dames, ou Entretiens sur la lumière, sur les couleurs, et sur L'attraction,* 2nd ed., tr. Mr. Du Perron de Castera (Amsterdam, 1741; 1st ed., Paris, 1738), I, vii.

26. "Il a pris les fleurs pour lui et m'a laissé les épines," wrote Voltaire, *Oeuvres,* XXXIV, 187.

27. According to Octavien de Guasco, Cantemir's biographer and friend (cf. *Satyres du Prince Cantemir* [London, 1749], the poet made his translation of the *Neutonianismo* "étant encore à Londres." But, in view of the book's length, it may be doubted whether Cantemir completed his translation before moving to Paris in 1738; yet he did apparently finish it. This much is confirmed by Domenico Michelessi, one of Algarotti's first biographers, according to whom the completed MS was sent to Algarotti, who then composed an epistle in verse in praise of the Russian Empress.

28. Cantemir to the Marquise de Monconseil, July 10, 1738, in Maikov, "Materialy dlia biografii kn. A. D. Kantemira," p. 110.

version remained unpublished. Instead, the St. Petersburg Academy brought out Cantemir's earlier translation of the *Pluralité des Mondes* at a time in his life when he no longer believed in the principles Fontenelle proclaimed.[29] And Algarotti had to content himself with a new and revised Italian version of the *Neutonianismo,* which he dedicated to Her Sacred Imperial Majesty of All the Russias (see fig. 37). Ironically enough, the preface to that edition ends on a warm note of gratitude to the Russian poet for turning his masterpiece into Russian. Thanks to him the true doctrine would be disseminated in the new world: "e la vera Dottrina fia ben tosto, mercé lui, sparsa in nuovi Mondi . . ." and Cantemir would be gratefully remembered by posterity as "il Propagatore del Newtonianismo nel vasto Imperio delle Russie'."[30]

In reality, of course, the St. Petersburg Academy was, as we have seen, far from receptive to Newtonian theories. This may have been largely responsible for the failure of Cantemir's *Razgovory o svete,* as he called his version of the *Neutonianismo,* to see the light of day. Its nonpublication must have been as disappointing to Cantemir as it was for Algarotti, who stood to gain much by the appearance of his work in a country which was just beginning to arouse the intense curiosity of Voltaire and the "philosophes." He may have been consoled, however, by the realization that the Cartesians who were so firmly entrenched in academies and universities throughout the Continent were

29. The book (Fontenelle, *Razgovory o mnozhestve mirov* . . ., tr. Antiokh Cantemir) was published in St. Petersburg in 1740, after the Holy Synod had tried to prevent publication. But Cantemir managed to convey that the work was in a sense an anachronism by insisting that the original date on which he completed the MS be stamped on the title page, i.e., 1730.

30. Cf. the preface to *Il Newtonianismo per le Dame, ovvero Dialoghi sopra la Luce, i Colori, e l'Attrazione,* rev. ed. (Naples, 1739), p. 11. This particular edition is scarce. A presentation copy, dedicated to Empress Anne, may be found in the Babson Collection at the Babson Institute, Wellesley, Massachusetts. (not listed in Gray; Babson catalogue, no. 146). The Empress died in 1740. Algarotti's work would have been known in St. Petersburg through other copies that he had sent there through Cantemir. Voltaire's *Elémens* was formally presented to the St. Petersburg Academy only in 1745; cf. *Protokoly zasedanii konferentsii Imperatorshoi Akademii Nauk s 1725 po 1803 goda* . . ., ed. Veselovsky, II, 78, for August 20, 1745: "Staehlinus tradidit Academiae librum Voltarii Gallice conscriptum, cui titulus: Elementa philosophiae Newtonianae, quem ab Imperii Cancellario acceptum retulit. Hanc librum Delilius secum sumsit, Bibliotheca Imperialis, cui destinatus est, brevi restituendum." As a result, Voltaire was made an honorary member of the Academy: "not only on account of his service to literature but also to science—through his work on *La Philosophie de Mr. Newton* mise à la portée de tout le monde." Cf. also M. S. Filippov, ed., *Istoriia Biblioteki Akademii nauk SSSR (1714–1964)* (Moscow and Leningrad: Izdatel'stvo "Nauka," 1964), p. 81, no. 361.

at last being forced to acknowledge defeat in an area where speculation was compelled to give way to demonstrable proof.

Through his own contacts in London and Paris (where Cantemir moved in 1738) the poet was particularly well placed to observe the resolution of the protracted controversy over the configuration of the earth. He knew Maupertuis, its main protagonist on the Newtonian side, as well as Clairaut, both of whom set out on their famous expedition to the Gulf of Bothnia in the summer of 1736. Algarotti announced his intention of accompanying them; Cantemir, who hoped to become the Russian Academy's next president, contented himself by doing all in his power to have it reward them for their monumental achievements in the field. Eventually he succeeded. By then the whole controversy had reached St. Petersburg, where the argument over the shape of the earth obliged Newton's opponents to re-examine their objections to the principle of universal gravitation.

14

The Shape of the Earth

It was only during the third decade of the eighteenth century that the Newtonian theory of universal gravitation began to make serious headway on the Continent, and, though the general character of the Cartesian and Leibnizian objections to it remained unchanged, the controversy over its validity became more specific. Newton had argued that the circumference of the earth was wider at the equator than across the poles owing to a diminution of gravity occasioned by the diurnal rotation of the earth (cf. *Principia,* Book III, Proposition XIX, Problem III). If this could be disproved, the validity of the theory of gravitation might be undermined and, with it, the whole Newtonian cosmology. This is why the problem of the shape of the earth assumed such significant proportions in the eyes both of Newton's disciples and of their opponents, which Voltaire so wittily summed up in his celebrated *Lettres Philosophique* (1734): "A Paris vous figurez la terre faite comme un melon: à Londres, elle est applâtie des deux côtés."

For the physical origins of this startling pronouncement, one must go back to the genesis of the revolutionary idea from which the *Principia* sprang. To prove that the action of gravity was as if the mass of the earth was concentrated at its center, Newton had to make his calculations from one mathematical point to another: for example, from the center of the moon to the center of the earth. In 1666, he had begun "to think of Gravity extending to ye orb of the Moon . . ."[1]

1. It is to this period that the famous story of Newton and the apple belongs. It had its origins with Stukely, while Voltaire passed it on (in the *Lettres Philosophiques* and elsewhere). See also Richard D. de Villamil's discussion of this in *Newton: The Man,* with a foreword by Albert Einstein (London [1931]), pp. 41–43.

Having found out how to estimate the force with which a globe revolving within a sphere presses the surface of the sphere from Kepler's rule, Newton deduced "that the forces which keep the Planets in their Orbs must [be] reciprocally as the square of their distance from the centers about which they revolve." Yet not till the 1680's did Newton return to this crucial problem. The reasons for the delay are still controversial, but the fact remains that, though in the middle of the sixties the data upon which Newton worked may not have been radically wrong, by 1684 he was able to make use of more accurate observations,[2] for, in 1672, a French expedition under Jean Picard had enabled simultaneous measurements of the altitude of Mars to be taken in Paris and in Cayenne. These results made it possible to secure a more accurate estimate of the sun's mean distance from the earth. It was worked out at eighty-seven million miles, coming near to the modern calculation of ninety-two million. The calculations of Picard were made known to Newton in 1684 and were the ones he used to obtain the results of the *Principia*.

By proving these results inconsistent with the true shape of the earth, Newton's opponents hoped to demonstrate the fallacy of his theory of universal gravitation. The French Academy selected this problem as peculiarly its own, and owing in part to the authority enjoyed by D. Cassini and J. Cassini, most of its members adopted the oblong form.[3] The dispute also found its distant echo at the very first meeting of the Academy in St. Petersburg on November 13, 1725. The problem chosen for the discussion on this occasion was taken from the Third Book of the *Principia* (Proposition XIX), and dealt

2. Brewster and others repeat the story that, "being away from his books" etc., Newton did not know the correct size of the earth and, therefore, the true distance of the moon. Hence, he may have found that his calculations gave incorrect results, i.e., results which did not agree with his revolutionary theory. But De Villamil (who unearthed the catalogue of a part of Newton's library) points out that R. Norwood had calculated the diameter of the earth by measuring the distance from York to London as early as 1636 (*Seaman's Practice*, 5th ed., 1662). "Is it conceivable," asks De Villamil in his *Newton*, "that Newton did not know it?" Yet it should be recalled that Newton hesitated for about a quarter of a century in publishing his three small papers on "Fluxions" and twenty-nine years elapsed between the time he made his optical discoveries and the time they were published in the *Opticks*.

3. Jacques Cassini (1671–1794) was the son of G. D. Cassini, who had been called to Paris by Louis XIV in 1669 and, till the year of his death in 1712, remained what was remarkable at that late date—a stunch anti-Copernican. The Cassini regime at the Paris observatory lasted for a century and a quarter (till 1794). Their conservative bias gradually weakened as the dynasty came to an end, but it was very injurious to French science.

precisely with this issue. Bilfinger apparently supported the Cassinian position, while Jakob Hermann argued in favor of Newton's: "the shape of the earth is a spherical figure, with the shorter axis passing through the poles—the shape demonstrated by Newton in the 'Principia philosophiae mathematicae.' "[4]

This is all the surviving record states, though this is enough, it might be argued, to suggest that a genuine difference of opinion existed. Nor would Bilfinger's opposition to the Newtonian view seem surprising in light of his support of Cartesian vortices. But was Jakob Hermann's defence of it more than a formality dictated by the requirements of academic discussion?[5] The answer is to be found in a work he devoted to the problem before his arrival in St. Petersburg. Hermann (1678–1733) came to the Russian Academy with an established reputation. The first foreigner to commit himself to serving the new institution, he was regarded as its most illustrious acquisition, a status duly recognized by his official sobriquet of "Professor primarius et Matheseos sublimioris." His prestige would easily have obtained for him a position of intellectual authority above and apart from the rivalry that formed within the Academy so soon after its opening. But he took part in the dispute over "vis viva,"[6] and in Bilfinger's quarrels with Daniel Bernoulli chose to ally himself firmly with the former. This was, so the secretary of the Academy suggested slyly in his report to Blumentrost, because the mild-tempered Hermann shared a house with Bilfinger and tended to fall under the influence of the younger man.[7]

A more convincing explanation, however, is the enmity Hermann felt toward Daniel's brother, Nicholas Bernoulli, who had harshly

4. I.e., Hermannus de figura telluris sphaeroide, cuius axis minor sit intra polos, a Newtono in Principiis philosophiae mathematicis synthetice demonstrata analytica methodo deduxit. Opposuit Bülfingerus has demonstrationes locum habere, si terra antequam circa axem rotaretur sphaerica fuisset, sed de hoc ipso dubitari posse. [Hermann proved synthetically by the analytic method that the shape of the earth is a spherical figure, with the shorter axis passing through the poles—the shape demonstrated by Newton in the 'Principia philosophiae mathematica.' Bilfinger opposed by maintaining that these proofs are relevant only, if, before revolving round its axis, the earth were spheroid, but this precisely may be doubted.]

5. Academician T. P. Kravets, who refers briefly to this episode in "N'iuton i izuchenie ego trudov v Rossii" (in *Isaak N'iuton, 1643–1727, Sbornik statei k trekhsotletiiu so dnia rozhdeniia*, ed. S. I. Vavilov, [Moscow and Leningrad: Akademiia nauk, 1943], p. 314), appears to have been under the impression that this discussion was evidence of the support Newton enjoyed as soon as the Academy opened. The contrary is in fact the case.

6. Cf. Hermann, "De mensura virium corporum."

7. Pekarskii, *Istoriia Imperatorskoi Akademii Nauk v Peterburge*, I, 70.

criticized one of his mathematical memoirs.[8] This would have predisposed him in Bilfinger's favor, with whom he had far more in common in spite of the fact that he, like the Bernoullis, came from Basel; and he had even studied mathematics under their uncle, Jacob. Yet he owed his career largely to Leibniz, whose protégé he became. He first gained Leibniz' gratitude when he defended Leibniz' differential calculus against a Dutch mathematician.[9] It was at Leibniz' recommendation that Hermann was made a member of the newly founded Berlin Academy in 1701, and it was also to him that Hermann owed his chair of mathematics at Padua, which he later exchanged, again on Leibniz' recommendation, for a similar position at Frankfurt on the Oder, where he completed the *Phonoromia, seu de viribus et motibus corporum solidorum et fluidorum* (Amsterdam, 1713). This was Hermann's most important work, highly thought of by Leibniz.[10] From it Hermann's real attitude to the Newtonian proposition concerning the earth's shape may easily be ascertained.

The subject of the treatise—the forces and motions of bodies—prompted Hermann to examine Huygens' problem of the relative equilibrium of rotating fluids under the action of a constant force directed to a point on the axis of rotation. His solution took two forms: one was based on Newton's principle of columns balancing at the center; the other, on Huygens' principle of the plumb line. But Hermann observed that his result did not agree with the one Newton had obtained for the ratio of the axes of the earth. Though he did not actually say that Newton was wrong, he did imply that his own was the correct result (the ratio he obtained was 577 to 578), thus lending support to the argument of D. Cassini and his friends in Paris, who used this discrepancy to pour scorn on the Newtonian theory of gravitation.

It is hardly likely, therefore, that Hermann could have changed his opinion on the matter when the meetings of the Russian Academy began in 1725, nor is there any real evidence that he did. He contributed nothing substantial to this controversy during his sojourn in the

---

8. Cf. J. F. Montucla, *Histoire des Mathématiques*, Vol. III, new ed. (Paris, 1802), p. 326. The offending review was placed in the *Acta Eruditorum* for 1720. "Ce géomètre," says Montucla of Hermann, "quoique élève distingué de Bernoulli, a en effet bien souvent encouru avec justice la reproche d'une certaine précipitation qui lui faisoit donner comme complettes des solutions imparfaites et même quelquefois erronées." (*Histoire*, p. 342.)
9. I.e., Jacob Hermann, *Responsio ad cl. Nieuwenteyt. Considerationes secundes circa calculi differentialis principia* (Basel, 1700).
10. Leibniz, *Virorum celeberr.*, II.

St. Petersburg Academy, and his association there came to an abrupt end in 1730 when he decided to accompany Bilfinger back to Germany. Yet the issue of the earth's configuration remained very much a live one and took a new turn in 1733 when a scheme was seriously proposed to settle the dispute between the Cassinians and the Newtonians by measuring an arc of the meridian near the equator in order to compare the corresponding length of a degree with that which had already been obtained by Picard and J. Cassini. If the degree proved to be longer in northern than in southern latitudes, then the earth was flattened in the polar regions, as Newton had reasoned from dynamical considerations.

While this project was being promulgated, each side tenaciously maintained its view, and the division of opinion was appropriately reflected in the double prize offered by the Paris Academy in 1734. The subject related to the inclination of the planes of the orbits of the planets to the planes of the Sun's equator. The prize was divided between John Bernoulli and his son Daniel, who had so recently left St. Petersburg. The elder Bernoulli used a system of vortices to arrive at his answer, and he went out of his way to depreciate Newton[11] while Daniel stoutly defended him. The central issue could only be validated in the field, however, and the French Academy finally decided to dispatch two expeditions. One, under Bouguer, La Condamine, and Godin, left Paris for Peru in May 1735 and did not return until 1743. Soon after it started, it was thought advisable to measure another arc as near as possible to the North Pole; another expedition was promptly organized under the direction of Maupertuis, who had pioneered the introduction of Newton's doctrines into France.[12]

Maupertuis, whose expedition had the support of Aléxis Clairaut, a brilliant mathematician who like him opposed the Cartesian majority within the French Academy, was long in doubt whether he should go to Iceland, Norway or the Gulf of Bothnia. Louis de l'Isle de la Croyère had earlier impressed astronomers through his pendulum

11. It ends, as Todhunter notes, on a note of premature triumph: "Après cette heureuse conformité de nôtre théorie, avec les observations célestes, peut-on plus long-temps refuser à la Terre la Figure de sphéroide oblong, fondé d'ailleurs sur la dimension des degrés de la méridienne, entreprise et exécutés par le même M. Cassini, avec une exactitude inconcevable?" (*A History of the Mathematical Theories of Attraction and the Figure of the Earth* . . ., 2 vols [London, 1875], I, 116; no. 221.

12. On Maupertuis, see Pierre Brunet's masterly study, *Maupertuis, étude biographique,* 2 vols. (Paris: A. Blanchard, 1929).

observations at Archangel on the White Sea,[13] but the difficulties of such work in northerly latitudes was well realized. When the Russian Academy met to consider the whole project, his brother, Joseph Nicolas de l'Isle, eloquently described some of the obstacles.[14] He was with the first contingent to arrive in St. Petersburg after the Academy's opening and was undoubtedly one of its most able members. Not only did he lay the foundations for systematic astronomy in Russia, but he was responsible for the first accurate maps of the country, which his own expeditions helped to complete. He stayed in St. Petersburg until 1747, but he appears to have lost faith in vortices at a remarkably early date and in the controversy over the earth's shape was predisposed against the Cassinian thesis. This is one of the reasons why Daniel Bernoulli wanted him to take part, with Maupertuis and Clairaut, in their enterprise to test Newton's theory. His participation, he wrote, would serve as a most praiseworthy token of international cooperation.[15]

But Fontainebleau and the Court of St. Petersburg were deeply suspicious of each other's political intentions; besides, Captain Bering's expedition to Kamchatka, then negotiating its arduous way through Siberia, was absorbing much of the Academy's resources and attention. Joseph de l'Isle therefore played no part in Maupertuis' venture and had to content himself with his own project "Sur les operations pour les mesures de la terre proposées en Russie."[16] This

13. Louis de l'Isle de la Croyère, "Observatio longitudinis penduli simplicis facta Archangelopoli . . .," *Commentarii* (St. Petersburg), IV (1729, pub. 1735), 322–328.

14. Cf. *Protokoly zasedanii konferentsii Imperatorskoi Akademii Nauk*, I, 351, for February 4, 1737. "Fürnehmlich zeigte Prof. De l'Isle, dass weder auf dem Wasser, noch auf dem Eise oder Lande, mit Schlitten oder Wagen oder zu Fuss so schlechthin der Vorschlag angehe, weil die Superficies terrae, wo die Linea tangens seyn soll, eine dergleichen Höhe nicht mache, dass das vorgeschlagene Licht wie ein Punct verschwinden und wieder erscheinen könne, maassen der geringste Hügel oder abfallende Tiefe auf dem Erdreich, Eise oder Wasser Welle solches behindere . . . etc."

15. Daniel Bernoulli to Leonard Euler, October 26, 1735, in Fuss, ed., *Correspondance Mathématique*, II, 428.

16. Cf. *Protokoly zasedanii konferentsii Imperatorskoi Akademii Nauk*, I, 346, for January 21, 1737: "Hr. Prof. De l'Isle hat hierauf eine Französische Piece produciret und zugleich aufgelesen, welche betitelt ist: Sur les operations de la terre proposées en Russie. Nachdem derselbe die Lection geschlossen, hat er das Praes. zwar darauf setzen und die Blätter paginiren, aber nicht dem Archiv gelassen, sondern wieder mit sich genommen." The Russian translation was promptly published by the Akademiia nauk in the same year: *Predlozhenie o merianii zemli v Rossii, chtennoe v konferentsii sanktpeterburgskiia imperatorskiia (sic) Akademii nauk, genvariia 21 dnia, 1737 goda chrez gospodina de l'Ilia, pervogo professora astornomii* (St. Petersburg, 1737).

paper, which was read before the Russian Academy and then promptly published in the *Philosophical Transactions,* has been praised by Todhunter, but it was written before the results of the two expeditions dispatched by the Paris Academy were known, and de l'Isle apparently hoped to engage the interest of various court officials in supporting a similar operation on Russian soil. This it failed to do, though the scheme attracted considerable attention abroad.[17] Maupertuis had finally decided to go to the islands along the shore of the Gulf of Bothnia, but they proved too low and too near the shore to form advantageous stations, and the expedition had to move inland up the river Tornea (which is crossed by the Arctic circle). It was a bold undertaking for the time. Tangled forests had to be penetrated, icy rapids navigated, and quadrants handled when the mercury had sunk far below zero—by gentlemen fresh from the drawing rooms of Paris. The exposure to extremes of heat and cold, excessive rains, want of proper food, and the havoc perpetrated by mosquitoes reaped its inevitable toll. Le Monnier, one of Maupertuis's colleagues, fell very ill, and Maupertuis's own health was permanently impaired. In his absence, Emily de Breteuil tirelessly waged the struggle on behalf of the Newtonians in the salon, while Voltaire and Francesco Algarotti dramatized their cause with their pen.

In the Russian Academy long discussions were held over the prospects of the expeditions's success. Andreas Celsius, the professor of astronomy at Uppsala University who was associated with the enterprise, joined these discussions by sending to St. Petersburg a paper defending the operations in Lapland from an objection urged against them by J. Cassini before the Paris Academy,[18] though the actual

---

17. De l'Isle's paper came out in German, Russian, and English translations, as well as in the French original: *Projet de la mesure de la Terre en Russie* (St. Petersburg, 1737). cf. "A Proposal for the Measurement of the Earth in Russia, read at a meeting of the Academy of Sciences of St. Petersburg, January 21, 1737, by Mr. Joseph Nicholas de l'Isle, first Professor of Astronomy, and F.R.S., Translated from the French printed at St. Petersburg 1737 4to. by T.S. M.D. F.R.S.," pp. 27–49 of No. 449 of the *Philosophical Transactions,* No. 449 (January–June 1737), pp. 27–49. It forms part of Volume XL (1738).

18. Cp. *Protokoly zasedanii konferentsii Imperatorskoi Akademii Nauk,* I, 349, for January 29, 1737: "Hr. Justiz-Rath Goldbach producirte 1) einen von dem Hrn Andreas Celsius in Schwedischer Sprache d. 5 Octobr. a praet. zum Druck gegebenen Brief von der *Figurae Terrae* 2) diesen Brief ins Lateinischer Sprache entworfenes Tentamen ad investigandum utrum axis terrae intra polos major sit an minor diametro aequotoris, er verlangte, dass diese dreyerley Schrifften in Continenti den Hrn Prof. De l'Isle zugestellet, und darüber vernommen werden soll ob es sein Wille dass auch die übrigen Hrn Professors gedachte Schriften de Figura Terrae zugleich mit seiner den 22 hujus einge-

measurements had by then already been made. The results, however, were at first kept secret, "tant pour se donner de loisir de la réflexion sur une chose peu inattendue, que pour avoir le plaisir d'en apporter à Paris la première nouvelle."[19] After suffering shipwreck in the Gulf of Bothnia, the expedition returned in August 1737; in the folowing year Maupertuis published its findings, which were subsequently published in most European languages. The ratio of the polar to the equatorial diameter of the earth was found to be 177/178, and critics of the Newtonian theory of the earth's shape, including the astronomer Jacques Cassini, were decisively confuted.

But the controversy was not really wholly settled, in spite of the triumphant welcome which Voltaire and his friends reserved for Maupertuis upon his return, the Lapland Hercules, who had "aplati les pôles et les Cassini." Newton's theory gave the ratio of flattening as 220/230, and this divergence provided the pretext some of its opponents required to continue to hold it in doubt. Indeed, Descartes's authority was still far from dead, as is indicated by the way the French Academy distributed its awards in 1740. Tides were chosen as the subject of the prize essay, and four essays were published as a result. One, by the Jesuit Cavalieri, fully retained the Cartesian system of vortices. He shared the prize with three other entries by Colin Maclaurin, one of Newton's outstanding Scottish disciples, Leonard Euler (then still at St. Petersburg), and Daniel Bernoulli, who, like Maclaurin, employed attraction for the purpose of his essay.[20] Euler did not, and it is curious to recall Bernoulli's courteous reprimand to him for failing to do so. He was glad, he wrote, to share the prize with Euler, but he knew "dass wenn Ihre pièce de aestu maris nicht so vollkommen schön ware befunden worden, wie ich sie auch befinde, Sie keinen Theil an dem praemio würden bekommen haben, da des

---

brachten Piece einsehen und sich bekannt machen mögen?" As is evident from the minutes of the meeting for February 4 (*ibid.*, p. 352), De l'Isle did not think highly of the paper by Andreas Celsius. The title is nowhere indicated, but it was probably the item noted by Joseph LaLande in his *Bibliographie Astronomique avec l'Histoire de l'Astronomie, depuis 1781 jusqu'a 1802* (Paris, 1803), p. 406: *De Observationibus pro figura telluris determinanda in Gallia habitis disquisitio* (Uppsala, 1738).

19. Paris *Mémoires* for 1737, on page 94 of the historical portion; quoted by Todhunter, *History of the Mathematical Theories* I, 98, no. 188. E.g., Maupertuis, *La Figure de la Terre déterminée par les observations . . . au cercle polaire* (Paris, 1738).

20. Cf. Todhunter, *History of the Mathematical Theories*, I, 327, no. 501 f. Bernoulli had this to say of Newton's method of determining the shape of the earth: "Quant à son raisonnement, il n'y a peut-être que lui pût y voir clair; car ce grand homme voyoit à travers d'un voile, ce qu'un autre ne distingue qu'à peine avec un microscope."

Newtons Reputation in Frankreich nunmehro so gross ist als in England selbst . . ."[21]

This was not perhaps strictly true, for the French Newtonians had still not conquered the Academy in Paris, but the certitude with which Euler rejected the theory of universal gravitation in his essay was nonetheless unusual, as Bernoulli went on to tell him: "Es hat mich auch Wunder genommen, dass Sie mit so grossen elogiis von den vorticibus reden, ja praetendiren demonstrirt zu haben, dass sich die Sach unmöglich anders verhalten könne . . ."[22] But if this was remarkable in 1740, Euler's continued opposition to Newton's doctrines proved even more extraordinary at a much later date. When Clairaut reinvestigated the Newtonian theory in 1743, he obtained a much better agreement with Newton's ratio, and he even produced a formula relating the earth's ellipticity to the gravitational attraction at any latitude which should have satisfied Newton's critics once and for all.[23] This too failed to shake Euler's convictions.[24]

Indeed, Euler, who never concealed his disagreement with Newton's doctrines, became the Russian Academy's most influential scientific figure in the course of the controversy over the earth's shape. Between 1733, when he assumed Daniel Bernoulli's chair of higher mathematics, and 1741, when he departed for Berlin, Euler's voluminous contributions to the *Commentarii* and the prizes awarded to him abroad lent him unprecedented authority in St. Petersburg, which he used to support the Cartesian system. At the very time that Maupertuis and his party were triumphantly vindicating Newton's calculations, Euler was preoccupied in composing his first treatise on light—*Sur la Nature et les Proprietés du Feu*—which rejected the Newtonian emission theory. In spite, or perhaps because, of this, the work was promptly crowned by the Académie des Sciences in Paris in 1738. Throughout that period, thanks in part to Euler, the St. Petersburg

21. Daniel Bernoulli to Leonard Euler, January 21, 1742, in Fuss, ed., *Correspondance Mathématique*, II, 481. He goes on: "Ich glaube in der hypothesi vorticum und derselben examine so weit gegangen zu seyn, als ein Anderer, und kommen mir doch dato noch ganz apocryphisch vor, ja, dass sie die gesunde Vernunft blassieren."
22. *Ibid.*
23. Cf. Pierre Brunet, "La vie et l'oeuvre de Clairaut," in *Revue d'Histoire des Sciences* (Paris), IV (1951), 105–132.
24. See Euler's comments on Clairaut's earlier investigations, expressed in a letter to Goldbach, November 6, 1744: "Die Solution ist . . . die *Newton*ianische und kann freilich nicht allgemein sein." (Iushkevich and Winter, eds., *Leonhard Euler und Christian Goldbach*, p. 208.) At the same time, Euler admired Clairaut and hoped he could be persuaded to come to St. Petersburg.

Academy did not try to make Newton's views better known in Russia. The only time his name is mentioned in the official record of its minutes during these years is in answer to a request for books from the Society of Jesus in China, to whom the Academy dispatched a copy of Newton's *Chronology*.[25] When Euler accepted Frederick the Great's call to Berlin in 1741, this changed. That same year Newton's optical discoveries became the subject of a series of public lectures given under the auspices of the Academy by Professor G.-W. Krafft.

25. Cf. *Protokoly zasedanii konferentsii Imperatorskoi Akademii Nauk,* I, 121, for November 25, 1734: "Der Brief, so an die Patres Societatis Jesu nach China von der Academie geschicket werden solle, wurde approbirt, und sollten Ihnen Kämpfer's Japan und Newton's Chronologie zugeschicket, auch, weilen es nicht in dem Büchladen vorhanden, aus der Bibliothek gegeben und dargegen diese Bücher ungesäumt wieder verschrieben und in die Bibliothek wieder zuruckgegeben werden."

15

Newton's Experiments with Light, and the Controversy over "vis viva"

Georg-Wolffgang Krafft (1701–1754) came to St. Petersburg at the same time as Bilfinger. He taught at first at the gymnasium, being appointed to the chair of mathematics at the Academy in 1730, and to that of natural philosophy in 1733. His marriage to the sister of one of Schumacher's intimate friends, Taubert, caused much merriment; Daniel Bernoulli ascribed to Krafft a new application of the doctrine of attraction at a distance—"on croit que Newton a aussi établi que la force de l'aimant agisse en raison triplée des distances . . ."[1] Krafft's connection by marriage was one reason why his pedagogic activities within the Academy were favorably regarded by those in authority. He also ingratiated himself at court by serving as Empress Anne's astrologer, though he did not himself apparently ascribe much significance to his work in this capacity; he was, in fact, a devoted man of science. Some of his scientific textbooks "for the profit of Russian youth" ("v pol'zu rossiiskago iunoshestva") were still being used in Russia more than a generation later, and he was proud to consider himself to be the first to write "Russian creations of that kind". He wrote in German and Latin, but most of his manuals were quickly translated into Russian at the Academy's expense.

The Academy fulfilled another of its educational functions by sponsoring public lectures of various kinds. In a surviving notice for one such series, delivered in the spring of 1741, we are told of Pro-

1. Daniel Bernoulli to Euler, March 7, 1739 (from Basel—Euler was still in St. Petersburg), Fuss, ed., *Correspondance Mathématique*, II, 453.

fessor Krafft's intention *"to demonstrate and explain all useful experiments, pertaining to the science of motion, air, heat and cold, and at the same time, those which were invented by the renowned Newton in England concerning the nature of light and colours."*[2] When Francesco Algarotti visited St. Petersburg two years earlier, he conducted similar demonstrations;[3] but Krafft prided himself on being the first member of the Academy to supplement a regular course of lectures on physics with experiments of this sort. Yet Krafft himself was not a Newtonian. Besides, when Schumacher was temporarily stripped of his powers in 1743, he wisely retired back to Tübingen. At the same time Newton's experiments with the prism lent themselves so dramatically and yet so simply to enactment before an audience (see fig. 38), that it would seem they were performed even by those who had little faith in the theories they were designed to support.[4]

Interest in these experiments was apparently sufficiently strong to justify an article on the camera obscura—for use in a carriage—in the popular scientific journal which the St. Petersburg Academy began to publish at this time.[5] Yet Newton's optical theories were not the only

2. *St. Peterburgskie Vedomosti*, I, 1741, p. 232 [emphasis mine].

3. Such public demonstrations acquired a considerable social success in London through Desaguliers, as Algarotti would no doubt have known. See Algarotti's *Saggio di Lettere sopra La Russia*, 2nd rev. ed. (Paris, 1763), p. 6: "Sperando di far quivi un corso do Fisica Sperimentale a quella Imperadrice che non so quanto avrà fantasia di vederlo. Onde Ella può ben credere, che non siamo senza un bello apparato di machine per dimonstrare a tutte le Russie il peso dell'aria, la forza centrifuga, le leggi del moto, la electricità, gl'inventi, e i giocolini della Filosofia." On Algarotti, see Chapter 13.

4. Even Lomonosov, who opposed Newton's doctrines on light (see below, pp. 185–199) refers to them in his "Letters on the Usefulness of Glass" (Pis'mo o pol'ze stekla . . .):

> Астроном весь свой век в бесплодном был труде
> Запутан циклами, пока восстал Коперник;
> Презритель зависти и варварству соперник;
> В средине всех Планет он солнце положил,
> Сугубое земли движение открыл:
> Одним круг центра путь вседневный совершает,
> Другим круг солнца год теченьем составляет.
> Он циклы истинной Системой растерзал
> И правду точностью явлений доказал.
> Потом Гугений, Кеплеры и Невтоны,
> Преломленных лучей в Стекле познав законы,
> Разумной подлинно уверили весь свет,
> Коперник что учил, сомнения в том нет.

Cf. *Polnoe sobranie*, VIII, 517, no. ii. 268–280.

5. Cf. "Opisanie osoboi kamery-obskury, kotoraia v karete upotrebliaemaia byt' mozhet," *Ezhemesiachnyia sochineniia k pol'ze i uveseleniiu*, V (1760), 462. The *Ezhemesiachnyia sochineniia*, which began to appear in 1755, ceased publication in 1764. About two thousand copies were printed to begin with, but the number decreased rather drastically as circulation shrank. See SK, IV, 130–131. Much of the material consisted of translations—from *The Spectator, The Tatler, The Guardian*, and other

ones to engage the attention of its members in this period. Within a year of Krafft's departure from the Russian capital, a fresh controversy broke out in St. Petersburg, which rasied again the familiar argument over Newton's dynamics and the validity of his doctrine of attraction at a distance.

The pretext for the new dispute was a small treatise published in London in the year of Cantemir's death. Entitled *De conservatione virium vivarum dissertatio,* it was sent to the Russian Academy by the author, who concealed his identity behind his "nom de plume"— "Phileleutherus Londinens." This, as it later transpired when Euler was asked to pass judgment on the treatise, was none other than James Jurin. By the time his *Dissertatio* appeared, Jurin had added to his polemical reputation by attacking Bishop Berkeley, who in the *Analyst* accused mathematicians, and Newtonians in particular, of infidelity.[6] Under a similar "nom de plume" he carried on a discussion with Dr. Pemberton in defence of Newton in "The Works of the Learned" for 1737–1739. His *Dissertatio* on the conservation of force was also inspired by a desire to protect Newton's reputation, this time against the "geometers of Basel," of whom Jurin held John Bernoulli to be the most nefarious.

The issue behind Jurin's attack was not new. As early as 1686 (in the *Acta Eruditorum*) Leibniz had produced an alternative to Descartes's method of measuring force (mv), and by differentiating "dead" forces from "live" ones, introduced mv^2 as the true formula. This opened a long-lasting controversy between the Cartesians and the Leibnizians, in which representatives of almost every European nation took part.[7] It was still going strong in the 1740's, as may be gathered

---

foreign journals, as extracts from the writings of D'Alembert, Buffon, Gellert, Pope, Voltaire, and other writers: see below, pp. 213, 218, 224.

6. Cf. *Geometry no Friend of Infidelity; or a Defence of Sir Isaac Newton and the British Mathematicians* . . . (London, 1734), and *The Minute Mathematician or the Free Thinker no Just Thinker* . . ., which came out, also in London, the following year. Both pamphlets appeared without Jurin's name, and were signed "Philalethes Cantabrigiensis." (The main brunt of Berkeley's attack had been aimed against Halley; Newton was too august a target for him to attack directly.)

7. Leibniz argued that since, by Galileo's law of falling bodies, the height to which a body rises is proportional to the square of the initial velocity, the effect of a force upon a body must be proportional to the product of the weight into the *square* of the velocity imparted, not into the simple velocity. Both sides were right except for Leibniz' error in taking the product mv^2 as the measure of the effect of the force instead of $\frac{1}{2}$ mv.

from the fact that the Marquise du Châtelet,[8] Kant,[9] d'Alembert,[10] and Lomonosov[11] all wrote on the subject at this time. Finally, d'Alembert in his *Traité de dynamique* of 1743 explained that the whole controversy was merely an empty dispute about words. According to him, it was equally legitimate to measure force by the "vis viva," which it imparts to a body upon which it acts through a certain distance, or by the momentum which it imparts to a body upon which it acts for a certain length of time (the formulation preferred by Leibniz and his followers). But before d'Alembert temporarily put an end to the discussion (which we cannot go into here), it was further complicated by the argument over the validity of Newtonian dynamics. Jurin had attacked Bilfinger in 1732 for his misinterpretation of Newton's position on the problem.[12] In 1744 the treatise Jurin sent to the Russian Academy was similarly provoked in the first instance by a work written in St. Petersburg—the *Hydrodynamica* of Daniel Bernoulli.

In this work, published only in 1738, Daniel Bernoulli's admiration for Newton, "Virum meritus suis immortalem," is stated clearly enough,[13] but he suggested that Propositions LI and LII of Book II in the *Principia* do not correspond to possible cases. It was this that made Daniel Bernoulli the object of censure in Jurin's *Dissertatio*, but most of his scorn was reserved for Daniel's father, John Bernoulli,

8. G. E. D. du Châtelet, *Réponse de Mme . . . à la lettre que M. de Mairan lui a écrite sur la question des forces vives* (Brussels, 1741). This was prompted by D'Ortous de Mairan's *Lettre à Mme . . . sur la question des forces vives, en réponse aux objections qu'elle lui fait sur ce sujet dans ses institutions de physique* (Paris, 1741), which was, in turn, a response to her *Institutions de Physique* (Paris, 1740).

9. Cf. Im. Kant, *Gedanken von der wahren Schätzung der lebendigen Kräfte und Beurteilung der Beweise* (1747) in Ernst Cassirer's edition of Kant's *Werke*, Vol. I (Berlin: B. Cassirer, 1912).

10. Jean Le Rond d'Alembert, *Traité de Dynamique, dans lequel les loix de l'équilibre et du mouvement des corps sont réduites au plus petit nombre possible* (Paris, 1743). See Carolyn Iltis, "D'Alembert and the *Vis Viva* Controversy" in *Studies in History and Philosophy of Science*, I (No. 2, 1970), 135–144. Some of the more recent literature on the subject is cited by Dr. Iltis. Lomonosov's contribution to the controversy cannot be examined here in a meaningful way without going far beyond the theme of the present study.

11. Lomonosov's letter to Euler, July 5, 1748, cf. *Polnoe sobranie*, II, 173.

12. Cf. James Jurin, *Dissertationes physico-mathematicae, partim antea in Actis philosophicis Londinensibus . . . partim nunc primarum impressae* (London, 1732).

13. Daniel Bernoulli, *Hydrodynamica, sive de viribus et motibus fluidorum commentarii*, p. 3. The work was begun in 1729 and published in Strasburg in 1738. As Lagrange noted: "Daniel Bernoulli a donné ensuite plus d'extension à ce principe et il en a déduit les lois du mouvement dans les vases, matière qui n'avait traité avant lui que d'une manière vague et arbitraire."

whose criticism of Newton was more fundamental. He still believed in a rectified Cartesian system, as his recent attempt to construct "une nouvelle Physique céleste" had shown, and in dynamics he followed Leibniz. Like him, he maintained that if Newton had understood the principle of conservation he would not have postulated two distinct principles of force: one to set bodies in motion; the second to maintain that motion. The philosophical implications behind this argument had already been enunciated by Leibniz in his controversy with Clarke.

For Leibniz, force bore the same relation to matter as Aristotelian form does; analogous to soul, its nature follows a perpetual law pre-established by God, which is responsible for all changes taking place spontaneously. Hence, unlike Newton, Leibniz assumed that the total amount of force in the universe suffers no dimunition. No body ever loses force without communicating an equal amount to other bodies, and it likewise shows no increase since no machine can ever generate force unless it receives an equivalent impulse from without. The Newtonian view—that from time to time it required divine interposition to restore the status quo—was one Leibniz found ridiculous.[14] Eventually, in 1750, Daniel Bernoulli was able to dispel the metaphysical mists which had gathered round this argument by formulating his great law of the conservation of energy.[15]

Until then, the controversy over this issue was fought over a range of specific cases. Jurin would proudly claim in his *Dissertatio* that the Leibnizians did not put forward one single problem which could not be solved by following Newton's method—whereas the Leibnizians had been quite unable to deal successfully with several put forward by

14. I.e., "If *active Force* should diminish in the Universe by the Natural Laws which God has established; so that there should be need for him to give a *new Impression* in order to restore that Force, like an Artist's Mending the Imperfections of his Machine; the Disorder would only be with respect to God himself." Cf. G. W. Leibnitz, *A Collection of Papers which passed between the late learned Mr. Leibnitz and Dr. Clarke in the Years 1715 and 1716* . . . (London, 1717), 3rd paper, pp. 65–67; for Leibniz' criticism of this, see *ibid.*, pp. 33–35.

15. Bernoulli came close to discovering the transition from molar into molecular motion and the equivalence of mechanical energy and heat, but his principle of the conservation of *vis viva* was initially restricted to mechanics and was extended to all branches of physics only in the middle of the nineteenth century—by Mayer, Joule, and Helmholtz. The widespread interest that discussion over this problem elicited was partly inspired by concern over the philosophical implications. Mme. du Châtelet, for example, expressed dread to Maupertuis that the solution of it would be a terrible blow to the doctrine of free will ("car enfin je me crois libre et je ne sais si cette quantité de forces toujours la même dans l'univers ne détruit point la liberté"). Cf. Ira O. Wade, *Voltaire and Mme du Châtelet: An Essay on the Intellectual Activity at Cirey* (Princeton, N.J.: Princeton University Press, 1941), p. 19.

the Newtonians.[16] When his treatise reached St. Petersburg, the dispute was at first maintained on the same technical level. Josiah Weitbrecht (1702–1747), who reviewed Jurin's work, agreed with his criticisms, while his colleague, Georg Wilhelm Richmann (1711–1753), supported the Leibnizian side. Their exchange of views became so acrimonious that, when it was decided by the Russian Academy on March 4, 1745, to seek Euler's opinion in Berlin, the papers he received included:

1. Weitbrecht's review of "Philaleutherus' " work:
2. Richmann's comments on his review, with additional comments ("notata ad dissertationem una cum additionibus");
3. Weitbrecht's "notes" on Richmann's "notes" ("notas ad notata");
4. Richmann's notes on Weitbrecht's notes on Richmann's notes.[17]

In an attempt to reconcile the disputants, these papers were expunged from the Academy's records, and it was only with the publication of the surviving documents in 1956 that some of the details of their polemic were revealed.[18]

In giving his verdict Euler was careful not to offend either Weitbrecht or Richmann, reserving his harshest words for Jurin and castigating his presumption and "ignorance of the subject of which he dealt."[19] Jurin's attempts to refute the principle of "vires vivae" were "weak and insubstantial." He was shocked, wrote Euler, at Jurin's lack of respect in questioning "the profound judgements of the renowned Bernoulli."[20] As to the issue between Weitbrecht and Richmann, these, in the last analysis, depended upon the interpretation of one crucial expression in Jurin's presentation—"in quibuscunque directionibus"—which they had chosen to understand differently. Neither could therefore be said to be wholly right.[21] Euler's hostility to Jurin, however, was clearly influenced by his own attitude toward the author of the *Principia,* as well as by his resentment of English

16. James Jurin, *De conservatione virium vivarum dissertatio* (London, 1744).
17. Cf. *Protokoly zasedanii konferentsii Imperatorskoi Akademii Nauk,* I, 52.
18. G. W. Richmann's works have been published, several for the first time, in *Trudy po fizike,* ed. Grigor'ian.
19. Euler's answer was read at the Academy on June 10, 1746; cf. Rikhman, *Trudy po fizike,* ed. Grigor'ian, pp. 494–496. (This letter has not before been published.)
20. *Ibid.,* p. 495.
21. *Ibid.*

mathematicians in general, as Daniel Bernoulli noted in a letter to Euler tactfully urging him to restrain the asperity of his criticisms of Newton ("Ich weiss zwar wohl wie wenig Ew. Ursach haben mit den Engländern zufrieden zu seyn, welche anstatt Sie als ein wahres ornamentum saeculi nostri zu veneriren, vielmehr alles verachten; aber ich bin versichert, dass wenn der grosse Newton noch lebte, er selbst ganz anders würde von Ihnen geredet haben").[22] This may seem ironic, for no one did more than Euler to extend the method of the calculus which Newton and Leibniz had discovered, but personal prejudice as well as a misplaced patriotism undoubtedly played a part in forming Euler's scientific orientation, as well as that of lesser men.

Nor can it be denied that Jurin provoked such sentiments. His Homeric allusions, his vituperative belittlement of his antagonists, his deification of Newton, and the tone of national condescension he managed so adroitly to convey in referring to the "geometers of Basel" —all of this was well designed to annoy and infuriate. As Richmann noted with hurt pride: "Great Leibniz . . . is of course just as much an adornment for Germany, as Newton is for England, and future generations will maintain his glory safe from the oblivion that his satirical writings threaten to reserve for Phileleutherus."[23] Euler felt the same respect for Leibniz, and though he rejected the doctrine of pre-established harmony, he completely accepted Leibniz' criticism of the concept of "attraction at a distance" as something "miraculous" and "occult."[24] For this reason, Euler was hardly the best choice to adjudicate in the dispute. At the same time, in a scientific sense the value of the whole controversy was very limited. As Euler himself remarked, the problems associated with running water moving from a wider into a narrower orifice, which had earlier prompted Jurin to criticize Bilfinger in the pages of the *Commentarii*,[25] were difficult, "and it is impossible to hope for an adequate solution."[26] Besides, as Diderot

22. Daniel Bernoulli to Euler, January 20, 1742, Fuss, ed., *Correspondance Mathématique*, II, 480.
23. "Quae observavi dum dissertationem perlegi cujusdam, qui sub nomine Phileleutheri Londinensis latere voluit, contra principium conservationis virium vivarum, sequentibus cum societate communicabo/ To chto ia zametil, chitaia rassuzhdenic nekoego avtora, kotoryi pozhelal skryt'sia pod imenem Fileleitera iz Londona," supplement to Rikhman, *Trudy po fizike*, ed. Grigor'ian, p. 443.
24. Leonard Euler, *Letters of Euler on different subjects in Physics and Philosophy addressed to a German Princess*, 2nd ed., 2 vols, tr. Henry Hunter (London, 1802), I, letter lxviii (dated October 18, 1760).
25. See Chapter 12.
26. Cf. Rikhman, *Trudy po fizike*, ed. Grigor'ian, p. 495.

noted later, what was lacking in this period were accurate numerical data for establishing the equivalence of energy and heat.

From a historical point of view, Weitbrecht's agreement with Jurin is of some interest. When Christian Wolff wrote to Schumacher about the author of the *Dissertatio*, he described him as "ein blinder Anhäger des N e w t o n s . . ."[27] With such people, Wolff added, "ist nicht viel auszurichten." And this, without any doubt, was the attitude toward the Newtonians to which most of the Russian Academy's men of science at this time subscribed. Why, then, did Weitbrecht dissent from the prevailing orthodoxy? His combative and argumentative nature is readily apparent from the Academy's proceedings, and a broken mirror in its Conference Hall bore witness to the passion with which he upheld his convictions.[28] Unfortunately, his intellectual development is less vividly documented, yet a curious parallel between his and Jurin's careers partly reveals some aspects of it.

Both studied philosophy as undergraduates and only later turned to medicine. Weitbrecht (1702–1747) came to St. Petersburg in 1725, after completing his studies at Tübingen University. At first he taught mathematics at the gymnasium attached to the Academy; in 1727 he became assistant in anatomy to Dr. Duvernois, the Academy's physiologist. By 1731 he was made a professor and himself gave lectures on physiology. His study, "De febrili constitutione petechizante" (then raging in St. Petersburg) gained him an M. D. from the University of Königsberg. He then established a medical practice in the Russian capital, and five years before his death in 1747 completed his classic work, the *Syndesmologia, sive historia ligamentorum corporis humani*.[29] It demonstrated, among other things, that the heart alone was not powerful enough to carry the blood around the body. In showing what an important part the smaller blood vessels played in this function, Weitbrecht became interested in some of the phenomena which James Jurin had studied earlier. Like him, Weitbrecht conducted experiments with running water, being similarly moved to attempt to ex-

27. Wolff to J. D. Schumacher, December 9, 1735, *Briefe von Christian Wolff aus den Jahren 1719–1735*, p. 91.
28. On Weitbrecht, see Pekarskii, *Istoriia Imperatorskoi Akademii Nauk v Peterburge*, I, 468–474.
29. Josiah Weitbrecht, *Syndesmologia, sive historia ligamentorum corporis humani* (St. Petersburg, 1742).

plain its suspension in capillary tubes.[30] He was also drawn into comparing its behavior with that of blood, to which he also applied the Newtonian principle of attraction.[31] Thus, Weitbrecht's interest in Newton was encouraged by the considerations that had earlier prompted Freind, Jurin, and others, to reconcile medical theory with Newtonian laws.

Richmann's concern with Newton, however, was prompted by entirely different scientific considerations. Having studied natural philosophy at Halle and Jena Universities, Richmann came to St. Petersburg in 1735 in order to teach physics in the gymnasium. He was born in 1711, in Pernau in Estonia, which had been acquired by Russia in the course of the Great Northern War with Sweden, but Richmann's upbringing was largely German. In St. Petersburg he became Krafft's protégé and eventual successor. When Richmann arrived in the Russian capital, Krafft was trying to derive experimentally an expression for the equilibrium temperature of homogeneous liquid mixtures, and it was natural for the younger man also to become interested in calorimetry; later he also became preoccupied with the phenomenon of luminescence and electricity, which, with Lomonosov, he was the first in Russia to explore.

Paradoxically, it was the empirical character of most of his work and not, as in Euler's case, his academic background that made him dubious about the Newtonian doctrine of "attraction at a distance." The lectures Richmann delivered on physics—he became professor of the subject in 1741—were based on L. F. Tümmig's *Institutiones philosophiae Wolffianae*, which Lomonosov translated into Russian in 1745. It is intriguing to record that one of Richmann's professors at Jena (G. E. Hamburger) supported Newton's gravitational theory.[32] Yet Richmann rejected it, as may be gathered from his *Tentamen stabiliendi leges cohaesiones*.[33]

Richmann's early philosophical views, stated in an essay he wrote in dialogue form some time between 1735 and 1742, remained unpublished and have only survived in incomplete form, but the essay is

30. Cf. Weitbrecht, "Tentamen theoriae, qua ascensus aquae in tubis capillaribus explicatur," *Commentarii* (St. Petersburg), VIII (1736), 261–309, and "Explicatio difficiliorum experimentum circa ascensum aquae in tubos capillares," *ibid.*, IX (1737), 275–309.

31. Cf. Weitbrecht, "Cogitationum physiologicarum de circulatione sanguinis. Caput de quantitate motus sanguinis," *ibid.*, VIII (1736), 334–340.

32. Cf. G. E. Hamburger, *Elementa physicae methodo mathematico in usum auditorum conscripta*, 2nd ed. (Jena, 1734), p. 81, no. 147.

33. Rikhmann, *Trudy po fizike*, ed. Grigor'ian p. 669; first published in 1956.

nonetheless noteworthy for being the first such attempt in Russia to deal with the metaphysical implications of scientific investigation. Originally written in German, the essay was intended for publication in the *Primechaniia na vedomosti,* the Russian periodical brought out by the Academy as a popular supplement to the *Commentarii.*

The dialogue is between a Plenist, a Vacuist, and a student, but for all its show of objectivity Richmann's own predilections are obvious enough. As the introductory comments imply, the essay was really aimed against the Newtonians, those "contemporary natural philosophers who on this question whether space is a vacuum or a plenum lapse into a most difficult position . . ."[34] In the second part of the dialogue, the student evidently finds the arguments of the Plenist more convincing than those of his antagonist, and indeed the Vacuist at this stage is there only to pose the necessary questions. The plenum is tacitly assumed to be the correct picture of the universe, and as the Vacuist wonders how motion is possible in a world full of matter, the Plenist replies: "The place deserted by a moving body would immediately be filled by some other matter—its place would be taken by still another body yielding its place, and thus everything you wish to say falls to pieces."[35] In this case, adds the Plenist, the cause does not precede the effect, but both occur simultaneously. To the Vacuist's dismay, the Plenist then adduces several other arguments against the existence of a Vacuum.

It is interesting to note, however, that Richmann goes out of his way to attack not only the assumptions on which Newtonians based the theory of universal gravitation but also the monadology of the Leibnizians and their principle of sufficient reason.[36] The argument of the Plenist is clearly taken from Descartes's *Principia Philosophiae.* Later in his career Richmann was to change his mind concerning the plenum and perform some of the first experiments on the effects of electricity in a vacuum, but the Cartesian approach to natural philosophy was the one with which Richmann appeared wholly sympathetic in his earlier scientific investigations.

This is suggested by his dispute with Georg-Wolffgang Krafft, who

34. "Beseda mezhdu dvumia filosofami o pustom i napolnennom prostranstve," *ibid.,* p. 399. The German title is not given by the editors, and the actual date of composition is uncertain.

35. *Ibid.,* p. 401.

36. *Ibid.,* p. 416.

also carried out various experiments on heat while he taught at St. Petersburg and tried to generalize Morin's formula for the resulting temperature of mixtures of water at different temperatures.[37] Prior to the publication of Krafft's results, Richmann had done some similar work on the same problem, which he resumed after reading Krafft's paper. His detailed criticisms are not as important as the method he employed. He set out from the explicit assumption that the heat of a substance, or at least of a liquid, is diffused evenly. Having put forward this principle, he then proceeded to argue that the intensity of a given quantity of heat will consequently vary inversely with the mass of the substance which it suffuses, but this formula was the result not of empirical observations or careful experiment but of abstract and a priori considerations.[38] In other words, his approach was entirely Cartesian in spirit, and he must have realized how differently Newton went about solving a related problem when, some years later, he came across one of Newton's papers on heat in the *Philosophical Transactions*.[39] He was so impressed that he wrote out Newton's paper in his own hand, but he was not yet prepared to accept the doctrine of attraction at a distance, as the *Tentamen stabiliendi leges cohaesiones* shows.

In this paper Richmann examined forces responsible for the cohesion of bodies, and, naturally enough, this problem led him to consider the Newtonian theory of gravitation, which he predictably rejected in favor of an ether hypothesis of the Cartesian type ("cohaerere possunt corpora ob pressionem fluidi substillissimi externam"). In another place, he says quite explicitly that the Newtonian theory is inadmissible because, and this charge we have heard before, it restores "qualitates occulta."[40]

Yet the corpuscular philosophy enunciated in the *Tentamen,* unpublished until 1956, is in a very significant respect different from the one adopted, for example, by Lomonosov. Richmann was prepared to make hypotheses, but he seemed to be well aware of their limitations

37. Cf. G.-W. Krafft, "De Calore et Frigore Experimenta Varia," in *Commentarii* (St. Petersburg), XIV (1744–1746), 218.

38. *Ibid.*, I (1747–1748), 168 ff. It was, in fact, no more than a tentative hypothesis awaiting verification. At first Richmann merely argued that Krafft's formula did not fit the results Krafft had observed as well as his own, but he later also carried out some experiments himself.

39. Newton, "Scala graduum caloris et frigoris," *Philosophical Transactions*, XXII (1701), 824. Richmann, it is claimed, reached the same results—independently. Cf. Rikhman, *Trudy po fizike*, ed. Grigor'ian, p. 571.

40. *Ibid.*, p. 605, "Primechaniia k trudam po elektrichestvu"; see also p. 669, n. 2.

when unchecked by experiment: "One might picture to oneself matter set into motion by various means and so explain all phenomena including cohesion. But no mortal has the ability to determine the distribution, fineness, and direction of the atmosphere necessary for the resulting phenomena, and even if one were able, then one would never have any assurance that this motion imagined by one actually exists."[41] This is why, Richmann concludes, it is "wiser to leave behind the internal toys of the mind to investigate only those laws" to which bodies are visibly subject. He therefore admonishes against the use of a corpuscular theory to *explain* the behavior of natural phenomena; it was more important to investigate observed effects.

Such caution on Richmann's part could well be attributed to Newton's influence. At any rate, he was certainly quite familiar with Newton's work by the time he came to write the *Tentamen,* as is indicated by the surviving notebooks in which Richmann kept a careful record of his reading. True, his knowledge of Newton's writings at that point (1743) was all secondhand: but he does refer to a number of well-known early textbooks by Newtonians, including those of John Freind,[42] Keill,[43] and 's Gravesande.[44] Clearly the idea of attraction continued to intrigue Richmann in the course of many years, and he notes with interest that Freind applies it "to chemistry, and accepts Keill's positions by deducing all phenomena of chemistry from it alone." But he goes on to add: "According to Keill and Freind, there are also similar powers of repulsion. In this manner the occult qualities are restored."[45] Yet Jurin's little tract may have prompted Richmann to think again. In 1744–1745, while he was still engaged in his dispute

41. *Ibid.,* p. 418.
42. John Freind (1675–1728) was one of Newton's early disciples at Oxford. His *Praelectiones chymicae,* first published in 1712, were widely admired, an extract of this Latin edition having earlier appeared in the *Philosophical Transactions* in 1709. The English original appeared as *Chymical lectures in which almost all the operations of chemistry are reduced to their true principles, and the laws of nature read in the museum of Oxford in 1704* (Oxford, 1704). Cf. Hélène Metzger, *Newton, Stahl, Boerhaave, et la Doctrine Chimique* (Paris: F. Alcan, 1930), pp. 45–50.
43. John Keill (1671–1721), an Oxford mathematician, was best known for his treatise on gravitation, first published in the *Philosophical Transactions* in 1703. Richmann is probably referring to the book which Diderot was still recommending to Catherine II for use in schools three-quarters of a century after publication; see Chapter 20, below.
44. Almost certainly, the *Philosophiae Newtonianae Institutiones, in usus academicos,* the first Newtonian textbook to be used at the St. Petersburg Academy; see Chapter 10, above.
45. I.e., "Datur secundum Keillium et Freindium etiam vis repulsiva. Sic qualitates occultae restituuntur." Cf. *Arkhiv Akademii Nauk SSSR,* Part I, p. 37, op. 94, no. 85. In the *Trudy po fizike,* ed. Grigor'ian, Richmann's writings are all given in translation.

with Weitbrecht, he did some experiments with light.[46] He consulted the *Opticks,* and again Richmann's reaction was negative, as is revealed by a note in his hand in which he freely paraphrases the thirty-first "Query" added to that great work.[47]

That Richmann should refer to this particular query is appropriate in view of his previously stated objection to the gravitation theory. This query was written by Newton in answer to critics who made precisely the same charge. In the advertisement to the second edition of the *Opticks,* which appeared a year after Leibniz' criticisms in 1716, he was anxious to repeat that the "cause" of gravity was unknown to him, "for he was not yet satisfied about it for want of Experiments." This statement was designed to invalidate the objection that there was anything "occult" about his theory. In the fourth edition of the *Opticks* (the one to which Richmann refers), Newton added a final "Query," in which he made his position clearer still: "what I call Attraction may be perform'd by impulse, or by some other means unknown to me."[48] Yet there were other phenomena, such as magnetism or electricity, the explanation of which had proved to be equally elusive. Far from undermining the gravitation theory, their existence, Newton argued, supports it. Indeed, he further implied that the theory might be applicable to several other fields, including chemistry. He even suggested an answer to the problem to which Richmann had devoted his *Tentamen:* Is not the combination or cohesion of hard atoms or particules, he asked, due to some form of attraction which is very strong when atoms come together, and by virtue of this great strength they stay together?

Richmann was not convinced, as further comments in his *Journal* indicate. "The cause of gravitation," he notes, "is unknown," as if to say that this was enough to condemn it.[49] In the same place he re-

46. Richmann, "Duo experimenta nova optica," in his *Trudy po fizike,* ed. Grigor'ian, pp. 500–501.

47. *Ibid.,* p. 418. Cf. Newton (*Opticks,* Query 31, p. 380): "This Analysis consists in making Experiments and Observations, and in drawing general Conclusions from them by Induction, and admitting of no Objections against the Conclusions, but such as are taken from Experiments, or other Certain Truths. For Hypotheses are not to be regarded in experimental philosophy." Richmann: "Since it is difficult to explain natural phenomena from their causes without admitting fictions and hypotheses, it seems reasonable to me not to look at a system timidly, in order that love for a theory may not deflect us from truth. This is the fundamental reason why I begin physical investigations with experiments."

48. *Opticks,* Query 31, p. 351.

49. Entry for April 11, 1752. Cf. Rikhman, *Trudy po fizike,* ed. Grigor'ian, p. 693.

veals his fidelity to Christian Wolff, an authority even Euler no longer admired. Several years later Richmann again appears to have become interested in Newton's work; in 1752 his own investigations prompted him to devote serious study to the *Principia* and the *Opticks,* and he borrowed the two books from the Academy library.[50] His reading of them was cut short by his tragic death a few months later while experimenting with electricity. Two other members of the St. Petersburg Academy, F. U. T. Aepinus and M. V. Lomonosov, took up where Richmann had left off, both with respect to his work on electricity and his study of Newton.

50. Entry for April 10, 1752. Cf. *ibid.* The edition of the *Principia* to which Richmann refers is that published by the Jesuits in Geneva in 1748. The *Opticks* he borrowed was also one of the late editions brought out by Clarke, "Will return in a month," Richmann's *Journal* notes.

It is the language used, and not the thing itself, that irritates the human mind. If Newton had not used the word *attraction* in his admirable philosophy, everyone in our Academy would have opened his eyes to the light; but unfortunately he used in London a word to which an idea of ridicule was attached in Paris; and on that alone he was judged adversely, with a rashness which will someday be regarded as doing very little honor to his opponents.—Voltaire (Letter to M. de la Condamine, June 22, 1734)

16

Electricity and Action at a Distance

After developing his doctrine of attraction at a distance with reference to celestial bodies, Newton began to apply that theory in several other fields, including chemistry, physiology, and pure physics. His disciples in Britain and on the Continent were to follow his lead, though inhibited at every step by the reluctance of the academies to come to terms with Newtonian theory. As Maupertuis noted, "il a fallu plus d'un demi-siècle pour l'apprivoiser les Académies du continent avec l'attraction . . ."[1] In this respect the Academy in St. Petersburg was no exception, as is illustrated by the way it sought to provide Newton's critics with an opportunity to expound their objections at its first public competition in 1750.[2]

1. "On s'applaudissoit tant d'avoir de la Philosophie les qualités occultes, en avoit tant de peur qu' elle revenissent, que tout ce qu'on croyait avoir avec elles la moindre ressemblance effrayoit: on étoit si charmé d'avoir introduit dans l'explication de la Nature une apparence de méchanisme, qu'on rejetoit sans l'écouter le méchanisme véritable, qui venoit s'offrir," P. L. M. de Maupertuis, *Oeuvres* rev. ed., 4 vols. (Lyons, 1756), II, 252. [Lettre XII: Sur l'attraction.]
2. The problem was spelled out in these words: "An omnes inaequalitates, quae in motu Lunae observantur, Theoriae Newtonianae sunt omnes consentaneae, et quaenam sit vera Theoria harum inaequalitatum unde locus Lunae ad quodvis tempus possit inveniri?" Though the topic of the concourse was only publicly announced in September 1750 at a session of the St. Petersburg Academy, it had been previously discussed in September 1749. Cf. *Torzhestvo Akademii nauk . . . , prazdnovannoe publichnym sobraniem sentiabria 6 dnia 1749 goda* (St. Petersburg, 1749).

Of these, the most serious had been the irregularities in the moon's motion, which did not seem to accord with the law of universal gravitation. To Euler's delight, it was this that the Russian Academy chose as the subject of its competition, for he was certain that "Newton's famous theory" could be "applicable to the heavens only to a certain degree . . ."[3] Yet the prize was won, to the acclaim of the Newtonians, by Clairaut, the very man whose election as a foreign associate of the St. Petersburg Academy had earlier been proposed by Cantemir and rejected.[4] Indeed, Clairaut's first work on lunar theory appeared in 1743, a year before Cantemir died,[5] but he was not then fully convinced that the irregularities of the moon's motion could be explained in terms of Newton's theory, and when he returned to the subject in 1747 he spoke of the difficulty and complexity of Newton's proofs.[6] In writing the treatise that was crowned by the Russian Academy, he decided to abandon Newton's methods of calculation, and to follow his own. He was surprised and elated to discover that his results corresponded so well to those of Newton.[7]

On the strength of his brilliant treatise Clairaut's merits were at last recognized in St. Petersburg,[8] and, with his election to the Russian Academy in 1754, it too in a sense formally accepted the victory of the Newtonian cause, which was now being even more triumphantly vindicated by the persuasive application of the principle of attraction at a distance to other phenomena. Thus, unknown to Richmann, John Michell (1724–1793) had applied it to magnetism at the very moment

3. Euler's letter to Razumovskii, congratulating him on the selection of so important a problem as the theme of the Academy's first public competition, is quoted by Pekarskii, in *Istoriia Imperatorskoi Akademii Nauk v Peterburge*, I, 269.

4. Chapter 13, above. Clairaut's immense ability had earlier been discerned by Daniel Bernoulli, who, in his role as adviser to the St. Petersburg Academy, recommended that Clairaut be invited to Russia. "Man sollte trachten," he wrote to Euler on December 18, 1734, "den jungen Mann Clairaut von Paris zu bekommen." Cf. Fuss, ed., *Correspondance Mathématique*, II, 415.

5. I.e., *De l'orbite de la Lune dans le système de Newton* (cf. *Histoire* [of the Académie Royale des Sciences] (Paris, pub. 1743; 1746), pp. 17–32. *Théorie de la Figure de la Terre, tirée des principes de l'hydrostatique*, referred to above (see p. 136), was written by Clairaut after this work.

6. Cf. *Histoire* [of the Académie Royale des Sciences], (Paris, 1745; pub. 1749), pp. 329–364 (this, like some earlier volumes of the *Histoire*, contains material of a later date than the given year of publication—material added afterward on the Academy's instructions).

7. Cf. Academician N. I. Idel'son's illuminating paper, "Zakon vsemirnogo tiagoteniia i teorii dvizheniia luny," in *Isaak N'iuton 1643–1727*, ed. Vavilov, p. 193.

8. Clairaut's work was published in St. Petersburg in 1752 with Razumovskii's imprimatur: *Théorie de la Lune déduite du seul principe de l'attraction réciproquement proportionelle aux quarrés des distances.*

that both he and Euler were prematurely celebrating its demise. By showing that "the Attraction and Repulsion of Magnets decreases, as the squares of the distances from the respective poles increase," Michell laid the foundations for the mathematical theory of magnetism.[9] Thus, the system of vortices was overthrown from yet another field in which it had until then prevailed. The same was to happen to electricity after a dramatic series of experiments in which scientists of several countries took part.

Of these perhaps the most spectacular was perpetrated by the Abbé Nollet, court electrician to Louis XV, who charged seven hundred monks holding hands in a circle, thus causing them all to leap into the air simultaneously. Not to be outdone by Western monarchs, Elizabeth ordered Richmann to begin the study of electricity so that she too might be amused by the drawing of sparks from ice, the electrification of monks, and other demonstrations of the kind that proved so popular at this time. Through Schumacher, Richmann was ordered (in a letter of March 6, 1745) "in accordance with the command of the Empress" to carry out electrical experiments and a special stone chamber was constructed in the Winter Palace where these could be performed before Elizabeth and her court.

Richmann was convinced that the phenomenon of heat, with which he had been preoccupied until then, was related to electricity. Both were the result of corpuscular motion, the electrical *tremulus* being imparted to particles surrounding the surface of an electrified body while heat is conducted through the motion of particles *within* the body. The other parallel which visibly impressed Richmann had been made as early as 1746 when Professor Winkler of Leipzig University suggested that "electrical matter" and lightning were one and the same thing. Musschenbroek and others denied it. After Benjamin Franklin was able to demonstrate that this was indeed the case, Richmann, who carefully followed the reports of Franklin's experiments, was tempted to repeat one of them and was electrocuted in the process. News of the tragedy[10] spread quickly through European capitals, and Richmann

9. Cf. John Michell *A Treatise of Artificial Magnets; in which is shewn an easy and expeditious method of making them superior to the best natural ones* (Cambridge, Eng., 1751), p. 17. Before his untimely death, Richmann became aware of Michell's discovery, and in fact he refers to this work in his posthumously published paper on magnetism: cf. "De virtute magnetica absque magnete communicata experimenta," *Novi commentarii* (St. Petersburg), IV (1758), 235. Yet Richmann obviously did not realize its significance, and spoke of Michell's "lack of clarity."

10. Cf. April 21, 1752: *Protokoly zasedanii konferentsii Imperatorskoi Akademii Nauk,*

enjoyed a moment of posthumous fame and glory as a modern martyr for science. The main purpose of his investigations was to find an answer to the question which Newton, amongst others, had put forward at the end of the *Opticks:* to discover the physical cause of electricity. This problem also preoccupied Lomonosov, who shared some of Richmann's scientific interests.[11] Indeed, it was at his and Richmann's initiative that in 1753 the Academy devoted its prize competition to this theme. A hundred gold rubles were offered to the most successful paper to disclose "the cause of electricity, and to give an exact theory of it."[12]

By the time the competition closed (June 1, 1755), Richmann was no longer alive. Though Lomonosov continued his investigations, his membership in the Russian Academy made him ineligible to compete. However, Leonard Euler, then in Berlin and whose status was in any case different from Lomonosov's in that he was considered a "foreign" member, was able to circumvent the provision that kept Lomonosov from contributing. He sent an entry in the name of his son, Johann Albrecht Euler,[13] and this was the essay that was awarded the prize.

Its conclusions should not surprise anyone familiar with Euler's objections to the Newtonian theory of gravitation. Since Gilbert (1540–1603) had advanced his emission hypothesis, most investigators of electricity agreed that it was some sort of fluid. This view only began to lose its credibility in Euler's lifetime, more especially in the fourth decade of the eighteenth century, when, after the invention of the Leyden jar, "electrical matter" began to reveal new and unsuspected attributes. Particularly astonishing, so it appeared, was its ability to set off sparks and induce combustion. Perhaps, it was now argued,

II, 275–276. An idea of the impression conveyed by Richmann's death may be gathered from Charles Rabiqueau's *Lettre élèctrique sur la mort de M. Richmann* (Paris, 1784; reprinted from the author's *Spectacle de la Nature*).

11. Cf. Lomonosov's letter to Shuvalov, dated May 31, 1753, to which he attached a description of Richmann's experiments with a Leyden jar ("Mushenbrekov opyt s sil'nym udarom") and a drawing, neither of which have survived. See M. V. Lomonosov's *Polnoe sobranie*, X, 482. Richmann and Lomonosov were intending to give an account of their investigations before the Academy ("Richmann will present his experiments, and I—the theory and uses arising from them"), but Richmann was killed on July 26, a few days before the announced meeting.

12. Lomonosov presented a special "Programme" before the Academy, suggesting the scope and purposes of the competition *Protokoly zasedanii konferentsii Imperatorskoi Akademii Nauk*, II, 292.

13. Cf. S. Ia. Lur'ie, ed., "Neopublikovannaia nauchnaia perepiska Leonarda Eilera," in *Leonard Eiler, 1707–1783. Sbornik statei i materialov k 150-letiiu so dnia smerti*, ed. A. M. Deborin (Moscow and Leningrad: Akademiia nauk, 1935), p. 156.

electricity was not a "fluid" after all. Instead, it began to be freely compared not only with gravitation or magnetism as Newton had done but with fire, which in several ways it seemed to resemble. This is why Lomonosov, who thought this parallel fallacious, suggested, in drawing up the terms of the competition, that candidates state "unequivocally and in detail" why this comparison was really false.

Light was the other phenomenon with which electricity was thought to bear a credible resemblance. Euler believed their physical source to be identical, and he elaborated an ether hypothesis that sought to explain their properties in terms of the same doctrine. His views on electricity developed from his ideas on light, which he had first put forward almost two decades earlier in opposition to the Newtonian theory in a treatise he wrote in St. Petersburg.[14] His main argument against it was twofold. In the first place, even if light consisted, as Newton supposed, of rays emitted from the sun, how was it possible to explain that "such floods of luminous matter, with a velocity so prodigious" did not speedily exhaust their solar source? The other difficulty, "equally insuperable" as Euler believed, was that

the sun is not the only body which emits rays, but that all the stars have the same quality; and as everywhere the rays of the sun must be crossing the rays of the stars, their colission [*sic*] must be violent in the extreme. How must their direction be changed by such a colission! This colission must take place with respect to all luminous bodies visible at the same time. Each, however, appears distinctly, without suffering the slightest derangement from any other: a certain proof that many rays may pass through the same point without disturbing each other, which seems irreconcileable to the system of emanation.[15]

His own theory was suggested to him by an interesting parallel to which Euler often refers in the course of his writings:

As the vibrations of the air produce *sound,* what will be the effect of those of ether? You will undoubtedly guess at once *light.* It appears in truth abundantly certain, that light is with respect to ether, what sound is with respect to air; and that the rays of light are nothing else but the shake-up or vibrations transmitted by the ether, as sound consists in the shakings or vibrations transmitted by air.[16]

14. For further comments concerning Euler's scientific views, see Chapter 20.
15. *Letters of Euler . . . to a German Princess,* ed. Hunter, I, 69 (letter xvii: "Of Light, and the Systems of Descartes and Newton," dated June 7, 1760).
16. *Ibid.,* pp. 77–78 (letter xix: "A different System respecting the Nature of Rays and of light proposed," dated June 14, 1760).

This seductive notion may well have been a real incentive for Euler's rejection of the corpuscular theory. Newton, he suggests with almost disingenuous simplicity, turned against the Cartesian system of vortices, "lest a subtle matter, such as *Descartes* imagined, should disturb the motions of the planets"; in fact this was "quite contradictory to his own intentions."[17] For if the theory of emanation is true, surely space "instead of remaining a vacuum, must be filled with the rays, not only of the sun, but likewise of all the other stars which are continually passing through it, from every quarter, and in all directions, with incredible rapidity."[18]

This prompted Euler to conclude that space, far from being a vacuum, was in reality filled with "a fluid" resembling air, "but incomparably finer and more subtile . . . [which] insinuates itself by the pores of all bodies, and passes irresistibly through them."[19] The same phenomenon, Euler further believed, explained the action of electricity: in his words, electricity depended for its activation on the withdrawal or decrease of the complex of ether, with which its pores are normally suffused. That is to say, a body remains electrified until the ether liberated from the surrounding pores replaces this loss. That is how the full equilibrium in the elasticity of the ether in that body is restored. "Bodies may be said to be devoid of electricity when the ether in its intersteces possesses the same elasticity as the rest of the ether diffused everywhere [around it], and when full equilibrium exists between [the ether in the body and that outside it]."[20]

This doctrine explained electrical phenomena without introducing action at a distance by supposing that something which forms an essential part of the electrical system is present at the spot where any electric action takes place. Electricity was thus made subject, in Euler's system, to the same laws of motion as light, the velocity of which Euler explained by the extreme elasticity of his "fluid." It expanded in much the same way that, Euler believed, the explosion of gunpowder was brought about by the air which it contains in a state of violent compression. It also followed from this hypothesis that if light did in

17. *Ibid.*, p. 72 (letter xviii: "Difficulties attending the System of Emanation," dated June 10, 1760).
18. *Ibid.*
19. *Ibid.*, pp. 75–76 (letter xix).
20. J. A. Euler, *Disquisitio de causa physica Electricitatis ab Academia Scientiarum Imp. Petropolitanae praemio coronata* (St. Petersburg, 1755), p. 14.

fact travel "as the sound of a bell," it could hardly do so in a Newtonian vacuum: "for if the air, intervening between the bell and our ear, were to be annihilated, we should absolutely hear nothing, let the bell be struck ever so violently."[21]

This theory of Euler's led him to question the dynamic aspects of the theory of universal gravitation from much the same motives as Huygens had earlier done, and these motives were equally pertinent, as it turned out, to Euler's doctrine of electricity.[22] The chief novelty of his work on light was his explanation of the manner in which material bodies appear colored when viewed in white light, and, in particular, of the way in which the colors of thin plates are produced. He rejected Newton's explanation that such colors are due to a more copious reflection of light of certain particular periods and instead supposed that they represent the vibrations generated within the body itself under the stimulus of the incident light. Moreover, he anticipated Maxwell in asserting that the source of all electrical phenomena is the same ether that propagates light: elasticity is nothing but a derangement of the equilibrium of the ether, and this derangement might explain not only electrical, but also gravitational phenomena.[23] Newton's unpardonable "error" in explaining the latter lay in his insistence that bodies "were endowed with a secret or occult quality, by which they are mutually attracted." Gravitation could much more plausibly be accounted for with reference to the ether's action on the bodies around it. "Bodies may be put in motion by the ether, just as we see that a body plunged into a fluid receives several impressions from it."[24] In other words, motion was the product not of attraction, as Newton had maintained, but of impulsion. How precisely ether exercises this power, Euler did not always make clear. "Its manner of acting," he admits in one passage, "may be unknown," but it seemed "more reasonable" than "to have recourse to an unintelligible property."[25] For attraction "in so far as it is given for a property essential

21. *Letters of Euler . . . to a German Princess,* ed. Hunter, I, 87 (letter xviii).

22. As Academician S. I. Vavilov has pointed out in his excellent essay on Euler's optics, "Fizicheskaia optika Leonarda Eilera," in *Leonard Eiler, 1707–1783,* ed. Deborin, his theory does differ in certain details from the ondular theory associated with Huygens.

23. Leonard Euler, *Opera omnia sub auspiciis Societatis Scientiarum Naturalium Helveticae . . .,* Ser. 3, "Sur la Nature des Moindres Parties de la Matière," (Leipzig and Berlin: B. G. Teubner, 1911–   ; imprint varies), I, 149–150.

24. *Letters of Euler . . . to a German Princess,* ed. Hunter, I, 210 (letter lxviii: "More particular Account of the dispute respecting Universal Gravitation," dated October 18, 1760).

25. *Ibid.,* 262.

to all bodies" was "an occult quality," and occult qualities were now banished from philosophy, which is why "attraction ought not to be considered in this sense."[26]

The implications of Euler's own hypothesis, however, were clear. He supposed that the pressure of the ether increases with the distance from the center of the earth—so that the force pressing a body toward the earth is stronger than that directed away from it, the balance of these forces being the weight of the body. In this view, the force on each atom would be proportional to the volume of the atom; therefore, the weight of the atom must be proportional to its volume, that is, the densities of all atoms are equal. The fact that the densities of the bodies differ from each other is accounted for by assuming that the atoms are not in actual contact. Curiously enough, though Euler appeared to be so tenaciously opposed to the irrational and "miraculous," he turned a blind eye to the call of Newton's and Locke's disciples who produced a similar rationale for wanting "to rid natural philosophy of all fine fluids."[27] The prevailing theories of electricity were all based on the "one " or "two fluid" hypothesis, but who could actually say that they had seen such a substance in operation? Had Euler been consistent to his own premises, he would have been better able to appreciate the insistent demand of the Newtonians to limit scientific hypotheses to phenomena actually perceptible to our senses. This argument is unceremoniously swept aside in the *Disquisitio*. But there were other objections of an empirical nature that could be offered against Euler's theory. Why, for example, was the action of the ether that supposedly transmitted the vibrations by which light was propagated so limited? This could easily be shown by repeating some of Newton's own experiments. Rays of sunlight allowed through a small opening into a darkened room failed to lighten it, as might have been expected if Euler's 'ether' had the contagious "elasticity" with which he endowed it.

The same objection could be made against his theory of electricity. What prevented the electric matter from being disseminated through certain substances such as glass, for example? Nor did his theory give a credible explanation for certain newly discovered and very puzzling properties of electricity such as the mutual repulsion of resinously electrified bodies. As is well known, this phenomenon also

26. *Ibid.*, p. 263.
27. J. A. Euler, *Disquisitio de causa physica Electricitatis*, p. 3.

caused considerable perplexity to Benjamin Franklin, and he eventually adopted a rather ambivalent solution. He arrived at what was really a theory of action at a distance by assigning repellent and attractive powers to the actual (vitreous) electric fluid, but, as to the physical nature of electricity, Franklin refused to abandon the doctrine of effluvia.[28] This ambiguity was finally overcome by one of his continental followers, Franz Ulrich Theodor Aepinus (1724–1802), who came to St. Petersburg in 1757.

After taking his doctor's degree in 1747 at the university of his native city of Rostock, Aepinus was appointed a professor of astronomy in the Berlin Academy of Sciences. In 1751 a young Swede named Johan Carl Wilcke (1732–1796) came to Rostock and attended Aepinus' lectures there, following him to Berlin in 1755, where they met Euler for whose scientific views both of them had, at that time, much respect. Two years after Aepinus was invited to Russia, he published his *Tentamen Theoriae Electricitatis et Magnetismi* in St. Petersburg, which contained the results of his and Wilcke's work in Berlin.[29] Overnight it made Euler's ether hypothesis redundant and removed the incongruities from Franklin's theory. In the first place, it generalized the doctrine that glass is impermeable to electricity—which had formed the basis of Franklin's theory of the Leyden vial—into the law that all nonconductors are impermeable to the electric fluid. Secondly, the *Tentamen* proved that this applies even to air—which Aepinus and Wilcke demonstrated by constructing a machine analogous to the Leyden jar. In their version of it, however, air took the place of glass as the medium between two oppositely charged surfaces.

The success of this experiment led Aepinus to deny altogether the existence of electric effluvia surrounding charged bodies. In accord with Newtonian dynamics, he concluded that attractions and repulsions observed between bodies compel us to believe that electricity acts at a distance across the intervening air. He further argued that, since two vitreously charged bodies repel each other, the force between two particles of the electric fluid must (on Franklin's one-fluid theory, which Aepinus adopted) be repulsive, and, since there is an attraction

28. He describes this electric fluid as a substance consisting of "particles extremely subtile, since it can permeate common matter, even the densest metals, with such ease and freedom as not to receive any perceptible resistance."

29. Cf. F. U. T. Aepinus, *Tentamen theoriae electricitatis et Magnetismi* . . . (St. Petersburg [1759]). A Russian translation of this work was published by the Soviet Academy of Sciences in 1951.

between oppositely charged bodies, the force between electricity and ordinary matter must be attractive. Franklin had made the same assumptions, but he was not able to account for the repulsion between resinously charged bodies. Euler's ether hypothesis was similarly unable to explain this. Aepinus introduced a new supposition: that the particles of ordinary matter repel each other.

This was something his contemporaries found difficult to believe, though Aepinus' explanation of the relation of "electricity" and "ordinary matter" is easier to understand in the light of modern atomic physics, where the relationship of "electrons" to "atomic nuclei" is closely comparable. "Unelectrified" matter, in other words, is really matter saturated with its natural quantity of the electric fluid, and the forces due to the matter and fluid balance each other. Thus, instead of assuming, as Euler did, that the difference between an electrified and "unelectrified" body lay in the equilibrium of ether that each maintained with respect to the ether around it, Aepinus suggested an equilibrium of a different sort—that between matter and the electrified fluid. Further, he suggested that a slight want of equilibrium between these two forces might give, as a residual, the force of gravitation.

Assuming that the attractive and repellent forces increase as the distance between the acting charges decreases, Aepinus applied his theory to explain a phenomenon which had been more or less indefinitely observed by his predecessors: namely, that if a conductor is brought close to an excited body but without actually touching it, the remoter part of the conductor acquires an electric charge of the same kind as the excited body, while the nearer part acquires a charge of the opposite kind. Aepinus was able to show that this effect followed naturally from the theory of action at a distance.

But Euler remained strangely unmoved by the mounting evidence amassed in favor of the Newtonian theory. This is not, however, as surprising as it may seem. Euler's *Disquisitio* on electricity, which was not richly informed by experiment, formed only a minor addition to his accomplished work on motion, light, and color. And it was not of sufficient importance for Euler to want to revise his scientific theories to which he had been so long attached. For this is what acceptance of the doctrine of attraction at a distance at this late point would have required. He preferred to stay faithful to the ideas he had imbibed in his youth (see the oration he delivered when he received his Master of Arts degree at Basel University, in which he compared the philosophies of Newton and Descartes—to the latter's advantage). Lomonosov's case was altogether different. Though he was only four years younger than Euler, he matured intellectually much later. When Euler published his first opus on coming to Russia in 1727, Lomonosov had not yet left his native village; by the time he began writing his first work more than a decade later, Euler had already become one of the major mathematicians of the age.

The theory of ether Lomonosov later developed was not unlike Euler's, and his disagreements with Newton, both with regard to universal gravitation and to his optical theories, were also similar. It would, however, be misleading to assume that Euler played any large part in shaping his scientific views. If any man had a lasting influence on Lomonosov, that man was Christian Wolff, who, for three years, supervised his studies. They met in Marburg in 1736.[30] Twenty years later Lomonosov was still under Wolff's spell. This much is clear from the title of the work he began in 1756, when he belatedly decided to take up the challenge posed by the Academy's competition of the previous year by elaborating a "Theory of electricity according to the mathematical method" (a titular form characteristic of Wolff). Yet Lomonosov never completed this opus. Only the first chapter and part of the second have come down to us, together with a synopsis of the contents.

The work was to consist of eight chapters, including a final one suggesting what directions "future successes" in the investigation of electricity might take. How he wished to approach the problem is clear enough. Bodies exposed to electrical power in similar conditions, he says, do not react in the same way: this means that "in constructing a theory of electricity it is essential to investigate their nature in order to determine what brings this difference about."[31] In other words, that mysterious substance called electricity would only yield its secrets if we investigate the "inner structure of bodies . . . through chemistry." Without it, "it would be difficult and even impossible" to discover "the real cause of electricity."[32]

One of the puzzles of Lomonosov's intellectual biography is why

30. Cf. A. V. Topchiev *et al.*, *Letopis' zhizni i tvorchestva M. V. Lomonosova* (Moscow and Leningrad: Akademiia nauk, 1961), p. 34.
31. Cf. "Theoria electricitatis methodo mathematica concinnata . . . ," *Polnoe sobranie*, III, 280, Section 18.
32. *Ibid*, p. 282, Section 20.

he never refers to this particular work in his other writings. Indeed, it is even ignored in the bibliographies Lomonosov compiled of his own compositions. B. G. Kuznetsov suggests that the reason may be found in the fact that Lomonosov liked to base his hypotheses on empirical demonstrations. Before committing himself to his chemical theory on the source of "electric power," he wished to conduct additional experiments.[33] What Kuznetsov leaves unanswered is why Lomonosov waited so long to conduct them, preferring to leave his "Electrical Theory" in abeyance if not oblivion. Yet the answer seems perfectly plain. When Aepinus came to St. Petersburg in 1757 to publish his *Tentamen theoriae electricitatis*, he made Lomonosov's "chemical hypothesis" redundant. There was surely no purpose, as Lomonosov would have realized, to insist on an alternative to a theory which was instantly accepted everywhere in Europe. Besides, Aepinus was soon to become his harshest critic within the Academy. Their quarrels were to last until Aepinus eventually withdrew from scientific activity to become director of the Imperial Cadet School. He survived most of his contemporaries in the Academy and lived to see the dawn of the Alexandrian era, when he died at the ripe age of seventy-eight.

Yet Lomonosov's "hypothesis" was not without its merits. In another work (on light) he rejected the view that electricity could be explained with reference to the theory of "fluids."[34] He even vaguely implies that electricity may be identified with a state of stress in the ether, an idea which bears some resemblance to the theory introduced nearly a century later by Faraday. Unfortunately Lomonosov could not develop this brilliant insight; his feet were bound by his refusal to accept the doctrine of attraction at a distance, without which Aepinus would have been equally helpless. Indeed, Aepinus owed much both to Newton and to Benjamin Franklin,[35] and he acknowledges his generous debt to the former in the opening pages of his *Tentamen*. In the florid dedication to the president of the St. Petersburg Academy, Count K. G. Razumovskii, he compares Newton's successes in making nature yield her secrets to the progress of an enormous river,

33. B. G. Kuznetsov, *Tvorcheskii put' Lomonosova*, 2nd ed. (Moscow: Akademiia nauk, 1961), p. 200.

34. Lomonosov, "127 Zametok k teori sveta i elektrichestva," in *Polnoe sobranie*, III, 237–263. It was written in the same year as the "Theory of electricity"—see below, Chapter 18.

35. Franklin's debt to Newton was no less considerable; see I. B. Cohen's *Franklin and Newton* (Philadelphia: the American Philosophical Society, 1956).

flooding its banks and inundating the surrrounding fields, covering them in all directions as far as the human eye could see. Lomonosov, however, remained largely uninfluenced by Newton's physical theories. This was Wolff's worst legacy to his Russian disciple, and it was to have unfortunate consequences for his scientific work as a whole, in spite of the fact that he successfully emancipated himself from much that he thought was "mystical" and trivial in the German philosopher's thought.

39. Mikhail Lomonosov (1711–1765).

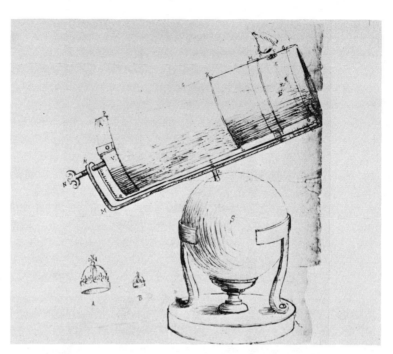

40. Drawing of Newton's reflecting telescope.

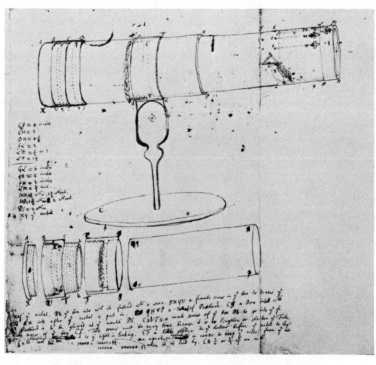

41. Newton's own drawing of a telescope and its parts.

42. Lomonosov's drawing of the "Newton-Lomonosov" telescope, showing the inclination of the speculum and the position of the eyepiece.

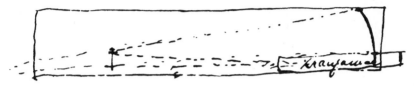

43. Lomonosov's drawing of the "Newton-Lomonosov" telescope, indicating the path of light.

44. Lomonosov's initial sketch of the "Newton-Lomonosov" telescope with the eyepiece in the same position as in Newton's reflector telescope.

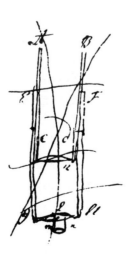

45. The "Tubus Nyctopticus modo Lomonosov-Newton," indicating a return to the refracting principle.

россійская универсальная

ГРАММАТИКА,

или

ВСЕОБЩЕЕ ПИСМОСЛОВІЕ,

Предлагающее

Легчайщїй способъ основательнаго ученїя рускому языку съ седмью присовокупле-нїями разныхъ учебныхъ и полезно-забавныхъ вещей.

Clericus es! legito hœc. Laicus! legito ista libenter. Crede mihi, invenies hic quod uter-que voles.
$\qquad$ D. Collis

ИЗДАНО ВО ГРАДѢ СВЯТАГО ПЕТРА
1769 года.

46. Title page from Kurganov's *Pismovnik*, the first edition (St. Petersburg, 1769).

Что может собственных Платонов
И быстрых разумом Невтонов
Российская земля рождать.

Lomonosov

17

Newton, Lomonosov, and Christian Wolff

LOMONOSOV AND WOLFF

The trials and tribulations that brought Lomonosov to Marburg University were even more remarkable than those that brought Cantemir from Constantinoble to Georgian London. A polymath of genius, Lomonosov was born in 1711, near Kholmogory, where the Northern Dvina flows into the White Sea. Though cut off from cultural centers such as Moscow and Kiev, the region's geographical isolation had saved it from the Mongol yoke and serfdom. Having long served as a refuge for fugitives from political oppression, it was colonized almost entirely by runaway peasants, religious dissenters, and Old Believers. Foreign trade by way of Archangel further stimulated its cultural development. Here, in the Antonievo-Siiskii monastery, about fifty miles from Kholmogory, one of Russia's earliest typographies was founded about 1670; and the monastery became an early focus of enlightenment. At the beginning of the eighteenth century a theological seminary was also built there, and books began to circulate among the peasantry. A school of icon painters flourished, and in 1692 Archbishop Athanasii funded an astronomical observatory near Kholmogory, about which little is known.[1]

Athanasii was hardly a typical representative of the clergy of his time, but neither was Kholmogory typical of the Russian provinces. Lomonosov's father, Vasilii Dorofeev, owned several merchant vessels,

1. See V. Veriuzhskii, *Afanasii, arkhiepiskop kholmogorskii; . . .* (St. Petersburg: Típografiia T. Leont'eva, 1908).

and was by local standards a wealthy man, a circumstance sometimes forgotten owing to his son's peripatetic and eventful youth. But his father shared none of Lomonosov's intellectual interests, which were probably aroused through his association with local religious dissenters. As far as is known, it was only in 1725 that he became familiar with secular literature that included Smotritskii's *Slavic Grammar,* a well-known literary manual of the period, and Magnitskii's *Arithmetic,* published in Moscow in 1703 "for the instruction of wisdom-loving boys and girls." For half a century Magnitskii's work was the mathematical primer Russians generally used, and for Lomonosov it was probably his earliest introduction to the "mathematical sciences." It had sections on geometry, forty pages on algebra, and an account of Franciscus Vieta's method for solving equations,[2] which by this time had been superseded by the methods of James Gregory and Newton. Still, these two books may have had something to do with Lomonosov's early love for natural philosophy.

"Although I had a father who was kind by nature," Lomonosov would later write, "he was brought up in utmost ignorance, and my stepmother was malicious and envious and tried to rouse anger in my father in every way she could. She made out that I was idly wasting time reading books, and I had to read and study as much as I could in solitude and in desolate places, suffering cold and hunger . . ."[3] He decided to leave home and obtain an education by his own efforts elsewhere. In 1730 he set out on foot for Moscow, and, after covering a distance of about 620 miles, he arrived in the city, where he applied for admission to the Slavo-Greco-Latin Academy. To be enrolled, Lomonosov had to conceal his peasant origin.

The Slavo-Greco-Latin Academy had been founded in 1684, and the curriculum was still in the Scholastic mould. Grammar, syntax, rhetoric, poetry, arithmetic, Aristotelian philosophy and theology were the principal subjects. The type of problem studied in the senior classes has been described by V. A. Steklov: "How do [angels] communicate their thoughts? . . . Do roses in paradise have thorns?"[4] A more recent study suggests, however, that Lomonosov's time was not

2. Cf. A. K. Sushkevich, "Materialy dlia istorii algebry v Rossii," in *Istoriko-matematicheskie issledavaniia,* Vol. **IV** (Moscow: Gosudarstvennoe izdatel'stvo tekniko-teoreticheskoi literatury, 1951), p. 239.
3. Lomonosov to I. V. Shuvalov, May 10, 1753, in M. V. Lomonosov's *Polnoe sobranie,* **X,** 482; this edition cited throughout.
4. V. A. Steklov, *Mikhailo Vasil'evich Lomonosov* (Berlin: Z. I. Grzhebin, 1921), pp. 17–18.

entirely wasted, though his own allusions to the Slavo-Greco-Latin Academy were always dismal:[5]

When studying at the Spasskii school there were exceedingly strong influences on all sides, hindering me from learning *(nauk)*, and in those years these had a power which was well nigh irresistible. On the one hand, my father, having no other children than myself, said that I, being the only son, had abandoned all the fortune (measured by local conditions) which he had amassed for me by bloody sweat, and which, after his death, others would plunder. On the other hand, there was inexpressible poverty: since I had only an *altyn* a day for pocket money, I could not have more food each day than half a *kopek*'s worth of *kvas,* and the rest was spent on paper, shoes, and other necessities. In such a manner I lived for five years and never abandoned learning. On the one hand they wrote that, knowing my father's means, the good people at home set their daughters' caps for me—they had proposed even while I was yet there. On the other hand, the schoolboys, little children *(rebiata)* used to cry and point at me with their fingers. 'Look, what a blockhead *(bolvan)* to be studying Latin at twenty': but all this did not turn me away from learning . . ."[6]

Lomonosov's education was almost cut short when his humble origins were discovered. At the end of 1735, however, Baron Korff, who had been appointed president of the Academy of Sciences the previous year, petitioned the Senate to issue a decree requiring that the best students in the monastic schools be enrolled in the Academy's gymnasium. Only twelve students could be found, but Lomonosov was included. Soon afterward he was sent abroad, where his instruction was supervised by Christian Wolff.

Wolff, in turn, had a considerable influence on Lomonosov— exactly how much is still a matter of dispute. "It was not deep enough to make him a Wolffian," one biographer has suggested.[7] Another has asserted that Lomonosov's view of nature "was wholly Cartesian."[8] Others have insisted that Lomonosov owed most to Leibniz.[9] Since Wolff was in some respects a Cartesian and in others a Leibnizian, the particulars of this debate are not of consequence here. In both

5. See V. P. Zubov's excellent article, "Lomonosov i Slaviano-Greko-Latinskaia Akademiia," in TIIE-T, I (1954), 5–52.

6. Lomonosov to Shuvalov, May 10, 1735, *Polnoe sobranie,* X, 479.

7. Cf. M. I. Sukhomlinov, "Lomonosov—student Marburgskogo universiteta," *Russkii vestnik* (No. 1, 1861), pp. 141–142, 152–153.

8. Cf. N. A. Liubimov, "Lomonosov kak fizik," in *V vospominanie 12 ianvaria 1855 goda* (Moscow, 1855), p. 12.

9. Cf. Vladimir Tukalevskii, "Glavnye cherty mirosozertsaniia Lomonosova (Leibnits, Lomonosov)," in *M. V. Lomonosov, 1711–1911; Sbornik statei,* ed. V. V. Sipovskii (St. Petersburg: Akademiia nauk, 1911), pp. 13–32.

capacities, Wolff's reaction to Newtonian philosophy was equally hostile. He had played a rather disreputable part in the squalid controversy between Newton's friends and the champions of Leibniz over the invention of the calculus,[10] and, in the controversy over gravitation, he firmly took the side of the Cartesians as his polemic with Keill in the *Acta Eruditorum* (February 1710) shows. Yet in his *Vernuenfttigen Gedancken von den Wuerckungen der Natur,* which closely follows the courses on physics Wolff delivered at Marburg in the period that Lomonosov was a student there, he pronounces himself an adherent of the inductive method. One of his rules, says Wolff, is "to desist from all fanciful hypotheses," a maxim reminiscent of Newton's "hypotheses non fingo."[11] But his claim, that he based all his conclusions "on experiments alone," is, as Kant pointed out, quite untrue; Wolff, though perhaps he did not realize it himself, was quite opposed to the empirical bias of Newton's method.

Nor did Wolff have any of Leibniz' mathematical genius. He was, therefore, ill equipped to appreciate the revolutionary nature of the *Principia.* One important defect in Wolff's philosophy is his almost complete failure to distinguish the physical from the metaphysical, and, by deriving the principle of ground and consequent from that of contradiction, he established a hard-and-fast rationalism in respect of the world of nature which Leibniz would certainly have rejected. According to Gottsched, Wolff's most faithful hagiographer, however, his services to humanity were of another order, and, cumulatively, they were far more important than the discoveries of a Torricelli, a Kepler, a Huygens, or a Leibniz. Wolff was able "to encompass all of philosophy in his gaze . . ." as only Aristotle before him had done.[12] He may have been inferior to Newton as a physicist, "but in a dozen other ways Wolff stood head and shoulders above him." Newton, Gottsched goes on to say, was a mere scientist; Wolff was "a universal philosopher" superior even to Descartes, who had only paved the way for him.[13]

10. "Another German mathematician of feeble morality," says David Brewster in *Memoirs of . . . Sir Isaac Newton,* II, 57; see also p. 67.

11. Christian Wolff, *Vernuenfttige Gedancken von den Wuerckungen der Natur, den Liebhabern der Wahrheit mitgetheilet von Christian Wolffen,* 5th ed. (Halle im Magdeburgschen, 1746; 1st ed. dedicated to Peter the Great, Halle, 1723), Section 5, no. 2.

12. J.-Ch. Gottsched, *Historische Lobschrift des weiland hoch- und wohlgebohrnen Herrn Herrn Christians Freiherrn von Wolf, . . .* (Halle, 1755), p. 142. Gottsched, of course, was one of Wolff's pupils.

13. *Ibid.,* p. 142.

Incredibly enough, this view of Wolff's achievement was taken quite seriously at many German universities. Even Frederick the Great, a man not given to praising his fellow creatures, seems to have shared it. Wolffian philosophy was seen, not least by Wolff himself, as a cosmic antidote to the materialism and skepticism that, "together with the principles of the now well-known Englishmen," were ravaging France and Italy.[14] Voltaire was to show how tawdry this antidote was—in *Candide, Micromégas,* and elsewhere—and he offered the opinion that if Leibniz had been brought into this world to bring eternal muddle into philosophy, Wolff's function was to propagate that mass of confusion "in fifteen volumes in quarto" so that the passion of the Germans "to read much and understand little" was duly satisfied.[15] Lomonosov was at Marburg before the inevitable reaction against Wolff began. As A. A. Morozov's brilliant study of Lomonosov's early years has shown, he took much from Wolff—more especially in physics and mechanics.[16] At times he even adopted his style of exposition, that is, Wolff's celebrated "mathematical method" which the German philosopher applied indiscriminately to physics as well as to ancient Hebrew.[17] Did Lomonosov also imbibe Wolff's prejudices against Newton?

With all of his faults, Wolff was a man of vast erudition; from the copious bibliography attached to one of his major works, almost all of which were dutifully acquired by Lomonosov, a diligent student may have learned much.[18] Wolff thought highly of several English mathematicians and men of science—particularly John Wallis, John Ray, Robert Boyle ("Paradoxa Hydrostatica"), Christopher Wren, Edmond Halley and Robert Hooke (the "Micrographia"). He also recommends the *Principia* (in a nonexistent edition of 1682!) as a

14. Wolff to Manteuffel, June 7, 1739, in *Ch. Wolff's Eigene Lebensbeschreibung,* ed. Wolff von Heinrich Wattke (Leipzig: Weidmann'sche Buchhandlung, 1841), p. 177.

15. Quoted by Hans Droysen, "Marquise du Châtelet, Voltaire und der Philosoph Christian Wolff," in *Zeitschrift für französische Sprache und Literatur,* XXXV (1909), 226–248.

16. Cf. A. A. Morozov, *M. V. Lomonosov—put' k zrelosti, 1711–1741* (Moscow and Leningrad: Akademiia nauk, 1962), pp. 274–290.

17. *Ibid.,* p. 264. Curiously enough, Wolff's reputation in Russia during the eighteenth century rested on his mathematical writings (some of which were translated), though Wolff contributed nothing original to the subject.

18. Cf. Christian Wolff, *Kurzer Unterricht von den vornehmsten Mathematischen Schriften aufgesetzet von Christian Wolff* (Halle, 1717), which contains more than a hundred pages of items on mathematics, mechanics, architecture, artillery, optics, astronomy, and so on. The annotated list first appeared by way of supplement to Wolff's *Anfangsgründe aller mathematischen Wissenschaften,* Pt. II (Halle, 1710).

work which contains "many ingenious discoveries on motion," though he warns his students that it makes few concessions to the reader, who must know "higher geometry and the two calculi in order to understand it."[19] How enticing Lomonosov would have found this commendation we can only conjecture, but he was probably well aware of the stir being caused by Newtonian philosophy through Voltaire's *Lettres Philosophiques,* which he may have read at this time.[20] Moreover, he attended the lectures of Professor Duising, who thought very highly of Freind's *Praelectiones Chymicae,* a text used at Oxford and elsewhere.[21] Indeed, this was the work Richmann had earlier criticized for wantonly introducing Newton's doctrine of attraction at a distance into chemistry.[22]

Lomonosov, as it turns out (unknown to Richmann but for much the same reasons), appears to have been opposed to it already at this early point in his career. This transpires from one of the "specimens" of his work which, as a student, Lomonosov was asked to send to the St. Petersburg Academy of Sciences in 1739 from Marburg. Echoing Wolff, he describes the Newtonian doctrine as "an occult" theory[23] and when Lomonosov began his study of the *Principia* four years later, after returning to his homeland, nothing he found there appears to have changed this view.

THE PRINCIPIA

On more than one occasion Lomonosov implies that Newton himself did not believe in gravitation, that it was invented by his followers and successors.[24] So wrongheaded does this claim appear that it is tempting to resort to Guhrauer's ingenious explanation in his study of Leibniz, where he suggests that one reason why Leibniz did not

19. *Ibid.*

20. This work ("Lettres sur les Anglois") is included in the list of books acquired by one of the two students with whom Lomonosov was sent to Marburg. Cf. Morozov, *M. V. Lomonosov,* p. 464, n. 137. The books the three of them acquired were shared.

21. *Ibid.,* pp. 265, 464.

22. See Chapter 15, above.

23. I.e., Lomonosov, "Fizicheskaia dissertatsiia o razlichii smeshannykh tel . . . ," in *Polnoe sobranie,* I, 24–63.

24. A similar point is made in *Letters of Euler . . . to a German Princess,* ed. Hunter (I, letter xxiv, dated July 1, 1760), where he says that "the great Newton himself strongly felt [the] force" of "the difficulties and contradictions" of his theory. The *Letters* were published only after Lomonosov's death.

follow Newton's lead is that he never took the trouble to read the *Principia*.[25] It can be proved beyond doubt, however, that Lomonosov was in this instance more conscientious than Leibniz.[26]

On May 28, 1743, Lomonosov was put under house arrest for interrupting Academy meetings and for general insubordination. He was not released till January 1744. And, shortly after his arrest, he wrote to the Chancellery of the Academy requesting copies of the *Principia* and the *Universal Arithmetick*.[27] References in Lomonosov's writings further show that it was the second edition of the *Principia* that he received, conceivably the same copy which (as noted above) passed from the library of J. D. Bruce into the collection of the Academy in 1736.

This second edition (1713) contains the celebrated preface by Roger Cotes, with its vehement riposte to Newton's critics. This, as it turns out, is the key to Lomonosov's arbitary distinction between Newton and his followers. For the polemical purpose of Cotes's preface gave rise to some careless expressions which Newton himself would not have used. Above all, Newton had never explicitly stated that attraction takes place between bodies *at a distance without any intervening medium*.[28] He makes it clear in both the *Opticks* and the *Principia* that his mathematical formulation is neither the metaphysical nor the "ultimate," underlying cause of gravitation. Indeed, this is stated by Newton at the very opening of the *Principia*, as James Jurin had pointedly tried to impress upon Bilfinger.[29] Bilfinger's main objection to the Newtonian theory was similar to Leibniz': Newton had failed to give a purely physical explanation for action at a distance and

25. G. E. Guhrauer, *Gottfried Wilhelm, Freiherr von Leibnitz: eine Biographie* (Breslau, 1846), I, 279.
26. This slur on Leibniz' reputation may now at last be removed owing to Emil Fellman's recent recovery of a copy of the first edition of the *Principia* containing marginalia in Leibniz' own hand. Fellman described the circumstances of his discovery at the XIII International Congress of the History of Science, in Moscow (Section V,B, August 19, 1971): "Leibniz' Marginalnoten in Newton's *Principia Mathematica*." The marginalia are to be published.
27. *Polnoe sobranie*, IX, 322.
28. Cf. statement by Samuel Clarke (whom Newton accepted as his advocate in the celebrated controversy with Leibniz): "That one body should attract another without any intermediate means, is indeed not a miracle but a contradiction. For it is supposing something to act where it is not." See *A Collection of Papers, which passed between the late learned Mr. Leibniz and Dr. Clarke, in the years 1715 & 1716, relating to the principles of Natural Philosophy and Religion* . . . (London, 1717), p. 51. Clarke denied the use of any "occult qualities" and also denied the existence of a rare ether; see especially his fourth reply.
29. See Chapter 12, above.

disguised this failure by introducing a discredited Scholastic concept. Jurin's insistence that Newton considered attraction "not physically, but mathematically" was regarded by Bilfinger as an evasion, but Cotes's preface to the second edition made it appear something far worse.

There Cotes does not actually use the phrase "action at a distance," nor does he explicitly affirm that celestial spaces are void—doctrines which to any believer in the impulse theory would seem by definition implausible. He does, however, suggest *the existence of an ethereal matter*. This notion made the theory of gravitation seem patently inconsistent, and it is clearly the reason why Lomonosov, who was in any case predisposed against gravitation by virtue of his dynamic theory, could conclude that Newton had not actually meant what his less cautious followers ascribed to him. Had not Newton, in the same passage where he attributes forces only to "mathematical points," denied giving any "explanation" of gravity? (Book I, Definition VIII.) By admitting that interplanetary space is not a vacuum, but filled with "an aethereal matter," Roger Cotes merely added another possible objection to the Newtonian theory. There was no conceivable reason why gravitation should be made to apply to this ethereal substance, and, once this is granted, it would seem suspicious that Newton made no allowance for it in his calculations.

In fact, Cotes did offer an explanation: he argued that his ethereal matter "has no inertia, because it has no resisting force."[30] But it was not an explanation that could satisfy Lomonosov, and in a treatise on light he notes ironically:

Can one conceive of any finer ethereal matter than Newton's, anything faster in motion, and which is stronger in attraction to the sun than any other matter? What an extraordinary miracle! Why does not the force of these particles obey the laws of gravity?[31]

The ethereal matter postulated by Cotes (and *not* by Newton) would also have to account for the velocity of light, as Lomonosov here suggests. In his preface Roger Cotes also attacks the Aristotelians who "affirm that the several effects of bodies arise from the particular nature of these bodies."[32] Since, however, "they don't tell us . . .

---

30. *Principia*, Preface.
31. Lomonosov, "127 Zametok k teorii sveta i elektrichestva," *Polnoe sobranie*, III, 254, Section 87.
32. *Principia*, Preface.

whence it is that bodies derive those natures . . . they tell us nothing." But, Lomonosov could counter, was not Cotes guilty of the same offense as the Aristotelians?

It is a historical fact that Roger Cotes, more than anyone else, contributed to this alleged misunderstanding of gravitation. In his preface he writes of "the nature of gravity in earthly bodies," for example, and affirms that "gravity must have a place among the primary qualities of all bodies."[33] And even Bentley, the admirer of Newton's who published the second edition, shared these mistaken notions. Newton said nothing in the *Principia* itself of the qualities of the "ether" separating the planetary masses. This was Cotes. Nor did Newton ascribe a primary quality or essence to elementary particles.

Lomonosov drew the appropriate conclusion. In a rough note to an incomplete manuscript, also written in 1743–1744, he writes that "attraction must be refuted. There is no such thing as pure attraction . . . Newton regarded it as a phenomenon, in the same way that astronomers suppose, in order to explain [certain phenomena] that celestial bodies rotate every twenty-four hours."[34] That is to say, Newton's laws of gravity were an approximation, a description of empirical reality. Further investigations would show what gravity *really* was. And to show that what Newton meant by gravitation was not the same as the interpretation offered by his followers, he cites the authority of Newton himself. "We do not here impugn the opinion of men of great merit in the sciences [literally, de republica litteraria], who suppose that the force appearing as attraction is a phenomenon which [really] explains other phenomena . . . Illustrious Newton, who established the laws of gravity, did in no way have pure attraction in mind (nullam meram attractionem supposuit)."[35] To support this startling conclusion he refers to two passages from Book I of the

33. *Ibid.*
34. Lomonosov, "De Particulis Physicis Insensibilibus Corpora Naturalia Constituentibus, in quibus qualitatum particularium ratio sufficiens continetur," in *Polnoe sobranie,* I, 286 f. He makes the same reference to "quo astronomi motum diurnum siderum circa tellurem supponunt, etc." in the "Tentamen theoriae de Particulis Insensibilibus Corporum Deque Causis Qualitatum Particularium in Genere," in *ibid.,* p. 190, Section 44, the point being that this is what they assume: precise calculations give different results. It is true, of course, that Newton could not *prove* that gravitation was the key to all celestial motions—for not all of them were known in his time. And the irregularities of the moon's motions were enough to warrant reasonable doubt.
35. Lomonosov, "Tentamen Theoriae . . .," *ibid.*

Principia, the most pertinent being the scholium to the sixty-ninth proposition:

I here use the word *attraction* in general for any endeavour whatever, made by bodies to approach each other, whether that endeavour arise from the action of the bodies themselves, as tending to each other by spirits emitted; or whether it arises from the action of the ether or of the air, or of any medium whatever, whether corporeal or incorporeal, in any manner impelling bodies placed therein towards each other . . .[36]

Secondly, Lomonosov invokes the authority of John Bernoulli, who "explains many cases of attraction as impulse, which makes all cases quite suspect."[37] This "suspicion" of Newtonian gravitation is one Lomonosov harbored to the end of his life. It did not prevent him from expressing the utmost admiration for Newton, for his ingenuity in devising experiments, and so on, but he continued to criticize his followers. Thus, in one of his most accomplished scientific writings, his *Observations on the density and fluidity of bodies* (1760), in which the freezing of mercury is described for the first time, Lomonosov states quite explicitly that "Newton did not accept the force of attraction in his own lifetime, but became its involuntary champion owing to the excessive zeal of his followers."[38] To confirm the paradox he cites another section of the *Principia,* where Newton writes,

I have hitherto been treating of the attractions of bodies towards an immovable centre, though very probably there is no such thing existent in nature . . . I shall therefore at present go on to treat the motion of bodies attracting each other; considering the centripetal forces as attractions though perhaps in physical strictness they may more strictly be called impulses.[39]

These references prove, if more proof were needed, with what attention Lomonosov read the *Principia,* though the possession of any but Cotes's second edition would have prevented him from making so

36. *Principia,* Book II, Proposition LXIX. Scholium.
37. Lomonosov, "Tentamen Theoriae . . . ." *ibid.,* p. 186, Section 40 ("Celeberrimus Bernoulli plerasque attractiones per impulsionem explicat, quo omnis attractio adeo suspecta redditur.").
38. The *Rassuzhdenie o tverdosti i zhidkosti tel* was written simultaneously in Latin and Russian (and published in St. Petersburg in 1760 in both languages), but the two versions have slight variants. At this point in the Latin, for example, Lomonosov is less emphatic: "quam Newtonus invitus post fata attractionis patronus, in ipso initio Philosophiae naturalis Mathematica inculcat, etc." (*Polnoe sobranie,* III, 380, Section 3.)
39. *Principia,* Book I, Section XI ("Of the motions of bodies tending to each other with centripetal forces").

convenient a dichotomy between Newton and his "followers." It was convenient because the textual inconsistencies between Cotes's preface and the *Principia* itself made it easier for Lomonosov to reject gravitation; they were not the cause. The character of his objections had already been determined by his studies in Marburg, as is shown by the *Dissertatio Physica de Corporum Mixtorum Differentia,* which Lomonosov wrote as one of his undergraduate exercises in 1739. He attacks Newtonian gravitation and announces his refusal to "recognize the power of attraction or any other occult quality (qualitas occulta)."[40] His reason is that gravitation is incompatible with the impulse theory of motion. As he states elsewhere, "without impulse bodies can neither act nor react"; it followed that "every attraction is suspect and for the most part may be explained by impulse."[41]

This is precisely the type of criticism one would have expected from a disciple of Christian Wolff, for Wolff followed Leibniz in his critique of Newtonian dynamics.[42] In his *Vernuenfttige Gedancken von den Wuerckungen der Natur,* which Lomonosov knew well, Wolff carefully defines the limits of his acceptance of the theory of universal gravitation. He approved of it as applied to the planetary bodies, *but he regarded it solely as an empirical fact;* Newton had not stated *what* gravitation was, merely *how* it behaved:

Es ist wohl wahr das CARTESIUS durch eine Druck die Sache erklaeret: hingegen Kepler und Newton durch ein Anziehen, Nun wissen wir, das keine anziehende Krafft in der Natur als nur dem blossen Scheine nach Platz findet. Allein unerachtet die meisten Anhaenger Herrn *Newton*'s die Schweere als eine eigenthuemliche Krafft der Materie ansehen und keine natuerliche Ursache derselben verlangen; so haben wir doch nicht noethig hierinnen ihrer Meinung beyzupflichten. Wir nehmen diese

40. Lomonosov, "Dissertatio Physica de Corporum Mixtorum Differentia . . . ," in *Polnoe sobranie,* I, 42, Section 45.

41. Lomonosov, "Tentamen theoriae . . . ," in *ibid.,* p. 188, Section 41: "verum omnis attractio est suspecta et plerumque per impulsionem explicatur." Lomonosov returns to this objection on other occasions: cf. Theorem III, in *ibid.,* p. 186 ("Corpora in motum excitantur per solam impulsionem"), and note "d" to chapter i. Section 7 of "De Particulis Physicis Insensibilibus . . . ," in *ibid.,* p. 288 ("Gravitas est qualitas particularis, quia est motus, qui est qualitas particularis . . .").

42. Russell believed that Leibniz' objection to action at a distance was nothing but prejudice—there is no reason, he says, why monads (which do not in any case interact) cannot be made to function according to the Newtonian doctrine. The latter could be just as much a cause for their action as that which takes place near them (impact). "There seems to be no metaphysical reason, and in the time of Newton no dynamical reason for such a reaction." Cf. Bertrand Russell, *A Critical Exposition of the Philosophy of Leibniz with an Appendix of Leading Passages* (London: G. Allen and Unwin, 1937), p. 91. The same may be said of Wolff's system.

Schweere der Planeten gegen einander, oder ihre magnetische Krafft an als eine Sache, die in der Erfahrung gegruendet ist, aber eine uns zur Zeit noch nicht voellig bekandte Ursache hat, die wir zu weiterer Untersuchung ausgesetzet seyn lassen, als wie man die Schweere der Coerper gegen die Erde annimmet als eine Sache, die Vermoege der Erfahrung in der Natur statt findet, und andere Dinge daraus erklaeret die von ihr herruehren, auch wenn mann gleich nicht eigentlich weiss, wie die Natur die Coerper gegen die Erde Schweer machet.[43]

In this criticism lies the source for young Lomonosov's objections to Newtonian gravitation. As he grew more mature, he was able to emancipate himself from Wolff's more blatant faults. Not only did he jettison his style of exposition, but he disassociated himself (in private) from Wolff's monadology, which stood at the heart of Leibnizian metaphysics. Nonetheless, what continued to divide Lomonosov from Newton most fundamentally was not the question of dynamics, but the approach to natural philosophy—to which each brought some wholly different series of assumptions.

EMPIRICISM VERSUS RATIONALISM

The idea of centripetal forces acting at a distance, which both Wolff and Lomonosov found so unpalatable, was for Newton a natural law, a scientific description of natural relations expressed in mathematical terms. It was a concept that science could use to good advantage. Having proved centers of force attracting at a distance to be "experimentally" and mathematically correct, Newton felt no obligation to add supplementary hypotheses within the framework of the *Principia* to explain *why* gravitation behaved as it did. He would do so later, more especially in the queries added to the second edition of the *Opticks* (1717), but not without drawing a firm distinction between explanations which were "physical," and empirically proven and those expressly intended as hypotheses or conjectures.[44] In other words, to Newton a mathematical expression in experimental philosophy deals with physical relationships or the laws which are in nature, while the primary notions in general deal with the underlying "substratum" or "ethereal" and "ultimate" causes of nature. Gravitation

43. Wolff, *Vernuenfttige Gedancken von den Wuerckungen der Natur*, p. 556, Section 359.
44. See below, pp. 192–193.

for him did not imply any metaphysical explanation of cause or motion. He did not identify the mechanical or secondary causes with theological or primary causes of a metaphysical type, and it was because, when he did try to suggest a first cause for gravitation (in the *Opticks*), he did so by suggesting considerations *outside* the realm of physics, that Leibniz accused him of invoking "qualitates occulta."

Lomonosov, as we have seen, makes the same charge, and his objections to gravitation have this in common with Leibniz': they are in the last analysis metaphysical and could be made without reference to the mathematical demonstrations offered in the *Principia*. In fact, Lomonosov never refers to them.

Both Leibniz and Wolff believed that matter is infinitely divisible. However small the "modes" or the manifestations of extended substances may be, they are still divisible, for where can we stop dividing matter when material substance is synonymous with extension? To put it in another way: for Wolff the primary and absolute qualities of matter are extension, divisibility, and motion.[45] From these primary qualities, it follows that ultimate atoms do not exist. Matter being synonymous with extension, there is no vacuum. The material world is infinite, for it is impossible to limit extension, and changes in nature must be explained according to "ultimate" natural laws, logically consistent—for example, the mechanical laws of impact.

Newton's treatment of matter as a series of mathematical mass points was open to a number of logical objections on these counts, and Lomonosov offers the same type of argument that is found in the *Vernuenfttige Gedancken von den Wuerckungen der Natur*. "Centripetal force (vis centrifuga) can only be ascribed in the nature of things to that which turns around some center. Since unextended parts can have no surface separately from their centers, it follows that they cannot turn around their center, nor, therefore, exert any centripetal force on other particles."[46] What this approach illustrates above all is the difference in scientific method between Lomonosov and Newton. The former was a rationalist; logical deductions, even though physically unverifiable (such as the one above) had for him an existential value.

For Newton, on the other hand, logical deductions were useful providing only they were not physically contradictory and were physically verifiable. Cotes, stung by the rejection of the inverse-square

---

45. Wolff, *Vernuenfttige Gedancken von den Wuerckungen der Natur*, Section 6.
46. Lomonosov, "De Particularis Physicis Insensibilibus," in *Polnoe sobranie*, I, 294.

law of gravitation by the French Cartesians, boldly stated the implications of Newton's scientific method in his preface to the second edition of the *Principia* by saying that the aim of theoretical physics is simply the prediction of future events. Everything not strictly needed for this purpose and not directly deducible from observable facts should be eliminated. What he meant in modern scientific terms is clear enough. Investigations of the theoretical physicist are concerned partly with events which can actually be observed—these are called phenomena—and partly also with events which cannot be detected in themselves, but which are assumed to exist in order to set up continuity between separated events that are in fact observed. These hypothetical events have been called *interphenomena* by modern writers.[47] An example from optics may illustrate this distinction. As is well known, light cannot be perceived except when it impinges on matter. These impacts are phenomena. However, since the velocity of light is finite, there are intervals of time between successive encounters with matter. If the belief is adopted that the luminous disturbance continues to exist in some form between its leaving one point and appearing at another, then during this intermediate stage it is classified as interphenomena.[48]

Gravity provides another illustration. Assuming, as has in fact now been established, that its speed of propagation is finite, the questions as to what is happening while it is in the course of propagation belongs to the realm of interphenomena. Cotes would have held that these interphenomena should be ignored. They were not susceptible to observation, and the formulas of prediction could be established without using them. They belonged to metaphysics rather than to physics, though their juxtaposition was still partly disguised by the way in which Newton's and even Lomonosov's contemporaries used the term "natural philosophy." The school of which Descartes and Huygens were the first illustrious members regarded the understanding of interphenomena as one of the primary aims of the natural philosopher. Leibniz, Wolff, and almost all the physicists, as they came to be called, of the next two centuries belonged to the same tradition. It is, therefore, hardly surprising that Lomonosov also failed to realize what Newton was trying to do. Like all the continental rationalists,

47. E. H. Reichenbach.
48. In Lomonosov's day the propagation of gravitation was assumed to be instantaneous, and consequently there was no span of time to be bridged by interphenomena.

178

he thought of the forces formulated by Newton as metaphysical or final explanations of gravity, and he appropriately advanced other metaphysical arguments to show that Newton's solution was inconsistent. How irrelevant this type of objection seemed to Newton may be gathered from the fact that he freely admitted in the *Principia* that he had not yet discovered the *cause* of gravity. And he suggested that this ignorance was due to a lack of suitable experiments which the future would doubtless supply.

If this failed to check Lomonosov's criticisms, it is easy to see why. Like Christian Wolff, he failed to make any distinction between physical or mathematical descriptions of natural relations and the explanatory causes of philosophy in general. This confusion is of enormous importance historically. It was responsible for the divergent attitudes the Newtonians and the continental rationalists adopted toward induction. Newton says we must find the laws first "before we may speak about their causes" (that is, of gravitation).[49] But Lomonosov starts with the antithesis—we must find the causes before we speak of physical laws. In his *Prolegomena to Natural Philosophy,* for example, which was written shortly before his death, he writes: "How difficult it is to establish first principles: but, whatever obstacles there might be, we must encompass in one view the sum totality of all matter (*sovokupnost'*)."[50] Next to this he adds, "I want to base our understanding of nature on some definite principle . . ."

The philosophical consequences of this difference of emphasis were significant. Fontenelle's celebrated *Éloge* of Newton drew a similar contrast between him and Descartes.[51] "One cannot argue from the particular to the general," wrote Lomonosov.[52] This emphasis on deduction characterized most of Lomonosov's scientific work and dominated his quest for a systematic set of principles on which chemistry

49. *Opticks,* cf. Book III, Query 31.

50. Quoted by S. I. Vavilov, "Lomonosov i russkaia nauka," in *Sobranie sochinenii,* III, 572.

51. "Both of them founded their Physicks upon discoveries in Geometry . . . But one of them taking a bold flight, thought at once to reach the Fountain of All Things, and by the clear and fundamental ideas to make himself master of the first principles; that he might have nothing more to do than to descend to the phenomena of Nature as to necessary consequences: the other more cautious, or rather more modest, began by taking hold of the known phenomena to climb to unknown principles . . . The former sets out from what he clearly understands, to find the Causes of what he sees; the latter sets out from what he sees . . . ," in *An Account of the Life and Writings of Sir I. Newton,* 2nd ed., tr. from the *Éloge* of M. Fontenelle (London, 1728), p. 18.

52. Lomonosov to Euler, July 1748, *Polnoe sobranie,* X, 440.

might be based.[53] It meant that, in the last analysis, experiment took second place, and, at the same time, it implied a certain condescension for the empirical method. "True," he wrote, "many people deny the possibility of founding chemistry on mechanics so that it may be made an exact science . . . But if only those who darken their days with smoke and soot, with chaos reigning in their brains and teeming with undigested experiments (*neprodumannykh*), would only learn the sacred laws of Geometers, once strictly followed by Euclid, and perfected in our day by the illustrious Wolff, they could penetrate more deeply into the secrets of nature . . ."[54]

The rationalists generally shared this attitude: "geniuses such as he," wrote Leonard Euler of Lomonosov, "are very rare: for the most part they go no further than their experiments and refuse even to reason from them, or when they do—stumble into such absurdities as to contradict all principles of sound philosophy."[55] At issue was not only a different emphasis on induction, but a different attitude toward mathematics.

When Lomonosov calls his dissertations a "Theory of electricity according to mathematical method" or "Elements of mathematical chemistry," all he means is the use of quantitative definitions. It is one of the traits he borrowed from Wolff; his expositions are clearly and arithmetically ordered, but it was a purely formal characteristic that had nothing in common with mathematical analysis in the Newtonian sense. Indeed, he says at one point: "for chemical investigations it is not necessary to have the kind of mathematician who is adept only at difficult calculations; but one rather who, having become familiar with the rigours of mathematical proofs and demonstrations, can educe concealed secrets from Nature in precise order rather than feeble impulses."[56] Mathematics, in other words, is important, but important merely as a formal method.

This is a weakness in Lomonosov's scientific armory, which N. A. Liubimov blamed on Christian Wolff, who "for all his vast erudition, was backward as a mathematician, and could never bring himself to

53. Cf. Vavilov, "Lomonosov i russkaia nauka," in *Sobranie sochinenii*, III, 573.
54. *Ibid.*
55. Quoted by S. I. Vavilov, "Osnovopolozhnik russkoi nauki M. V. Lomonosov," in *ibid.*, p. 578.
56. *Trudy M. V. Lomonosova po fizike i khimii*, ed. B. N. Menshutkin (Leningrad: Akademiia nauk, 1936), pp. 379–380.

appreciate Newton's great achievement."[57] This is not quite accurate because Lomonosov actually took lessons in mathematics in Marburg not from Wolff but from Professor Duising; the criticism is nevertheless sound. Lomonosov's metaphysical assumptions prevented him here also from assigning the same importance to mathematics as Newton had. "A mathematical division," he writes, "is defined arbitrarily, and it cannot be proved that bodies really have parts that are divisible from each other, as assigned mathematically (*mathematice assignatas*)."[58] What Lomonosov does is to draw a sharp distinction between mathematics and the phenomenal world. "Bodies consisting of matter can be divided into the minutest parts," he writes, "this division can be considered from mathematical and from physical reason."[59] And hence, the following crucial two sentences (emphasis Lomonosov's):

Corpus mathematice dividi dico, quando in data ejus extensione partes pro lubitu per calculum assignantur. Physice corpus dividitur, quando partes ejus actu a se invicem sejunguntur.[60]

Thus, having asserted the truth of Wolff's metaphysical position—that a finite thing could contain an actually infinite number of real parts—Lomonosov had to reject Newton's objective idea of space occupied by a determinate number of centers of force. By rejecting that, he also had to reject the synthetic or apodictic character of mathematics. This was a quandary Lomonosov shared with most pre-Kantian rationalists, who failed to draw any clear distinction between physical and metaphysical conceptions.

By denying the a priori validity of applied mathematics, Lomonosov could with an easy conscience accept gravitation as merely a description of *how* certain natural phenomena appear to operate. Like Wolff he could regard it as mere "Erscheinung." He believed, however, that the more ultimate question—*why* it acted the way it did—had to be both asked and answered. This forlorn quest largely accounts for the long-lasting refusal of continental rationalists to accept the universal

57. Quoted by V. I. Lamanskii, *Mikhail Vasil'evich Lomonosov* (St. Petersburg, 1883), p. 96. N. A. Liubimov wrote a life of Lomonosov (*Zhizn' i trudy Lomonosova* [Moscow: V Universitetskoi tipografii, 1872]) and what was long a standard work in Russia, *Istoriia fiziki*, 3 vols. (St. Petersburg, 1892); the third volume (pp. 318–354) contains a discussion of Newton.
58. Lomonosov, "De Particularis Physicis Insensibilibus," in *Polnoe sobraniie*, I, 294.
59. *Ibid.*, p. 288.
60. *Ibid.*

theory of gravitation. It also explains how he could pass over the mathematical demonstrations contained in the *Principia*. It is possible, and even likely, that most of them were beyond his reach. They were, it would seem, for Wolff, who preferred to cite Edmond Halley's easier commentaries:

"Weil aber der Newton die Sache unter schweren mathematischen Demonstrationen vorgetragen; so hat Herr Haley zu besserem Verstande derer die in der Mathematick nicht geuebet sind, oder doch nicht weit kommen, daß sie das tiefsinnige Werk Herrn Newtons verstehen koennten, dieselbe eine leichtere Weise fuergetragen."[61]

Similarly, this indifference to mathematical analysis, may partly account for the cavalier fashion in which Lomonosov dismisses a priori Newton's demonstrations in support of another vital principle put forward in the *Principia*—the proportionality of mass and weight.

This issue was first discussed by Lomonosov in an interesting letter he sent to Euler in July 1748, and again in a dissertation *On the relationship of matter and weight*, written in 1757, where he repeats almost word for word the points made earlier in his letter.[62] That "the quantity of matter is the measure of the same, arising from its density and bulk conjointly" is the first definition with which the *Principia* opens, and Newton returns to its proof in the section on the "Motion and resistance of pendulous bodies."[63] No purpose is served by enumerating Lomonosov's objections in detail. They are largely irrelevant and vitiated by the fact that Lomonosov never apparently took the trouble to duplicate Newton's experiments. A brief summary will suffice to show the rationalist nature of his scientific approach.

He compares gold and water. The specific gravity of the latter is twenty times lighter than that of gold. Since it is impossible to compress water to a smaller volume, Lomonosov argues that therefore the "particles" of water must be immediately in contact with one another. Why, then, is water lighter than gold? Lomonosov supposes that it is because the configuration and size of the "particles" are different. Gold, he deduces, must have particles of such a shape as to cause the greatest density; they must therefore be cube-like and pressed tightly against each other. Water, on the other hand, Lomonosov concludes, must have particles that are spheroid. The interstices between these

61. Wolff, *Vernuenfttige Gedancken von den Wuerckungen der Natur*, p. 544.
62. Cf. *Polnoe sobranie*, X, 439–450.
63. *Ibid.*, p. 440; cf. *Principia*, Book II, Section VI.

particles cannot be larger than the particles themselves. It follows that the "quantity of matter" in gold will be approximately twice as large as in water, not twenty as Newton had proved.

On several occasions Lomonosov submitted the problem to the Academy of Sciences for consideration in a public contest. Each time it was turned down on the ground that no experiments could be devised to prove the accuracy of Lomonosov's criticisms. There was no way of testing the molecular structure of matter in his day. Like his objections to gravitation, therefore, Lomonosov's disagreement with the opening definition of the *Principia* was purely hypothetical. There was one central "idea" in the *Principia* which Lomonosov very much admired, however, and it deserves special emphasis because to many people in the eighteenth century it represented Newton's essential achievement: his mathematical conception of nature. At the beginning of the third Book Newton lays down the renowned "rules of reasoning in philosophy."[64] "In the preceding Books I have laid down the principles of philosophy," he writes, "principles not philosophical but mathematical: such namely, as we may build our reasoning upon in philosophical inquiries . . . It remains that, from the same principles, I now demonstrate the System of the World."[65] The demonstrations are preceded by the following rules:

I. *We are to admit no more causes of natural things than such as are both true and sufficient to explain their appearances.* To this purpose the philosophers say that Nature does nothing in vain when less will serve; for Nature is pleased with simplicity, and affects not the pomp of superfluous causes.

II. *Therefore to the same natural effects we must, as far as possible, assign the same causes.* As to respiration in a man and in a beast; the descent of stone in Europe and in *America;* the light of our culinary fire and of the sun; the reflection of light in the earth, and in the planets.[66]

In Lomonosov's *Tentamen Theoriae de Particulis Insensibilibus . . .* these words are almost literally transcribed:

Axiom III
Section 22. Like effects are brought about by like causes.

Section 23. Scholium: Thus respiration in man and beast, the fall of a stone in Europe and America, the light in the kitchen fire, and in the sun,

64. In the first edition they are called "hypotheses."
65. *Principia*, Book III, "The Rules of Reasoning in Philosophy," Rules I and II.
66. *Polnoe sobranie*, I, 178.

the reflection of light on earth and in the planets—so says illustrious Newton.

This mathematical conception of nature was an idea Newton shared with all the great philosophers of his age, from Descartes to Spinoza and Christian Huygens, but it became associated in the course of the eighteenth century primarily with Newton's name. Stripped of its religious and magical associations, this radical view of nature lay at the very core of the so-called "scientific revolution." It led ultimately to revolutions of another order, for it fostered philosophical material-ism and a universe without God. In Russia this was not realized until the generation after Lomonosov's.

Сами свой разум употребляйте. Меня за Аристотеля, Картезия, Невтона не почитайте. Ежели вы мне их имя дадите, то знайте, что вы холопы, а моя слава падет и с вашею. —Lomonosov

18

Lomonosov and the Opticks

Lomonosov was the only Russian scientist of the period to make original contributions toward explaining the behavior of light, a subject to which he brought the eye of an artist, the inspiration of a poet, and the gifts of a natural philosopher and inventor. Whether it was in making mosaics, designing a telescope, or observing the aurora borealis, light in its chameleon manifestations provided almost a tributary source for Lomonosov's voracious interests. He had become interested in optics very early in his scientific career and had even hoped to start certain experiments in that area after returning from Marburg in 1741. Other investigations, primarily in chemistry, prevented Lomonosov from doing so, and only after 1750 did he began to concentrate his attention on light.[1]

This was the period that he began his optical experiments in the new laboratory especially designed and equipped by him for the Academy of Sciences, bringing him into a series of interesting encounters with Newton's work in the same field. From a theoretical

1. The first Russian theoretical work on light was, in fact, completed one year before Lomonosov's crucial "Slovo o proiskhozhdenii sveta," i.e., in 1755. Its author, Litken, taught at the "gymnasium" attached to the newly opened university in Moscow. His dissertation on fire and light, "Ob ogne i svete," which he submitted for his Master of Arts degree, is Litken's only known work, and little is known about his life. According to Dr. Chenakal, his theory of light was Cartesian in character; unlike Lomonosov, however, it would seem Litken went out of his way to make no attacks on Newton. Cf. Chenakal's most useful survey, "Priroda sveta v vozzreniiakh russkikh estestvoispytatelei XVIII i nachala XIX veka," in TIIE, III (1949), 179–180.

point of view, the whole problem of light was inseparable from that of matter. As we have seen, Lomonosov's atomic theory was closer to the Cartesian than to the Newtonian one. Like Descartes, he considered space to be filled with ether, which he conceived to be composed of separate but extremely small particles. These particles were perpetually in contact with each other, accounting for phenomena such as light, heat, and electricity. "Imagine," he wrote, "the extent of the structure of the universe, consisting as it does of insensible globules of various sizes, their surface indentured by particles and fine inequalities by which these particles can cohere with each other as teeth do on a wheel. It is well known from mechanics that those wheels are connected and move each other harmoniously when the teeth are of equal size and like arrangement. Those which are different in size do not connect and therefore do not move together in accord. This I find to be the case in the primary insensible particles, making up all bodies, designed by the all wise Architect and almighty Mechanician in accord with immutable natural laws. I call these particles which are connected to each other *compatible (sovmestnye)* and those which are not *incompatible (nesovmestnye.)*"[2] It is by this analogy that Lomonosov describes the behavior of light.

He divides his fine particles of ether into three sizes, all spherical in shape:

The first type of particle is the largest and by continuous mutual contact has a square arrangement. Therefore, considering a cubic body as against a sphere of the same diameter doubled, there will remain almost as much empty space between these particles as the spheres occupy. In these spaces I assume ether particles of a second type, which, being much finer, are packed in considerable number against each other in a square position and by continuous mutual contact similarly occupy half the space in the interstices, hence the amount of material is half as compared to the first. Also I assume a third sort of the finest ether particles in the interstices of the particles of the second sort. This third sort of particle is of the same order of arrangement and by the above mentioned geometrical dimensions will have an amount of material in proportion to the amount of material of the second sort as one to two, to the amount of the first sort as one to four. There is no reason to divide the particles further nor do I see any necessity for this. These three sorts of ether particles are each connected with the other sorts and are incompatible with different sorts

2. Lomonosov, "Slovo o proiskhozhdenii sveta, novuiu teoriiu o tsvetakh predstavliaiushchee, v publichnom sobranii imperatorskoi Akademii Nauk iiulia i dnia 1756 goda govorennoe Mikhailom Lomonosovym," *Polnoe sobranie,* III, 329.

of particles so that when the particles of the first type turn by a rotary motion, joining with the others of their type, a large number in the circle around them are moved by the force of cohesion.[3]

While one kind of particle rotates and brings other particles of the same size into motion, the other two kinds may remain stationary. The copestone to his entire theory is his belief that these three types of ether correspond to three basic chemical substances, that is, acid, mercury, and sulphur. The ingenious part of Lomonosov's theory is his belief that "the color red comes from the first kind of ether, yellow from the second, blue from the third. Other colors come into being as a result of the mixture of these primary ones."[4]

So far the entire system is entirely fanciful and shows how different Lomonosov's approach was from Newton's. The three elements are taken from alchemy, for "sensible bodies," writes Lomonosov, "are divisible according to the most illustrious chemists into primary matter, active and passive." The former consisted of acid, mercury, and sulphur; the latter, of earth and water. How did Lomonosov know that the composition of primary matter coincided with the three types of ether? "I observed this over the years with many earlier conjectures," writes Lomonosov. In one respect, however, his "conjectures" turned out to be surprisingly correct.

The notion that there are three basic colors was actually Mariotte's, whom Lomonosov describes as that "renowned and diligent investigator . . . who tried not to refute Newton's doctrine of the separation of light by the refraction of beams, but only to correct it [to the extent] that there are three and not seven main colors . . ."[5] Moreover, Lomonosov's own experiments in mixing pigments at his glass factory seemed to confirm Mariotte's views.

In modified form this idea gained general acceptance a generation later. Thomas Young (1773–1829) proved that white light may be obtained by mixing *any* three basic colors, though Lomonosov's notion that his three colors corresponded objectively to three physical elements turned out, of course, to be mistaken. The truly prophetic part of his theory was in his belief that ether was a conductor of both

3. *Ibid.,* p. 331. I have used Henry M. Leicester's excellent translation of this passage: see his *Mikhail Vasil'evich Lomonosov on the Corpuscular Theory* (Cambridge, Mass.: Harvard University Press, 1970), p. 259.

4. *Polnoe sobranie,* III, 332.

5. Mariotte first set down this theory in his book, *De la Nature des Couleurs* (Paris, 1681).

light and electricity, and that, somehow, the two phenomena were interrelated. He had begun his experiments with electricity (with his colleague G. V. Richmann, who was killed as a result of one of them) at about the same time as Benjamin Franklin (1752), and he became convinced that the nature of light might ultimately be explained in terms of electricity. This idea was developed only in the nineteenth century, when Maxwell's electromagnetic theory of light gave an elegant explanation of the relation between the velocity of light and the ratio of electromagnetic units.

Lomonosov's theory brought him into conflict with what he conceived to be Newton's corpuscular theory of light. He had already decided in 1743, as one rough jotting ("lux per oscillatorium") shows, that the ondular theory best explained light phenomena. B. N. Menshutkin, who "rediscovered" Lomonosov in 1907, even affirms that here the Russian scientist was far ahead of his contemporaries, "in spite of the fact that almost all physicists of the eighteenth century considered Newton's authority infallible and held strongly to the 'matter of light' and the theory of effluxion."[6] This claim is often made on Lomonosov's behalf: how true is it?

As is well known, many natural philosophers of the seventeenth and eighteenth centuries sought to describe light either in terms of moving particles or of waves. To them these forms of moving energy were sharply differentiated. Particle energy is highly localized, and the kinetic energy of a rifle bullet, for example, travels from one well-defined, small region of space to another and does not spread during transit. If, however, a wave is started by dropping a stone into a pond, the energy quickly disperses over the whole surface, and usually no small region receives a very high proportion of it. This wave theory was first expounded clearly and thoroughly by Christian Huygens, though it was not till the nineteenth century that really detailed experiments on the reflection and refraction of "polarized" light became available as a result of the work of Thomas Young, Etienne Louis Malus (1775–1812), and Auguste Jean Fresnel (1788–1827). Their experiments with the phenomena of interference and diffraction ap-

---

6. I.e., emission theory. Cf. B. N. Menshutkin, *Russia's Lomonosov* (Princeton, N.J.: Princeton University Press, 1952), pp. 66–67. This is a translation of an earlier Russian work, and an incompetent one (i.e., Young=Jong, Fresnel=Frenelle, etc.). Menshutkin's writings on Lomonosov, however, are still indispensable: see notes to the bibliography.

peared to give a clear decision in favor of a wave theory. Consequently, in the nineteenth century the ether-wave theory completely super-seded the corpuscular theory.

What Menshutkin ignored, however, is that the quantum theory propounded by Max Planck in 1900 involved something like a return to a corpuscular structure of light.[7] By this reckoning, it was Newton's and not Lomonosov's hypothesis that acquired a curiously prophetic character. But a more serious objection lies in supposing that New-ton was wedded to the corpuscular theory. This interpretation was really thrust upon him by his successors. Netwon was always careful not to commit himself to any specific theory of the structure of light, as a letter intended for Robert Hooke (who supported the wave theory) shows:

And first of the *Hypothesis* wch Mr. Hook hath assigned me in these words: *But grant his first supposition that light is a body, & that as many colours or degrees thereof as there may be*[,] *so many bodies there may be, all wch compounded together would make white &c.* This it seemes Mr. Hook takes for my Hypothesis. 'Tis true that from my Theory I argue the corporeity of light, but I doe it without any absolute positiveness, as the word *perhaps* intimates, & make it at most but a very plausible consequence of the Doctrine, and not a fundamentall supposition, nor so much as any part of it, wch was wholly comprehended in the precedent Propositions. And I wonder how Mr. Hook could imagin that when I had asserted the Theory with the greatest rigor, I should be so forgetfull as afterwards to assert ye fundamentall supposition it selfe with no more than a *perhaps*.[8]

In attacking the Newtonian theory of light, therefore, Lomonosov greatly oversimplified it. This was not because he knew the *Opticks* only at second hand, as T. I. Rainov suggests. Nor is it true that "there are no direct references to it" in any of Lomonosov's writings.[9] In fact, Lomonosov refers to Newton's great work in the *127 Observations on the theories of light and electricity,* where he cites the twenty-ninth "query" of the *Opticks.* This is an important point, for it explains why

7. Cf. D. C. Miller, "Newton and Optics," in *Sir Isaac Newton, 1727–1927, A Bicentenary Evaluation of His Work* (Baltimore: Williams & Wilkins Co., 1928), p. 44 f. The modern quantum mechanics incorporates the appropriate parts of the electromagnetic wave theory, the quantum theory, and the relativity theory. It is therefore a matter of con-jecture, *ceteris paribus,* whether a corpuscular theory or a wave theory will prevail a hundred years from now.

8. Newton to Oldenburg, June 11, 1672, *Correspondence,* I, 173–174.

9. T. I. Rainov, "N'iuton i russkoe estestvoznanie," in *Isaak Niuton 1643–1727,* ed. Vavilov, p. 332. Rainov's three pages on Lomonosov were written before the publication of the latter's collected works and are, therefore, valueless. See notes to bibliography.

Lomonosov so readily ascribes the corpuscular theory to its author.[10] The "queries" were the fruit of Newton's reflections in his old age, in which he abandoned the severe, almost Euclidean style of the earlier sections of his work and the twenty-ninth "query" is the one place where he comes close to suggesting that a corpuscular theory best explained the behavior of light:

Are not the Rays of Light very small Bodies emitted from the shining substances? For such Bodies will pass through uniform Mediums in right lines without bending into the shadow which is the Nature of the Rays of Light . . .[11] I only say that whatever it be, it's difficult to conceive how the rays of Light, unless they be Bodies, can have a permanent Virtue in two of their Sides, which is not in their other Sides . . .[12]

From this Lomonosov could conclude that Newton believed light to consist of corpuscles, and from another passage in the same "Query" he could conclude that Newton ascribed "the difference between colors . . . to the different mass of parts ether":[13]

Nothing more is requisite for producing all the variety of Colours, and degrees of Refrangibility, than that the Rays of Light be Bodies of different Sizes, the least of which may take violet the weakest and darkest of Colours, and be more easily diverted by refracting surfaces from the Right Course; and the rest as they are bigger and bigger, may make the stronger and more lucid Colours, blue, green, yellow, and red, and be more difficultly diverted . . .[14]

By this Newton meant to account for the colors of natural bodies. He discusses this problem, which had long baffled natural philosophers, in two different parts of the *Opticks,* where "by the discovered properties of light" he explains "the permanent colours of natural bodies"[15] and points out the "analogy between such colours, and the colours of thin transparent plates."[16]

After showing that all bodies, whatever their colors, exhibited these

10. Lomonosov, "127 Zametok k teorii sveta i elektrichestva," p. 246, Section 51, which was written in 1756 in preparation for the *Oration on Light.* Lomonosov does not indicate which edition of the *Opticks* he used.

11. *Opticks,* p. 349.

12. *Ibid.,* p. 348.

13. Lomonosov, "127 Zametok k teorii sveta i elektrichestva," p. 246, Section 51. "Newton statuit differentiam colorum pendere a diversa mole particularum aetheris. Fortiorum particulas esse majores, debiliorum minores."

14. *Opticks,* p. 347.

15. *Ibid.,* p. 156.

16. *Ibid.,* p. 219.

colors best in white light (or in light containing their peculiar color), he proves by experiment that when colored bodies are illuminated with homogeneous *red* light, they appear *red, blue* with homogeneous *blue* light, and so on, "their colours being most brisk and vivid under the influence of their own daylight colours." This can be demonstrated in a variety of ways. For example, the leaf of a plant appears *green* in daylight, because it has the property of reflecting *green* light to a greater degree than any other. But if the leaf is placed in homogeneous *red* light, it no longer appears *green* because there are no green rays in red; it does reflect red light in a small degree because there are some red rays in the compound green which it has the property of reflecting. It follows that the colors of natural bodies are due to the property they possess enabling them either to stop or absorb certain rays of white light, while reflecting or transmitting to the eye the other rays of which the light is composed. The green leaf, for instance, stops or absorbs the red, blue, and violet rays of the white light falling upon it and reflects and transmits only those which compose its peculiar green.

To this extent Newton's views are demonstrable and have been universally adopted. When, however, he tries to explain the manner in which the color of any body is insulated from the other colors falling on it, whereby these other colors are stopped or lost—in other words, the physical constitution of natural bodies by which these processes are effected—Newton was far less successful. It was the part of his theory which Robert Hooke, P. Pardies, and others, found least convincing. As followers of the wave-ether hypothesis, they rejected the conclusions Newton drew from the connection between refrangibility and the physical constitution of bodies, though here again they tended to ascribe to him what he had not actually said:

I thank you for giving me notice of ye objection wch some have made. If I understand it right they meane that colour may proceed from the different velocities wch aethereall pulses or rays of light may have as they come immediately from ye Sun. But if this be their meaning, they propound not an objection, but an Hypothesis to explain my Theory. For ye better understanding of this I shall desire you to consider that I put not ye different refrangibility rays to be ye internal or essential cause of colours, but only the means whereby rays of different colours are separated. Neither do I say what is that cause, either of colour or of different refrangibility, but leave these to be explained by Hypotheses, & only say that rays wch differ in colour, differ also in refrangibility, & yt different refrangibility conduces to ye production of colour no other way then by causing a dif-

191

ferent refraction and thereby a separation of those rays wch had different colours before, but could not appear in their own colours till they were separated.[17]

The whole problem was debated at a meeting of the Royal Society on February 3, 1675/76, when it was discussed "whether the rays of light, which, though alike incident in the same medium, yet exhibit different colours, may not reasonably be said to owe that exhibition of different colours to the several degrees of the velocity of pulses, rather than, as Mr. Newton thought, to the several connate degrees of the refrangibility in the rays themselves. *Mr. Hooke* was of the opinion that the former of these was sufficient to give a good account of the diversity of colours."[18] Did Lomonosov know of Hooke's objections? Whether he did or not, his argument against the Newtonian theory is curiously similar to the one adopted by Robert Hooke. He had decided that refraction "non est causa colorum, quia colores proxime ad corpus videntur."[19] Similarly, "colores conspiciuntur ab omni parte sive puncto, ergo non fiunt per refractionem."[20]

The crucial issue here was the relationship Newton tried to establish between refractive power and the physical structure of a body. Having supposed, as we have seen from his twenty-ninth "Query," that the different colors of rays are due in part to the size of the component light particles, Newton deduced that the size of the "primary" parts composing natural bodies might also be inferred from their colors. He concluded that the "least parts" of almost all natural bodies are to some extent transparent, also that between the parts of opaque and colored bodies in any receptor there are many spaces or pores, either empty or filled with media of other densities. Two concurrent principles were: the parts of bodies and their interstices or pores must be not less than of some definite bigness to render them colored; and the transparent parts of bodies, according to their several sizes, reflect rays of one color and transmit those of another on the same ground that thin plates reflect or transmit these rays.

To each of these propositions Lomonosov objects. "In pellucilis radii propagantur per linea rectam," he writes, "unde sequitur lucem

17. Newton to Oldenburg, February 15, 1675/76, *Correspondence,* I, 417–418.
18. Thomas Birch, *The History of the Royal Society of London . . .,* 4 vols. (London, 1756–1757), III, 280, 295.
19. Lomonosov, "276 Zametok po fizike," in *Polnoe sobranie,* I, 138, Section 181.
20. *Ibid.,* p. 116, Section 64.

non per sinuosos poros propagari."[21] And again: "colores non sunt effluvia e corporibus emanantia, nam in tenebris nulli colores conspiciuntur, sequitur eos a luce sive aethere dependere."[22] These are rough notes that Lomonosov did not intend for publication, and it may therefore not be entirely fair to judge his theory of light on their basis, but in a sense they are more revealing than the *Oration on the Origin of Light,* which he delivered before the Academy of Sciences in July 1756, because they show from what assumptions he began. And they show that, like Hooke, Lomonosov misunderstood the Newtonian theory. He writes, for example, "[d]emonstrandum hoc primo, quod per refractionem corpora colorata esse non possunt"; this, he adds, he must show to be wrong before presenting his own theory.[23] But Newton stated expressly that "I put not ye different refrangibility of rays to be ye internal or essential cause of colours, but only the means whereby rays of different colours are separated." Robert Hooke and Christian Huygens had both misunderstood this, and Newton was considerably distressed both at the doubt which lesser men cast on his experiments and at the difficulties which great men, like Huygens, had in understanding his point of view. Typical of Huygens' attitude is the remark that "De plus quand il seroit vray que les rayons de lumiere, des leur origine fussent les uns rouges les autres bleus, & etc., il resteroit encore la grande difficulté d'expliquer par la physique mechanique en quoy consiste cette diversitè de couleurs."[24] It was useless for Newton to protest that "to examin how colours may be thus explained Hypothetically is besides my purpose. I never intended to show wherein consists the nature and difference of colours, but onely to show that *de facto* they are originally & immutable qualities of the rays wch exhibit them, & to leave it to others to explicate by Mechanicall Hypotheses the nature & difference of those qualities; wch I take to be no very difficult matter."[25]

The same objections continued to be made, and Lomonosov's are in the main equally irrelevant as when he states: "non dependunt colores a foraminulis seu poris, et a solo et unico situ diverso

21. *Ibid.*, pp. 114–116, Section 62: "Neque lux perfluit per poros: nam si perflueret, tum omni ex parte seu in omnes directiones corporis translucidi forent foramina, adeoque totum corpus illud esset foramen id est nihil."
22. *Ibid.*, p. 156, Section 257.
23. *Ibid.*, p. 138, Section 181.
24. Huygens to Oldenburg, September 27, 1672, [N.S.], *Correspondence*, I, 235–236.
25. Newton to Oldenburg, April 3, 1673, *ibid.*, p. 264.

corpusculorum."[26] Here he does not cite Newton, but he evidently had his theory in mind because he develops the same arguments later in his *Oration on Light*, where he limits himself to discussing the two "most important theories": "the first—that of Descartes, confirmed and expounded by Huygens, the second, which originated with Gassendi, and acquired significance by virtue of Newton's agreement and exposition of it."[27] It seemed to Lomonosov that the only positive proof of the latter's corpuscular theory, as he describes it, was, first, "his arbitrary postulate of the force of attraction of bodies which to-day physicists rightly reject as an occult quality resurrected from the ancient Aristotelian school, [and] contrary to good sense. That is why, though [it] supports the author's intelligence *(ostroumie)*, does not at all confirm his view. The sensible but very short time it takes for light from the sun to reach the earth no more confirms the flowing motion [i.e., the corpuscular theory] of ether than does the duration of time it takes for a voice to travel a certain distance show the motion of air."[28]

This criticism of gravitation was hardly original. Fontenelle had made it in his *Eloge* of Newton, and it became the main buttress of the Cartesian attack on Newton, though it is rather extraordinary that the Russian scientist should refer to "physicists *now*." [1756!], since by that time there were few Cartesians left to withstand the general acceptance of the Newtonian theory. More significant is his second objection, motivated as it was by the same reasons which led Huygens and Hooke to disagree with the conclusions they thought Newton drew from his experiments with refracted light. To understand Lomonosov's criticism we must return once more to the twenty-ninth "Query," where Newton concludes his observations on passing a beam of light through Iceland spar by suggesting that:

This virtue of refracting the perpendicular Rays is greater in Island-Crystal, less in Crystal of the Rock, and is not yet found in other Bodies. I do not say that this Virtue is magnetical: It seems to be of another kind. I only say, that whatever it be, it's difficult to conceive how the Rays of Light, unless they be Bodies, can have a permanent Virtue in two of their Sides,

26. Lomonosov, "276 Zametok po fizike," *Polnoe sobranie*, I, 138, Section 181.
27. Lomonosov, "Slovo o proiskhozhdenii sveta" *ibid.*, III, 319.
28. *Ibid.*, pp. 319–320. Professor Leicester's translation (*Lomonosov on the Corpuscular Theory*, p. 250) of the allusion to Newton's theory of gravity is erroneously rendered. Where the English text reads that it "has been supported by intelligent authors," Lomonosov did not concede this much. The Russian actually states: "supports the author's intelligence."

and this without any regard to their Position to the Space or Medium through which they pass.[29]

Newton assumed that his particles of light had *sides* to account for the phenomenon called *polarization* (the misleading title given it by Malus). That is to say, he assumed that his particles of light were different in different directions *across* the direction of travel. For example, an oblong sheet of cardboard traveling sideways (that is, in a direction perpendicular to the surface) in a horizontal direction can be said to have a direction of tilt, to be different in different directions of travel. A circular disc of cardboard, on the other hand, is the same in all directions across the directions of travel. According to Newton, the varying sizes of these particles in turn accounted for the seven colors comprising white light.

To prove this wrong, Lomonosov devised a series of experiments which the corpuscular theory seemed incapable of explaining. If light consisted of particles traveling in straight lines in accord with Newton's Laws of Motion, it should be possible to make these particles block each other's transit through the narrow interstices of a given receptor. One of his experiments involved a diamond "which is transparent throughout from every point of its surface and in all the interior of the body extend the straight line pores of the whole diamond; these pores carry the material of light." On either side Lomonosov placed a candle. The simple fact that "beams from both sides pass through the diamond with equal force, and one candle can be seen from one side through the diamond at the same time as the candle on the other side," seemed to show the inadequacy of Isaac Newton's explanation.[30]

Newton, as we have seen, spoke in his twenty-ninth "Query" of "a Virtue of refracting the perpendicular rays." This was a conjecture and nothing more, as in fact Newton elsewhere admits.[31] He never completed his investigations on the problem of diffraction, and the questions raised by Lomonosov (to which his own answers were quite inadequate) were only answered convincingly by Thomas Young in 1801. His *general law of the interference of light* solves the issues

29. *Opticks*, p. 348.

30. Lomonosov, "Slovo o proiskhozhdenii sveta," *Polnoe sobranie*, III, 323.

31. Cf. *Opticks*, end of Book II, p. 313: "When I made the foregoing Observations, I design'd to repeat most of them with more care and exactness . . . But I was then interrupted, and cannot now think of taking these things into further Consideration. And . . . I shall conclude with proposing only some Queries, in order that a farther search be made by others . . ."

raised by the experiment with the candle and diamond. Young suggested the model of a stagnant lake, with a number of equal waves moving on the surface at a constant velocity toward a narrow channel leading out of the lake. Imagine another equal series of waves arriving at the channel at the same time. One series of waves will not destroy the other. Their effects will be combined. If they enter the channel in such a way that the elevations of the one series coincide with those of the other one, the result will be a series of greater joint elevations, but, if the elevations of one series correspond to the depressions of the other, they will fill up those depressions, and the surface of the water will remain smooth. Young maintained that this is precisely what happens whenever two portions of light are mixed in this way.

Thus, Newton's chief objection to the wave theory—"that if light consisted in motion, *it would bend into the shadows*"—was overcome. Young proved that light *does* bend, though the bend is extremely small owing to the minuteness and immense speed of the waves. Newton's assumption that his light particles had *sides* could therefore be abandoned, and Fresnel, who was in correspondence with Young, gave the wave theory its modern form by explaining polarization and the celebrated experiment with Iceland spar in terms of vibrations which take place *across* the direction in which the light is traveling, what is commonly called today a *transverse* wave.

In sound waves, from which Young had drawn his picture of light waves, the vibrating particles move in a direction parallel to the propagation of the wave. This is "longitudinal" vibration. In water waves the water particles move up and down at *right angles* to the direction of the wave. This is *transverse* vibration as is found in light. The action of the Iceland spar upon light waves impinging on it may therefore be compared to a set of railings with vertical chinks. Vibrations parallel to the rails will pass on between its chinks, but the remainder will be stopped. The light that passes through is said to be polarized. This much was realized by Newton, and he gave to light periodic, that is, fluctuating, properties just where the wave theory has to invoke them.[32] A ray of light, however, has this complication: the plane of vibration is only partly restricted. A ray of light may therefore consist of waves vibrating in any plane passing through the direction of the ray. This complex phenomenon is largely responsible for the con-

32. Newton's theory thus had a considerable periodic or wave element in it.

fusions and controversies the whole problem of light engendered in the eighteenth century.

The immense prestige enjoyed by the *Principia* did much to support the authority of Newton's *Opticks,* and Lomonosov's criticisms were not, it appears, taken seriously either within the Academy in St. Petersburg or outside it, in spite of the fact that the *Oration on Light,* the fruit of fifteen years labor, was translated from Russian, and published in Latin.[33] To some extent, Lomonosov's attempt to correct Newton was thought rather brazen. An anonymous satirical poem (in German) which circulated at the time, has Lomonosov in a fit of "Grossenwahn" comparing himself to Newton. Even his own pupil, Stepan Rumovskii, ignores it, from which Lomonosov inferred that Rumovskii was bitterly "hostile" to him, "as it appears from the ending of his paper on optics."[34] In this paper, delivered before the Academy seven years after Lomonosov's *Oration on Light,* the latter is not even mentioned. What Rumovskii does is to devote a perfunctory passage to the problem "of the manner whereby light reaches us from the sun." After briefly summarizing the views of Descartes-Euler on the one hand, and of Newton on the other, he comes to the spineless conclusion that though both "opinions" have their merits, "it was difficult or even impossible to decide which of them should be considered true."[35]

Much stronger doubts about Newton's optical views would be expressed by scientific authorities in Russia later in the century. It should therefore seem hardly surprising that Lomonosov tried to challenge them with such conviction at an earlier date. It can hardly be denied, however, that his differences with Newton—in optics as in physics— handicapped the entire range of Lomonosov's scientific work. "The Science of Colours," wrote Newton, "is most properly a Mathematical Science."[36] Lomonosov disagreed, as did Robert Hooke and others in Newton's own lifetime and Goethe and the followers of his *Farbenlehre* long after Lomonosov's death. But it was a crucial weakness, affecting, as we have seen, Lomonosov's whole approach to natural philosophy.

33. *Oratio de origine lucis sistens novam theoriam colorum . . . a Michaele Lomonosov, consiliario academico,* tr. Gregorio Kositzki (St. Petersburg, 1759). Cf. Sk, no. 3781: *Slovo o proizkhozhdenii sveta . . .* (St. Petersburg, 1756).
34. Cf. "Kratkaia istoriia o provedenii akademicheskoi kantseliarii v rassuzhdenii uchenykh liudei . . .," in *Polnoe sobranie,* X, 306–307, Section 59.
35. S. Ia. Rumovskii, *Rech o nachale i prirashchenii optiki, do nyneshnikh vremen . . .* (St. Petersburg, 1763); *Polnoe sobranie,* X, 750–752, note to Section 59.
36. Newton to Oldenburg, April 13, 1672, *Correspondence,* I, 137.

It is only fair to add that Newton's theories of light and color were, for many of Lomonosov's contemporaries, equally unsympathetic. If no less a man than Euler disagreed with them, it is quite understandable why Lomonosov should feel, as he did in 1753, that among the foremost men of science they were no longer accepted. As it is, he showed an inventiveness and ingenuity in applying various ideas to basic principles already developed; he could advance them no further. This was not due to any deficiencies in Lomonosov's own gifts. No one who reads his scientific notebooks can fail to be impressed by the quality of his imagination, and the versatility of his mind. But his early training under Christian Wolff gave to his scientific methods a bias that ran counter to the tradition associated with Newton's name.

The future, in other words, belonged to the mathematical treatment of phenomena, not, as Lomonosov believed, to the qualitative analysis of physical properties. The triumph of the Newtonian world view led to the elevation of mathematics as the principal instrument for the solution of scientific problems, and it was Leonard Euler's emphasis on this aspect that largely accounts for the fact that the mathematical branches of science developed far more consistently in Russia during the second half of the eighteenth century than other fields such as biology and chemistry. Modern physics owed its progress largely to the extension of Newton's "fluxionary method" (though it was Leibniz' algebraic notation that came to be generally used). Without something like a calculus, no conception of instantaneous velocity could have been built up, nor the modern idea of momentum, ultimately the most important concepts in nineteenth-century physics. In this field Lomonosov contributed nothing of consequence, for he used little mathematics in his work.

This mattered less in chemistry, where his disagreements with Newton were mainly irrelevant. It was one of the last areas in natural philosophy to be affected by the so-called "scientific revolution." Natural phenomena continued to be investigated well into the nineteenth century in terms of esoteric fluids and substances (such as *phlogiston*) which had no basis in fact but which often served to impede acceptance of Newtonian theories, as in the case of Aleksandr Radishchev at the end of the century. To Lomonosov's credit, he eliminated some of them and advanced ideas about the nature of energy and electricity, some of which turned out to be prophetic. But he had no disciples, and his scientific work, though not wholly forgotten to the extent earlier scholars assumed, had little influence even within his own country. Many turned to Lomonosov's example in literature; few in science. With the historical revival of chemistry in the course of the nineteenth century—when its status in relationship to physics and biology was so dramatically reasserted—it was only natural that Menshutkin, who had himself devoted much study to it, should discover the relevance and pioneering character of this remarkable man's scientific work. During the post-Newtonian era, however, physics and mathematics were the fields in which the most arresting progress was made. Lomonosov's early neglect, which was related to his opposition to Newton's theories, is, therefore, not hard to explain.

19

A New Kind of Newtonian Telescope?

THE NEWTON-LOMONOSOV TELESCOPE

An item in one of Lomonosov's rough manuscripts refers to a *"Newtoniano-Gregor.-Lomon. tubus,"* and another runs: *"speculum ex Newtoniano, Gregoriano et meo compositum."*[1] Both notes refer to one of Lomonosov's proudest inventions, a new kind of Newtonian telescope, based on a principle usually associated with the great astronomer William Herschel, who suggested it some three decades later.[2] "The time has long passed," writes S. I. Vavilov, "that this optical system should be named after Lomonosov-Herschel."[3]

Before examining this claim in detail, it will be necessary briefly to see what in his telescope Lomonosov owed to Newton. In about 1668, Newton made a small reflecting telescope in which, as in the modern giants, the image was formed not by a lens but by a concave mirror. Such use of a concave mirror had earlier been proposed by James Gregory, but Newton was the first to make such an instrument.

The telescope aroused more interest at the time than Newton's

1. Lomonosov, "Khimicheskie i opticheskie zapiski," *Polnoe sobranie,* IV, 445, 423.
2. William Herschel (1738–1822) was born in Hannover—then a possession of the British Crown—and came to England as a musician in 1757. He then turned to astronomy, and acquired great skill in making instruments. He conducted four complete reviews of the heavens with telescopes of increasingly greater power; the second review revealed Uranus (1781), the first new planet to be discovered in historic time.
3. B. N. Menshutkin, *Zhisneopisanie Mikhaila Vasil'evicha Lomonosova, s dopolneniiami P. N. Berkova, S. I. Vavilova, i L. B. Modzalevskogo,* 3rd ed. (Moscow and Leningrad: Akademiia nauk, 1947), p. 159. On Vavilov and his commentary to Newton's *Opticks,* see bibliography.

lectures on optics at Cambridge to which, it appears, few people came. News of it reached the Royal Society. In response to urgent request, Newton made a second reflecting telescope, which is one of the Royal Society's greatest treasures today. "I was surprised," wrote Newton to Oldenburg in 1671/72, "to see so much care taken about securing an invention to mee, of wch I have hitherto had so little value. And therefore since the R. Society is pleased to think it worth patronizing, I must acknowledg it deserves much more of them for that, then of mee, who, had not the communication of it been desired, might have let it still remained in private as it hath already done some yeares."[4] The new invention "whereby long telescopes are considerably reduced in length without impairing the effectiveness of their use" was made public at the Royal Society in the same year.[5] The details of the new invention were transmitted to Huygens in Paris, who was at that time the foremost man of science abroad and an expert on the refracting telescope.[6]

The principle of construction was to form an image by means of the mirror at the base of the tube (see fig. 41), which was thrown to one side by a small mirror (later Newton used a right-angled prism) and viewed by a small lens. It gave a magnification of 38.1 times; the whole telescope was only about six and a half inches in length.[7] It performed so well and excited such interest at the Royal Society that Newton was promptly elected a Fellow. Encouraged by its reception, Newton promised to send the society an account of the discovery which had led him to make the telescope, this discovery being, of course, the nature of the decomposition of white light into colors by the prism, which showed that simple lenses must necessarily produce colored fringes.[8] Thus, with pre-Newtonian telescopes the picture was marred invariably by a colored ring due, as Newton discovered, to the different refrangibility of rays of different colors, blue rays at one end of the spectrum being more refracted than red at the other end. This phe-

4. Newton to Oldenburg, January 6, 1671/72, *Correspondence*, I, 79.

5. *Ibid.;* Henry Oldenburg (1626–1678) was Secretary of the Royal Society from 1663 to 1677. He came to England as consul for his native city of Bremen during the Long Parliament in Charles's reign: he became tutor to Lord William Cavendish, and was intimately acquainted with Milton.

6. Oldenburg to Huygens, January 15, 1672; also January 1, 1671/72, *Correspondence*, I, pp. 81, 72.

7. Cf. Newton's description, *ibid.*, pp. 74–75, and fig. 41.

8. Newton's discovery appeared in the *Philosophical Transactions* for February 19, 1671/72; it was Newton's first appearance in print.

nomenon of chromatic aberration, as it came to be called, was not generally understood until Newton explained it in 1672.

Lomonosov knew of Newton's discoveries, not least from his reading of the *Opticks,* which prompted him to write in 1754 that though he disagreed with "Newton's theory of colors more than Mr. Euler," he "was not, however, its enemy, though some think otherwise."[9] He had not yet formulated his own theoretical objections to the corpuscular theory of light, and his own passionate interest in the whole field of light and color was given further impetus by Empress Elizabeth's decree granting him permission to build a glass factory at *Ust Ruditsy.* He designed the equipment himself and initiated what was then a novel art form in Russia—the making of mosaics. A decade later his achievements earned him honorary membership of the Academy of Sciences in Bologna (May 1764). Throughout that period he retained a long-standing ambition to improve the Newtonian telescope. "I always desired these excellent celestial instruments (*nebesnye orudiia*), whose invention is the glory of Newton and Gregory, to increase not only the size, as has ordinarily come to pass, but that they should also be improved in other ways, in accordance with essential optical principles . . . Examining the construction of these telescopes containing the metallic mirror and the glass lenses, it occurred to me that they have a redundancy rather than a defect, which requires correction."[10]

The "redundancy" referred to by Lomonosov was actually noted when Newton's telescope was first presented to the Royal Society: "it was discoursed," wrote Oldenburg to Newton, "yt by this way of yours it was longsome, & difficult to find ye Object."[11] Newton replied, "that's the inconvenience of all Tubes that magnify much, & that after a little use the inconvenience will grow lesse. For I could readily enough grind any day objects by knowing wch way they were posited from other objects that I accidentally saw in it; but in the night to find starrs I confesse is troublesom enough."[12] As a solution he suggested attaching extra sights, and "if such sights bee not found a sufficient remedy, there may bee an ordinary perspective glasse fastened to ye

9. Cf. Lomonosov to G. V. Miller, May 7, 1754, *Polnoe sobranie,* X, 507:
Я больше, нежели г. Ейлер, в теории цветов с Невтоном не согласен, однако тем не неприятель, которые инако думают.
10. Lomonosov, "Ob usovershenstvovanii zritel'nykh trub," *Polnoe sobranie,* IV, 475.
11. Oldenburg to Newton, January 2, 1671/72, *Correspondence,* I, 73.
12. Newton to Oldenburg, January 6, 1671/72, *ibid.,* p. 79.

same frame wth ye tube, & directed towards the same object, . . ."[13]

This advice was, in fact, often taken. Though it is not certain whether Lomonosov knew of this auxiliary device because he writes of his own telescope that "although its position . . . may not seem important and is almost the same as Newton's, its advantage to the observer is that it frees him from the strained effort of craning his neck and (lowering his) eyes, which is necessary when using the Gregorian telescope, especially when focusing the mirrors before an observation. To do this he has to keep turning upwards, which is tiresome."[14] The new principle Lomonosov introduced was to incline his speculum (the metallic mirror) to 4° from the perpendicular dropped on the axis of the tube from its center. Newton had placed his speculum at right angles to the axis, and used a second mirror to reflect its image to the viewer through a plano-convex eye-piece in the side of the tube. Lomonosov's idea eliminated the necessity of a second mirror, and the eye-piece could therefore be placed next to the spectrum (as in the older telescopes; see fig. 42). The important advantage of the new method was that it obviated the necessity of adjusting the speculum, which, if done in the course of observation, tended to shake and distort the image. But there were other advantages which applied especially to Russian conditions, and it is interesting to note the sequence in which Lomonosov described them:

My newly invented "catadioptric" telescope is an improvement on the Newtonian and Gregorian (type) in that 1) there is less work, because there is no small mirror, and 2) it is also cheaper; 3) light is not obstructed from the large mirror, and is therefore not diminished, 4) cannot be spoiled so easily, as the heretofore mentioned, especially "en route," sun rays do not 'blunten' and mix (*putaiutsia*), and in this way clarity and definition are amplified.[15]

The increase in clarity and the facility with which magnification could be enhanced were the main reasons for Herschel's adoption of the same principle. The question weighing uppermost in Lomonosov's mind, however, was that of cost and ease of construction, because he had the greatest difficulties in obtaining skilled instrument makers for his own projects. Some were attached to the Academy, but they were not always qualified or expert enough to make Newtonian-type

13. *Ibid.*, p. 80.
14. Lomonosov, "Ob usovershenstvovanii zritel'nykh trub," *Polnoe sobranie,* IV, 483.
15. Lomonosov, "Khimicheskie i opticheskie zapiski, *ibid.*, p. 407, Section 1.

telescopes. It required considerable skill to match the concave speculum with its reflector. The extraordinary feature about Newton's invention was that he literally made it himself, that is, carried through the whole process. He even made the alloy from which the metallic mirror was formed and prescribed the addition of white arsenic, a formula still more or less used today. He also turned and polished the speculum, and made the entire mounting.

Lomonosov, however, used a speculum "prepared by Leutmann for making a Newtonian telescope."[16] Johann Georg Leutmann, first professor of mechanics at the Academy, had died in 1736. Presumably, the metallic mirror Lomonosov used was "borrowed" from an earlier telescope he had made. This is one of the few physical details about the construction of the telescope that Lomonosov gives, although rough notes show that he made various experiments with metallic alloys, calculations as to focal length, and so on. Some sketches show, too, that the initial idea from which the rest was derived involved inclining the mirror to one side. At first he may have considered having the eyepiece along the tube, like Newton (see fig. 44). The main disadvantage of the "Speculum ex Newtonia, Gregoriano et meo compositum" Lomonosov himself cites: the distortion in the image produced by the inclination of the mirror. This was slight, however, and by the time William Herschel constructed his telescope the optical techniques had advanced so far that this factor could be remedied in the eyepiece itself.

Unfortunately, there are no figures about the performance of Lomonosov's telescope. That it was made is beyond doubt, for on April 21, 1762, he turned to the Chancellery of the Academy with the announcement that he had "invented a new catadioptric telescope" and requested the services of two instrument makers, I. I. Beliaev and F. N. Tiriutin. He received no reply, and a few days later he wrote again: "many years now it has proved impossible to get these instrument makers. This is against all good sense . . ." If he were refused again, he would take the matter up elsewhere.[17] The threat had its desired effect and both Beliaev and Tiriutin came to work at Lomonosov's home. He also had Nikolai Chizhov, one of the best qualified Russian instrument makers of the period, working for him. He had

16. Lomonosov, "Ob usovershenstvovanii zritel'nykh trub," *ibid.*, p. 477; on Leutmann's correspondence with Bruce, see Chapter 6, above.

17. Cf. P. S. Biliarskii, *Materialy dlia biografii Lomonosova, sozdanny ekstraordinarnym akademikom Biliarskim* (St. Petersburg, 1865), pp. 503–504.

gone to England in 1759/60 to serve an apprenticeship with George Adams, the elder (1710–1773)—famous in his time for telescopes and other optical instruments.[18] Various craftsmen collaborated on the project during the summer, and it is evident from Lomonosov's jottings that the final design was quite impressive. The focal length of the speculum exceeded one meter: added features were a viewfinder (*navodnaia trubka*), a micrometer, a mechanism for moving the telescope and for turning the support on which it was mounted, and—a nice touch, this—a chair capable of mechanical elevation. On May 13, 1762, it was demonstrated before a session of the Academy.[19] Lomonosov also declared his intention to deliver a paper on the telescope at the next meeting, but, owing to the palace revolution on June 28 of that year which led to the assassination of Peter III and the succession of Catherine II, this meeting was never held. His paper "On the Improvement of Telescopes" thus remained unread.[20] His original manuscript was lost, and only the galley proofs survived to tell of Lomonosov's invention. It was published for the first time in 1827.[21]

Though it is doubtful whether Lomonosov lived to complete his telescope—he died in 1765—there can be no doubt as to the novelty of the principle involved; Vavilov's claim that Lomonosov anticipated a discovery generally attributed to William Herschel may, therefore, be accepted.[22] One significant point, however, deserves some emphasis because Soviet scholars have invariably neglected it.

Some of the best instruments in the possession of the Academy in St. Petersburg came from London, and Lomonosov was well acquainted with the work of celebrated instrument makers such as George Adams and James Short.[23] Better known, even, than either of

18. On Chizhov, see S. L. Sobol', *Istoriia mikroskopa i mikroskopicheskikh issledovanii v rossii v XVIII veke* (Moscow and Leningrad: Akademiia nauk, 1949), pp. 126, 127, 190, 191; a considerable number of instrument makers were employed by the Academy at this time, but for reasons of economy their services were used largely in making objects for public sale.

19. *Protokoly zasedanii konferentsii Imperatorskoi Akademii Nauk s 1725 po 1803 goda,* ed. Veselovsky, II, 483.

20. The surviving text has a note in G. V. Müller's hand (no friend of Lomonosov): "Dieses ist der Anfang einer Rede die Lomonossow wegen beständiger Trunkenheit nicht hat zu Ende bringen können." *Polnoe sobranie,* IV, 794.

21. By Ivan Dvigubskii, *Novyi magazin estestvennoi istorii* (No. 1, 1827), pp. 31–50; cf. *Polnoe sobranie,* IV, 793.

22. This has become a standard claim in Soviet historiography. See, for example, B. A. Vorontsov-Vel'iaminov, *Ocherki istorii astronomii v Rossii,* p. 76.

23. See Lomonosov, "Khimicheskie i opticheskie zapiski," *Polnoe sobranie,* IV, p. 430, Section 49: "pisat' k astronomu v London o Shartovoi trubke."

these was John Dollond, to whom Lomonosov refers briefly in a note above the sketch of his telescope, and again, in the same context, he cites "Dollond and Euler."[24] This is revealing in a rather important way, for the juxtaposition of these two names is enough to suggest why Lomonosov's good friend Euler remained totally uninterested in his discovery, and why, indeed, *it turned out to be just as untimely* as his theory of light.

In seeking to avoid or evade chromatic aberration of lenses, Newton tried differing contiguous media, hoping that the different refractions might correct each other. He failed to obtain any such solution. Hope of correcting chromatic aberration with lenses was thus generally abandoned, and successful telescopes were therefore built on Newton's model, until the middle of the eighteenth century when Leonard Euler returned to the problem. Basing his views on the supposed achromatism of the eye, itself a series of media of different refractive powers, Euler suggested in 1747 the possibility of correcting both spherical and chromatic aberration. At about the same time, a Swedish mathematician pointed out to John Dollond (1706–1761) that Newton's negative results were based on prisms of small apical angle and that the matter needed further inquiry. Dollond, having mastered the mathematical theory by way of Euler, succeeded, after years of experiment, in making an "achromatic" system which he completed in 1758 (five years before Lomonosov started building his telescope.) He computed, ground, and polished object lenses of crown glass with water in between, ultimately replacing the water with a lens of flint glass. In this way he reduced chromatic distortion by using lenses of two kinds, flint and crown, chosen so that the chromatic aberration of the one was equal to that of the other. Since that of the one, being convex, was in the opposite direction of that of the concave, their aberrations neutralized each other. As a result of Dollond's success, his principle rather than Newton's came to be followed in the construction of telescopes *in the second half of the eighteenth century*. On the whole and save for special reasons, refracting telescopes rather than reflectors of the Newtonian type held the field until the end of the nineteenth century.

For this reason, Lomonosov's modification of the Newton-Gregory telescope could not have been made at a less apposite moment in the history of optics. Moreover, Lomonosov knew of Euler's theoretical

24. *Ibid.*, p. 455, Section 147, item 10.

work,[25] and, as his brief allusions show, also of Dollond's; this may account for the fact that, in another type of telescope which it is claimed he invented, Lomonosov used Dollond's principle of construction rather than Newton's.

TUBUS NYCTOPTICUS MODO LOM. N.

"Tubus nyctopticus modo Lomonosov-Newton" is the grandiloquent way Lomonosov described his other telescope, which was designed expressly for use at night and in poor visibility.[26] The first official notification of its invention is contained in the minutes of the Academy for May 13, 1756, when Lomonosov intimated his desire to devote a paper to it. At that session, he is said to have

showed his machine for condensing light [*dlia sgushcheniia sveta*—another name by which his telescope was known] . . . which was made by the Academy's masters . . . Its tube is nearly two feet in length, and three-quarters of an inch in diameter. One lens is small [the eyepiece], and the other [the objective] is big, [and is intended] for gathering light. From all experiments it appears that an object placed in a dark room, may be distinguished more clearly with this tube than without it. But in so far as this is achieved only at small distances, it is too early to tell how it will work out at longer distances at sea. Lomonosov, however, believes that his invention can be brought to such a degree of perfection, that he can guarantee its undoubted usefulness at sea.[27]

But the professors of astronomy, A. I. Grishchov and N. I. Popov, who were present at the meeting, seemed to feel that apart from the novelty of its purpose, the new telescope gave "the same results as all other astronomical tubes."[28] And S. I. Rumovskii, who returned from abroad on August 24, 1756, wrote to Euler that "having had the honor of

25. Euler, then in Berlin, was in regular correspondence with members of the St. Petersburg Academy. See, for example, his letter to Goldbach on October 1, 1748: "und anjetzo bin ich bemühet, einen Einfall ins Werk zu richten, welchen ich gehabt, um solche Objektiv-Gläser zu verfertigen, welche eben den Dienst leisten sollen, als die Spiegel in den tubis *Neuton*ianis und *Gregor*ianis." See Iushkevich and Winter, eds., *Leonhard Euler und Christian Goldbach Briefwechsel,* p. 304.

26. He also calls it a "nochegliad" and then again "polesmocopium nocturnum": cf. Lomonosov, "Khimicheskie i opticheskie zapiski," *Polnoe sobranie,* IV, Sections 87, 147, 159, 105.

27. *Protokoly zasedanii konferentsii Imperatorskoi Akademii Nauk s 1725 po 1803 goda,* ed. Veselovsky, II, 350.

28. *Ibid.:*
Сказали, что иной новизны изобретения, кроме цели им назначенная, по сравнению с прочими трубами нет, и что все астрономические трубы дают то же самое.

being permitted to observe many objects through [Lomonosov's] telescope," he "noticed no difference from what is ordinarily seen through telescopes, excepting that all objects seemed highly colored. It shows the iridescent colors to the utmost degree of perfection, from which I conclude that the solution to this problem, according to Lomonosov, lies in placing the lenses in such a way that the iridescent colors may be seen through the telescope as clearly as possible."[29]

Undaunted by criticisms, Lomonosov continued to work on his "tubus nyctopticus," but on January 19, 1758, he presented the Academy with a Latin manuscript entitled *Problema physica de tubo nyctoptico* in which he declared that "he had no time for further improvement of the instrument"; he therefore commended it to the examination of his colleagues, "who may judge what usefulness to humanity may come of it in everyday life."[30]

There the matter would have rested were it not for Franz Ulrich Theodor Aepinus who had arrived at the Academy half a year earlier. Aepinus decided that Lomonosov's invention was unworthy of insertion in the Academy's records because, as he tried to show, the construction of the kind of telescope Lomonosov claimed to have invented was theoretically impossible. In support of his contention, he wrote a paper —"Demonstratio impossibilitatis nyctoptici Lomonossowiani."[31] The controversy flowered into an "affaire célèbre." On May 17, 1759, at another meeting of the Academy, Lomonosov again put forward his idea, suggesting that the construction of a night telescope be made the subject of a prize contest set by the Academy, but apparently many academicians shared Aepinus' doubts. Lomonosov offered to ask Euler to arbitrate the dispute, which was safe enough as the great mathematician was in Berlin. Aepinus in turn suggested that it should be decided by the "Académie des Sciences." A battle of memoirs ensued, though it is doubtful whether the controversy actually came to the hearing of the Paris Academy. It took another dramatic turn in June 1759, however, when Lomonosov received a new telescope from London, from his Maecenas, Prince I. I. Shuvalov.

Having examined and admired this telescope, Lomonosov declared its construction to be exactly like the new invention he had had in

29. Quoted by Pekarskii in *Istoriia Imperatorskoi Akademii Nauk v Peterburge*, II, 599–600.

30. Lomonosov, "Problema Physicum de tubo nyctoptico," *Polnoe sobranie*, IV, p. 119, No. 11.

31. Biliarskii, *Materialy dlia biografii Lomonosova*, p. 391.

mind! He convinced himself that the new English telescope made objects appear lighter at dusk, and he demonstrated it before the Academy later the same month in order to prove conclusively to his enemies that a "tubus nyctopticus" could be built.[32] But instead of subsiding, the controversy grew several degrees more intense. The immediate reaction of Lomonosov's colleagues to his artful counter-offensive is not known, but writing to Prince Shuvalov on July 8, 1759, Lomonosov declared that:

For the glory of the country (*otechestvo*) I must announce all my new inventions to the learned world at the earliest opportunity [in future] to avoid that which befell my night-watching telescope. This diminution in honor of my labors became doubly painful to me because those who reckoned this business impossible, are still bitterly quarreling . . . not seeing what they see, not hearing what they hear.[33]

The invention continued to occupy Lomonosov's mind, however. He refers to it repeatedly in his *Chemical and Optical Notes* for 1762/63 where, as we have seen, he also alludes to "John Dollond-Euler." Oddly enough, however, in spite of the assiduous attention Lomonosov's discoveries have inspired amongst Soviet scholars, none mention Dollond's name, an omission which appears all the more significant in view of the confident claims made on behalf of the "night telescope." For, sixteen years later, Johann Georg Lambert described a similar device, and, almost half a century afterward, Jerôme De Lalande also considered himself the first inventor of a telescope specially designed for night observation; Lomonosov, I maintain, anticipated them both.[34] The question always ignored is why Lomonosov should have insisted that the English telescope sent to him by Shuvalov was like his own 'tubus nyctopticus.'

The answer is suggested by the reference to Dollond. By 1758 Dollond had solved the problems involved in making telescopes according to his "achromatic" system. Lomonosov received his new English telescope the following year. Shuvalov would not have bothered to send him a model of the older Newtonian type, for these were already well known in Russia. It was probably a refracting telescope

32. *Protokoly zasedanii konferentsii Imperatorskoi Akademii Nauk s 1725 po 1803 goda*, ed. Veselovsky, II, 430.

33. Quoted in *Polnoe sobranie*, IV, 735: "vidia ne vidiat, i slysha ne slyshat."

34. Cf. *ibid.*, edited by T. P. Kravets and V. I. Chenakal. See J. H. Lambert, *Beyträge zum Gebrauche der Mathematick, und deren Anwendung* III (Berlin, 1772), 203–204; also, J. De Lalande, *Bibliographie astronomique avec l'histoire de l'astronomie* (Paris, 1803), no. 1811.

of the kind pioneered by Dollond. This is why Lomonosov could claim that the English telescope was like his own. His short description of the "tubus nyctopticus" in the paper read before the Academy in June 1758 shows clearly that it, too, was a refracting, not a reflector-type telescope of the Newtonian variety. The eyepiece was located at one end of the tube, in line with its horizontal axis, and there was a large convex lens "four London inches in diameter (*quatuor circiter pollices Londinienses*)" inside the second (moving) section of the tube. This arrangement served a simple purpose: by increasing the size of his objective lens opening as much as he could, he allowed more light into the tube, thus making it more effective in poor visibility. The eyepiece restored the definition of the image distorted by the large lens. All this was perfectly sound without being very radical; it only seemed to be so from Lomonosov's overly enthusiastic claims on its behalf, which Aepinus could justly ridicule (see fig. 45).

Thus it becomes clear why Rumovskii and others spoke of the "iridescent" colors seen through Lomonosov's telescope. With its arrangement of lenses, it managed to let in more light than the Newtonian type of telescope in which the mirrors invariably caused a loss of light due to the double reflection involved and the position of the smaller mirror, but it did nothing to solve the problems of chromatic aberration—hence the "iridescent" colors. When Lomonosov saw the instrument Shuvalov had sent him, he jumped to the conclusion that it was like his own, for so the arrangement of the lenses made it appear. And he could prove his point by demonstrating it before the Academy because a good telescope of the Dollond kind would, in fact, be more effective in poor visibility than a reflector type. What Lomonosov did not then know, or preferred to conceal, was that its effectiveness depended entirely on the grinding of the lenses and the complex calculations necessary to avoid chromatic and spherical aberration. Essentially, therefore, the "tubus nyctopticus" was a rather simple device which failed to grasp the more fundamental problems involved. The problem of constructing a telescope for night viewing could not be solved simply by increasing the aperture of the objective lens. Had Lomonosov wholly accepted Newton's laws of refraction, he would have realized this. There was no purely mechanical solution to the problem, and it was not until the beginning of the twentieth century that the physiological considerations involved began to be seriously investigated.

Whenever I hear a pretended Freethinker inveighing against the truths of Religion, and even sneering at it with the most arrogant self-sufficiency, I say to myself; poor weak mortal: how inexpressibly more noble and sublime are the subjects which you treat so lightly, than those respecting which the great *Newton* was so greatly mistaken!—Leonard Euler

20

Newton in the Catherinian Era

If Lomonosov's reluctance to accept Newtonian doctrines delayed their acceptance in the St. Petersburg Academy, his hostility toward them derived further authority from the support given to Newton's critics by Leonard Euler. Euler came back to Russia in 1766, a year after Lomonosov's death, and added his powerful voice to the waning band of the Englishman's detractors. His return to St. Petersburg was prompted by an invitation from Catherine the Great, whose wish it was that the great mathematician should once again add the luster of his presence to the Russian Academy.[1]

In this there is more than a little irony since Euler's most notable achievements lay behind him. His *Mechanica,* in which he applied the "analytical method" (that is, calculus) to mechanics, had been published thirty years earlier. Yet, as far as the Russian public was concerned, the immense authority he enjoyed in the Catherinian era was due neither to this renowned work nor to his copious contributions to the *Acta Eruditorum* and other learned journals, but, rather, to his *Letters to a German Princess.* This was a work with literary pretensions and was inspired by the same genre as Fontenelle's *Pluralité des Mondes.*

The letters, of which Euler wrote more than three hundred, represent the most exhaustive and authoritative treatment of natural

1. Cf. P. P. Pekarskii, "Ekaterina i Eiler," *Zapiski Imperatorskoi Akademii Nauk,* VI (1865), 59–92.

philosophy to be written by any major scientific figure in the eighteenth century. It was also the last work of that caliber to maintain a critical and hostile attitude to Newton and his doctrines. By the time the *Letters* appeared—they were begun in 1760 and completed within the next two years—the triumph of the Newtonians on the Continent was almost universally acknowledged: and it is thus appropriate that, while the *Letters* were being prepared for publication in St. Petersburg, a remarkable individual named Feodor Ivanovich Soimonov should at the same time launch his own bold attempt at giving a systematic exposition of Newton's cosmology. This he intended to do in a spirit very different from Euler's; indeed, the two undertakings were in a sense related in that success for one spelled failure for the other.

By the time Euler returned to Russia, Soimonov must have been one of the very few outstanding survivors of the Petrine era. He was born in a merchant's family in 1682 and attended the Navigation School Bruce had founded in Moscow after which he was sent to the Netherlands to practice what he had been taught. He spent three years abroad and, on coming back to Russia, served in the Caspian fleet, where he attracted the personal attention of Peter. Subsequently Soimonov published an excellent manual on navigation and several hydrographic surveys, among them a magnificent folio edition on the Varangian, that is, the Baltic, Sea.[2] At the end of Empress Anna's reign, Soimonov was appointed vice-president of the Admiralty "collegium," and it was here that his integrity brought him into collision with the mores of the times. He discovered that ammunition stores listed as full were in fact empty; in ferreting out these and other extravagant abuses, Soimonov made many enemies who awaited the opportunity for revenge. It came in 1740, when Artemii Volynskii, one of the powerful magnates in Anna's cabinet, quarrelled with Bühren, her favorite. After an unfair trial, Volynskii was executed; his friends suffered too, Soimonov among them. His refusal to sign a protocol of fabricated accusations against Volynskii led to his own incarceration. After being beaten with a knout, he was exiled to Siberia, where he toiled as a common laborer, forgotten by all. By then Soimonov was fifty-eight years old.

2. Soimonov, *Svetil'nik morskoi, to est' opisanie vostochnago ili variazhskago moria* . . . (St. Petersburg, 1738). This book is considered very rare. See G. N. Gennadi, *Ukazatel' bibliotek v Rossii* (St. Petersburg, 1864), no. 11; cf. SK, III, no. 6661. A second edition, dedicated to Empress Elisabeth, was published in St. Petersburg in 1743.

Two years later, on Elizabeth's accession, he was freed, but his rank was not restored. Though the next few years in his biography are obscure, we know that he remained in Siberia. In 1757 someone at Court was good enough to recall his existence, and Soimonov was promptly appointed governor of the whole vast region in which he had spent such thankless years. He became interested in Siberian exploration and also found time to write. He began a correspondence with Lomonosov, pressing him for the proper formula for manufacturing the glass he hoped to use in preparing lenses for telescopes. He also wrote to the Academy of Sciences, offering his services in furnishing it with information on the immense area under his control. At the same time, he began sending articles to the *Ezhemesiachnyia sochineniia,* the popular scientific journal published by the Academy. They were welcomed, and they marked the beginning of Soimonov's activities as a popularizer of the new astronomy. In 1768, after Catherine II came to the throne, he was called to Moscow to perform the sybaritic duties of a senator, which evidently took little of his time. In 1765 he published a most interesting anthology entitled: *Kratkoe iz'iasnenie o astronomii.*[3] As Soimonov says in his introduction, many readers were interested in astronomy, but lacked the opportunity to indulge their curiosity. His work was intended to do no more than compile in a useful way what others had written. The anthology consists largely of extracts taken from the *Pluralité des Mondes,* the *Primechanii k vedomostiam* for 1732 and 1738, and other such literature, including an eloquent article by Lomonosov in defense of the heliocentric theory and extracts from the *Kratkoe opisanie kommentariev.*

Even at this late date, however, there was only one work in Russian that Soimonov could draw on to describe the Newtonian theory of gravitation, and that work was almost half a century old: Bruce's translation of the *Kosmotheoros.*[4] Though its language must by now have seemed heavily archaic, Huygens' book was scientifically more advanced than Cantemir's translation of *Pluralité des Mondes* (which first appeared more than twenty years later). It was probably this that appealed to Soimonov, for, though Huygens rejected gravitation as an attribute of individual particles of matter, he accepted it as an attribute

3. I.e., Soimonov, *Kratkoe iz'iasnenie o astronomii v kotorom pokazany velichiny i razstoianiia nebesnykh tel, kupno s poriadkom v ikh razpolozhenii i dvizhenii po raznym sisteman i o velichine i dvizhenii zemnago globusa Vypisano iz raznykh astronomicheskikh i fizicheskikh avtorov* (Moscow, 1765).

4. See Chapter 5, above.

of the planetary masses. Unlike the *Pluralité des Mondes,* which did not mention Newton at all, the virtue of the *Kosmotheoros* was that at least it recognized the Newtonian system in part. Yet Soimonov's primary purpose in the anthology was to advocate the Copernican theory, which, as far as the public was concerned, appeared, even in 1765, to be a noteworthy undertaking. One of the twenty-eight engravings which luxuriously adorn the text contains a chart of the heavens illustrating the heliocentric system. Another is even more revealing; it shows the cosmology of Descartes.

Soimonov must have known that vortices were no longer tenable because in the following year he conceived the notion of publishing another book on the new astronomy, which Rumovskii agreed to write for him. It transpires from Rumovskii's correspondence with G. V. Müller,[5] that at least four chapters of the book were actually written. The first gives a general introduction explaining the structure of the universe. The second depicts the Copernican theory ("je prouve que le système de Copernic est la vraie système du monde"). The third chapter dealt in part with the controversy over the shape of the earth. The fourth one provided an explanation of celestial mechanics according to Newton. The book was never completed owing—so Rumovskii avowed—to other more "pressing work." What that other work comprised, Soimonov was fortunately never told, for it would no doubt have irked him that Rumovskii had been conscripted to translate a book of similar character in which Newton was exposed to detailed criticism—Euler's *Letters to a German Princess.*

The incentive for writing this famous work was provided by Princess Sophie Friederike Charlotte Leopoldine Luise, daughter of the Markgraf Friedrich Heinrich von Brandenburg and a distant cousin of the King of Prussia. The *Letters,* which have been compared to Burckhardt's "Weltgeschichtliche Betrachtungen," were soon translated into most European languages.

The book touches upon all the subjects that an undergraduate could then be expected to know, for not only does Euler discuss the main physical doctrines, but the philosophical implications as well. These take him far afield into theology and ethics. In epistemological questions he mainly follows Locke, emerging as an enemy of the

5. Extracts from Rumovskii's correspondence with G. V. Müller, who was Soimonov's intermediary, may be found in M. I. Sukhomlinov's *Istoriia Rossiiskoi Akademii,* 8 vols. (St. Petersburg, 1874–1888), II, 433 ff. See also V. V. Bobynin, "Rumovskii, Stepan Iakovlevich," in *Russkii biograficheskii slovar',* XVII (1918), 441–450.

idealism of Berkeley and his school, though this was not the main attraction of the *Letters*. What made them fashionable to many readers was the illustrious reputation of the author, a devout Calvinist, and the piety and orthodoxy of his personal convictions. Where scientific principle and revelation were out of harmony, Euler always tactfully avoided a confrontation, prompting the English translator to express his "mortification" at the fact that "the specious and seductive productions of a *Rousseau,* and the poisonous effusions of a *Voltaire,* should be in the hands of so many young men, not to say young women, to the perversion of the understanding and the corruption of the moral principle, while the simple and useful instructions of the virtuous *Euler* were hardly mentioned."[6]

In a "letter" entitled "Examination and Refutation of Newton's system," Euler recaptures the partisan tone of earlier encounters between the Newtonians and the Cartesians:

... It is besides rather hard and humiliating for a philosopher to acknowledge ignorance of any subject whatever. He would rather maintain the grossest absurdities; especially, if he possesses the secret of involving them in mysterious terms, which no one is capable of comprehending. For in this case the vulgar are the more disposed to admire the learned; taking it for granted that what is obscurity to others is perfectly clear to them. We ought always to exercise a little mistrust when very sublime knowledge is pretended to—knowledge too sublime to be rendered intelligible.[7]

Both the Russian and the French versions of the *Letters* began to be published in St. Petersburg under the auspices of the Academy of Sciences in 1768, and five Russian editions of Rumovskii's translation were to appear in the period that followed.[8] The influence of the book extended beyond its celebrated author's scientific views, and it is this that made the authority Euler enjoyed rather unfortunate. His reiterated attacks on "Freethinkers" and his fulsome piety gave those who wished to keep science subservient to religion the prestige and dignity

6. *Letters of Euler . . . to a German Princess,* tr. Hunter, I, xv–xvi.

7. *Ibid.,* I, 98–99 (letter xviii, dated July 1, 1760).

8. I. e., Euler, *Lettres à une princesse d'Allemagne sur divers sujets de physique et de philosophie,* 3 vols. (St. Petersburg, 1768–1772). Another French edition was published in Berne in 1778 (also in three volumes), followed by a Parisian edition edited by Condorcet and another in 1812 edited by J. B. Lakey. Also, *Pis'ma o raznykh fizicheskikh i filosoficheskikh materiiakh, pisannye k nekotoroi nemetskoi prinzesse, s franzuzskago iazyka perevedennye S. Rumovskim* appeared in 1768–1774; a second edition in 1785, a third in 1790–91, a fourth in 1796, and a fifth in 1808 (incomplete, with only the first two of the book's three parts actually appearing in print). All of the Russian editions were published in St. Petersburg.

of Euler's apparent support.[9] In England, where only the political reaction to the French Revolution saved the work from continued neglect, the philosophical orientation of the *Letters* found little favor; to the brilliant new generation of French mathematicians who were Newton's true heirs, Euler's attitudes and metaphysical opinions seemed quaint and even archaic. "Notre ami Euler est un grand analyste, mais un assez mauvais philosophe," wrote D'Alembert to Lagrange;[10] and Lagrange (who to D'Alembert seemed alone worthy among all of his contemporaries of taking Euler's place at the Berlin Academy) described the *Letters* as a major aberration of genius, comparable only in its absurdity to Newton's commentary on the Apocalypse.

In Euler's defence, it might be argued that the harmony he was never tired of discerning between revelation and the phenomena of nature was not a disingenuous device for evading the inconvenient issues of the day but, rather, a sincere attitude of mind which he owed to his upbringing in a Swiss republic, where the toleration and humanity of the confessional interest long delayed any direct confrontation between science and religion. In Russia, however, the situation was very different. From Antiokh Cantemir to Lomonosov and Aleksandr Radishchev, most of its prominent figures of the Enlightenment were bitterly anticlerical and with good reason. The obscurantism of the Russian church, which had not even yet come to terms with the Copernican theory, could only be fortified by the assurances of the great Euler that "in our researches into the phenomena of the visible world, which lies open to the examination of our senses . . . we [were subject] to weaknesses and inconsistencies so humiliating . . . [that] a Revelation was absolutely necessary to us; and we ought to avail ourselves of it, with the most powerful veneration."[11]

Condorcet tampered with such passages in his French edition of the *Letters*,[12] omitting some of Euler's stern rebukes to Newton, but they

9. As one of Euler's more recent admirers has put it, it cannot be denied that "die immer wieder kehrenden Ausfälle auf die Freigeister auf die Dauer lästig, ja zänkisch wirken." See Andreas Speiser's lecture, *Leonhard Euler und die Deutsche Philosophie* (Zürich, 1934), p. 14.

10. D'Alembert to Lagrange, June 16, 1769, quoted in *Journal des Savants*, January 1846, p. 57.

11. *Letters of Euler . . . to a German Princess*, tr. Hunter, I, 73–74 (letter xviii).

12. The French edition of Euler's *Letters*, with additions by Condorcet and Lacroix appeared in three volumes, and it was published in Paris between 1787 and 1789.

were left intact in the Russian translation. It is doubtful whether Rumovskii, who was a protégé of Euler, really shared these sentiments. They did, however, become fashionable later, in the wake of the French Revolution, when Newton's reputation was inevitably affected by the climate of reaction that event provoked in Russia. Since his name was linked so often with that of the philosophes, who were now proscribed, it suffered by association with them. Words Leonard Euler had infelicitously inserted in his *Letters to a German Princess,* were now commended as the best antidote to Voltaire: "Whenever I hear a pretended Freethinker inveighing against the truths of Religion, and even sneering at it with the most arrogant self-sufficiency, I say to myself: poor weak mortal; how inexpressibly more noble and sublime are the subjects which you treat so lightly, than those respecting which the great *Newton* was so greatly mistaken!"

This new attitude is most strikingly illustrated by one of Voltaire's more vehement Russian foes: Evgenii Bolkhovitinov, who decided to omit all references to Newton in his fresh translation of the *Essay on Man* in 1793.[13] The example is of some interest because Bolkhovitinov, who later became Metropolitan of Kiev, was regarded by contemporaries as a man with views so liberal that at one time he had even encouraged the study of the Newtonian system at the Voronezh seminary where he taught.[14]

In so far as the obscurantism of the church and the political response to the French Revolution affected Newton's standing, it was more than balanced by renewed public interest in English culture which was itself a reaction to events in France, having earlier been fostered by French men of letters such as Prévost, Montesquieu, Rousseau, Diderot, and, of course, Voltaire. Thus, when Count Mordvinov, the most noteworthy representative of the Anglomania affected by the Russian aristocracy at the end of Catherine II's reign, wrote to the brother of Jeremy Bentham to describe the "four geniuses" who had done most

13. With the exception of one factual reference on page 45, Bolkhovitinov's *Opyt o Cheloveke,* made from five different translations (including Du Resnel's and Silhouette's), appeared in 1806 in Moscow. It contains an introduction culled from the French translators Bolkhovitinov used, but it avoids carefully any allusion to Newton and the imputation of materialism that had caused the Pope so much anxiety.

14. Thus, according to E. F. Shmurlo (*Mitropolit Evgenii kak uchenyi. Rannie gody zhizni 1769–1840* [St. Petersburg, 1888]), Schmidt's *Kosmografiia,* which adopts the Newtonian cosmology, was translated by Gavrilo Uspenskii in 1801 at Bolkhovitinov's prompting.

for the happiness of mankind, he named Bacon, Newton, Smith, and Bentham, "each the founder of a new science."[15] And Karamzin, who visited England in 1789, paused in admiration before Newton's monument in Westminster Abbey, prompted by the thought that his achievement reflected the national character—the tendency of the English mind to probe for fundamental principles (rather than to soar upward, like the French).[16]

But the essentially novel idea concerning Newton to be reflected in Russian literature and writing in this period owed its origin to Voltaire. It was he who embellished, if he did not actually invent, a historical pedigree for Newtonian philosophy. In his brilliant polemic against the Cartesians Newton was, for Voltaire, not only the greatest philosopher who had ever lived, but heir to a tradition which it suited him to emphasize again and again, was fundamentally different from that which had prevailed in France or elsewhere. Newton's discoveries were due not only to his unrivaled genius, but to his expert use of the inductive method. This was the essential characteristic of the "philosophie anglaise," and was responsible for the fact that the English had attained more success in natural philosophy than the French. Russian writers making the same discovery owed it, paradoxically enough, to the philosophes rather than their own immersion in English thought, which was still known largely through French translation.

Thus, the first work to give Russian readers a vivid description of the ways in which gravitation affects space flight was one of Voltaire's best loved "contes"—*Micromégas,* which appeared initially in the columns of the *Ezhemesiachnyia sochineniia.*[17] The "contes" were a genre through which Voltaire set out to make the anthropocentric view of the universe, which should have vanished from human thinking somewhere between the discoveries of Copernicus and those of Newton but steadfastly refused to do so, totally ridiculous; they proved

15. Quoted by V. K. Ikonnikov, *Graf N. S. Mordvinov. Istoricheskaia monografiia* (St. Petersburg, 1873), p. 75.

16. N. M. Karamzin, "Pis'ma russkogo puteshestvennika," in *Izbrannye sochineniia,* 2 vols. (Moscow and Leningrad: Izdatel'stvo "Khudozhestvennaia literatura," 1964), I, 458, 491.

17. Publication of the *Ezhemesiachnyia sochineniia k pol'ze i uveseleniiu sluzhashchiia* was due largely to the instigation of Lomonosov, who managed to interest Shuvalov in the project. His idea was to provide a popular scientific journal, more accessible to the layman than the Russian supplement to the *Commentarii* (which the Academy had earlier launched as its first attempt at popularizing science). It consisted mainly of extracts or translations from foreign sources.

to be as popular with Russian readers as anywhere else in Europe, even though the immediate effect they often produced on the sensibility of the time was one of despair and melancholy at the insignificance of man's lot: "our earth contains nothing lasting."

The second reaction was hardly sufficient to absorb the force of the first, but at least it provided some comfort to know, as Voltaire reiterates throughout his later writings, that Nature is everywhere subject to the same laws of causality: "Newton dit que la Nature se ressemble partout: *Nature est ubique sibi consona.*" Elsewhere Voltaire cites the *Principia* more accurately—*Natura est semper sibi consona*—but the message itself stays the same and is repeated by him in several variations. For example, in *L'Ingenu,* the story of a Huron Indian's encounters with "civilized" society, and particularly in *Les Oreilles du Comte de Chesterfield.* Its implications were not merely scientific. Voltaire made it the point of departure for his attacks on every type of parochialism: religious, racial, and national. Humanity is clustered about one planet—Earth, which is only one of multitudes of inhabited globes filling the cosmos. How foolish to think that God makes any distinction between Turk, Chinese, Indian, or Russian; all belong to the same family of mankind, exposed to identical natural Laws.

It is not difficult to see why the implications of this view of nature were not to the liking of the Russian Holy Synod. Voltaire in any guise or shape was a natural target for its criticisms, but this was even truer with respect to the secular views propounded by the hero of *Micromégas,* an alien from outer space who appears to have studied the *Principia* and the *Opticks* with a devotion and understanding rare among earthlings. Indeed, the novel was only intended as a vehicle for expressing Newton's ideas. New methods of scientific thought, Voltaire believed, required new methods of presentation. The problem, as he became increasingly aware after writing the *Elémens,* was to reveal the Newtonian "catechism" to a wider audience.[18] Thus, "on a beau dire le siècle est philosophique," he wrote when Maupertuis published his treatise on Newton but it scarcely sold two hundred copies, "et si on montre si peu d'empressement pour un ouvrage écrit de main de maître, qu'arrivera-t-il aux faibles essais d'un écolier comme

18. "Un catéchiste annonce Dieu aux enfans, et Newton le démontre aux sages." Voltaire first made this statement in the *Lettres Philosophiques,* but he repeats it elsewhere, the most striking instance being in his letter on Spinoza (1767), in *Œuvres,* ed. Beuchot, XLIV, 388.

moi?"[19] Voltaire then gradually rejected the essay and the debating style he had earlier employed in argument with his opponents. Descartes and Leibniz, earlier treated as serious thinkers, became objects of ridicule. Similarly, metaphysics, to which Voltaire had devoted much study, became a synonym for futile enquiry. "Toute métaphysique ressemble assez à la coxigrue de Rabelais, bombillant dans la vide. Je n'ai parlé de ces sublimes billevesées que pour faire savoir les opinions de Newton."[20]

One outcome of this militant change of tone was Voltaire's emboldened use of satire in propagating the implications of Newton's natural philosophy, and *Micromégas* was the result. It is not a novel in the ordinary sense. The plot is no longer chained to this staid and solid earth, but revolves largely in outer space. The hero, who hails from the planet Sirius, is a Newtonian eight leagues in height, which is perfectly reasonable because where he lives far greater power is needed to overcome the tremendous gravitational pull. "Il n'avait pas encore deux cinquante ans, et il étudiait, selon la coutume, au collège de jesuites de sa planète, lorsqu'il devina, par la force de son esprit, plus de cinquante propositions d'Euclide."[21] In his 450th year, as he is emerging from childhood, Micromégas writes a scientific treatise that elicits the displeasure of the "muphti de son pays," who divines that it smacks of heresy. "Il s'agissait de savoir si la forme substantielle des puces de Sirius était de même nature que celle des colimaçons." After an extended trial the book is condemned, and the author is banned from court for 800 years. This was an allusion to Voltaire's own persecution for saying in the *Lettres Philosophiques* that the faculties of our soul develop in the same fashion as our other organs, and in the same way as the souls of animals. It could equally well have served as a prophetic anticipation of the fate of Voltaire's own book at the hands of the Russian Synod, which tried to have it suppressed.[22]

19. Voltaire, *Correspondance*, ed. Besterman, no. 596.
20. Quoted by Charles Rihs, *Voltaire; Recherches sur les origines du matérialisme historique* (Geneva: Droz, 1962), p. 202.
21. *Œuvres*, ed. Beuchot, LIX, 178–179.
22. The Holy Synod was concerned with the first Russian translation of Voltaire's *Micromégas*, the version which appeared in the January edition of the *Ezhemesiachnyia sochineniia* for 1756, but the attempt to have it banned failed. In August 1759 Sumarokov brought it out in his own translation in the *Trudoliubivaia Pchela*, the first private periodical in Russia (Book 8, pp. 455–476): under the title "Puteshestvie na nashu zemliu i prebyvanie na nei Mikromegasa." Andrei Reshetnikov published a translation in another journal in 1793. In book form, *Micromégas* was published in 1784, 1788, and 1789, as well as later. Cf. SK, I, no. 1119.

It did not succeed. Catherine the Great regarded herself, or so she professed, as Voltaire's "écolier"; she generously encouraged the translation and publication of his writings. Partly through Voltaire's works and partly through the works of the Encyclopedists, Newton's scientific prowess became inextricably associated with the method that had ostensibly enabled him to make his revolutionary discoveries. Thus, Baron Grimm, who visited Catherine in St. Petersburg at the same time as Diderot, would ironically suggest to her a parallel between his own extraordinary devotion to the Empress and the method that led to Newton's triumph: "Der welcher in seinem Leben nur einen Gedanken hat, der kann's weit bringen, und das Unerforschliche erforschen. So hat *Newton* die Welt oder das *Universum,* und der Grübler von Grimma ihre Zierde oder unsere Grossmutter erforschet."[23] More seriously, Diderot would tell Catherine that the future lay with the natural sciences and in systematically applying to them as well as to other human activities the method of investigation epitomized in Newton's credo—"hypotheses non fingo."

The sun of the system builders had set. Scholasticism was dead. It seemed to Diderot that "l'esprit humain semble avoir jeté sa gourme; la futilité des études scolastiques est reconnue; la fureur systématique est tombée; il n'est plus question d'aristotélisme, ni de cartésianisme, ni de malebranchisme, ni de leibnitzianisme; le goût de la vraie science règne de toutes part; les connaissances en tout genre ont été portées à très haut degré de perfection."[24] Such was Newton's authority that,

23. Cf. Ia. [K.] Grot, ed., *Pisma Imperatritsy Ekateriny II k Grimmu (1774–1796)*, 2nd ed., *Sbornik Imperatorskago Russkogo Istoricheskago Obshchestva,* Vol. XLIV (St. Petersburg, 1885), 245–246. Grimm to Catherine, July 15(26), 1782. Grimm imagines her grandsons, the future Emperor Alexander and his elder brother Constantine, visiting his grave, where they find the following inscription: "Ci-gît seul d'entre les mortels parvint à embrasser par la pensée ce que vaut ton aieule et ta maitresse d'école." To this Alexander exclaims: "De par le Dieu vivant . . . et par l'Impératrice mon aieule et très honorée Souveraine, voilà un insolent coquin! A quoi tient-il que je ne prenne mon sabre destiné à couper le noeud gordien, pour détruire cette tombe de fond en comble?" But Constantine restrains him: "Calme-toi, mon frère, und gönne seiner Asche die Ruhe. Der welcher in seinem Leben . . . etc."

24. Denis Diderot, *Œuvres complètes,* 20 vols. (Paris, 1875), III, 441. On Diderot's journey to St. Petersburg and his relations with Catherine, the classic study is *Didro v Peterburge* (St. Petersburg, 1884), by V. A. Bil'basov, author of an impressive but unfinished biography of the Empress. Through no fault of Bil'basov's, however, several of the points made in his book are no longer valid. For example, the *Essai sur les Études en Russie,* which is included in Vol. XIV of Diderot's *Œuvres complètes* was actually written by Grimm, as has since been established. Also, Diderot's unfinished memorandum on Russia was first published only after *Didro v Peterburge* appeared (Maurice Tourneux's *Diderot et Catherine II* [Paris, 1899]). Diderot's writings pertaining to Russia were conveniently gathered together and published in Volume X (1947) of the Soviet edition of

when Diderot drew up plans for a new Russian university at Catherine's invitation, he based its curriculum largely on works by Newton and the supplementary texts clarifying his achievement. He told the Empress that the principles of Newtonian physics were far more important to master than modern languages. And throughout Diderot's "Plan général de l'enseignement d'université" there is more emphasis on mathematics and physics than on any other subject.[25]

Diderot also thought that if arithmetic and "l'ésprit géometrique" were taught children at an early age, this would naturally turn them away from superstition. The culmination of such training would come in the student's third year, by which time he would be ready to apply himself to some "cent abrégés de *la Philosophie de Newton* [the *Principia*]," while "les maîtres" are enjoined to study Gregory's *Astronomiae physicae et geometricae Elementa* (Oxford, 1702).

The rest of the curriculum is no less Newtonian in character and shows dramatically how little progress astronomy and physics had made beyond the high point to which Newton had brought them at the beginning of the century. Thus, in astronomy Diderot recommends John Keill's *Introductio ad veram astronomiam*, one of the earliest expositions of the subject along Newtonian lines as the title indicates. It was first published in 1718, being translated "et pas toujours entendue"[26] by Le Monnier two decades later. In physics the best textbook Diderot could recommend was even older and by the same author, the *Introductio ad veram physicam*, which John Keill

---

his works (*Sobranie sochinenii Didro*). This volume also contains the *Plan d'une université pour le gouvernement de Russie*, which had first appeared in the *Œuvres* in 1826. The version it contains of Diderot's memorandum on Russia is superior to that in Maurice Tourneux's *Diderot et Catherine II*, where the original material was drastically rearranged. In the *Sobranie*, the Russian translation reflects the form and order in which Diderot left his notes. The first volume of the Soviet edition of Diderot's works appeared in 1935, with an introduction by J. K. Luppol, author of an excellent study on Diderot's aesthetics. Even Volume X (*Rossica*) does not, as it now appears, exhaust Diderot's work on Russia. New materials published elsewhere deal for example, with the questions Diderot put to the Academy of Sciences at its session in St. Petersburg on November 1, 1773 (which have been located) and with other matters. Cf. M. V. Krutikova and A. M. Chernikov, "Didro v Akademii Nauk," *Vestnik Akademii Nauk SSSR* (No. 6, 1947), 64–73; S. Kuz'min, "Zabytaia rukopis' Didro (Besedy Didro s Ekaterinoi II)," *Literaturnoe Nasledstvo* (Moscow), LVIII (1952), 927–948. See also M. P. Alekseev, "Didro i russkie pisateli ego vremeni," in *XVIII vek: Sbornik 3* (Moscow and Leningrad: Akademiia nauk, 1958), pp. 416–431.

25. "La théorie precise des temps des verbes," Diderot believed, "ne cède en difficulté aux propositions de la philosophie de Newton." Cf. his "Plan général de l'enseignement d'université," in his *Œuvres complètes*, III, 468.

26. *Ibid.*, II, 460.

published in 1702. This was the first Newtonian text on physics. In addition, Diderot recommends Musschenbroek's lectures on experimental physics, another Newtonian text by the Dutch instrument maker, which belongs to the same era as Keill's two works.[27] By his fourth year, Diderot assumed, a student could master Newton's *Opticks.*

The university Diderot proposed was never built. If it had been, Bruce would have found little in its scientific curriculum to surprise him, for his own library contained better manuals than Keill's and more advanced Newtonian texts on physics than those Diderot recommended.[28] But Bruce would have noticed a very dramatic change in Russian society, a change which occurred largely in Catherine's reign and one she wholeheartedly encouraged but did not in fact bring about: the vast growth in literacy duly reflected in the impressive rate at which new books and journals were published. Consequently, Newton's name is to be found with increasing frequency in periodicals of the Catherinian era, though, amid the popular articles explaining his ideas, only one appears to have been taken from the *Principia* itself.[29] The author, Pankratii Platonovich Sumarokov (1763–1814), was a cousin of the celebrated dramatist of the same name, and the provincial journal which he edited from Tobol'sk, was named appropriately: "Irtysh [being] transformed into Hippocrene." This was one of Russia's earliest provincial journals, published monthly and distributed free of charge by a local merchant, V. D. Kornil'ev, an ancestor of Mendeleev. It carried articles on several luminaries of the age, including Lomonosov, Voltaire, and Condorcet, and the style is surprisingly sophisticated, owing, no doubt, to the fact that some of Sumarokov's collaborators were teachers at the main school in Tobol'sk, the center of the province's educational system.[30]

27. Musschenbroek also became an early convert to Newtonianism. He was a friend of 's Gravesande, who was commissioned by Schumacher on his visit to Leyden to supervise Musschenbroek's construction of some scientific equipment for Peter the Great; see Chapter 7, above. (Cf. also C. A. Crommelin, *Descriptive catalogue of the Physical Instruments of the XVIIIth century, including the collection 'S Gravesande-Musschenbroek,* in the *Rijksmuseum voor de Geschiedenis der Naturwetenschappen at Leyden* ([Leiden], 1951).

28. Diderot was of course well aware of the contributions which his friend d'Alembert had made to certain problems advanced by Newton in the *Principia,* but he considered these too difficult for students to study (except "pour les maîtres").

29. "Kakim obrazom poznaem my rasstoianiia, velichiny, vidy i polozheniia predmetov," in *Irtysh, prevrashchaiushchiisia v Ipokrenu* (Tobol'sk), (September 1789), pp. 34–48. The article is signed P.S., with the notation "Vziato iz osnovanii N'iutonovoi filosofii."

30. Cf. M. K. Azadovskii, *Ocherki literatury i kul'tury v Sibiri* (Irkutsk, 1947), p. 64.

Maxims or principles from Newtonian philosophy appeared in Russian journals before Catherine II's reign.[31] Thus, not long after the publication of *Micromégas*, the *Ezhemesiachnyia sochineniia* gave a sober résumé of the law that enabled the Sirian to gravitate from one planet to another.[32] In another article, on meteors, readers could find Newton's definition of gravity—as well as the assurance that this "attractive force" (pritiagatel'naia sila) was to be found in all bodies.[33]

The most substantial of such articles was a translation from the work of P. Wargentin, the secretary of the Swedish Academy, which explained Newton's theory of tides after giving an instructive historical account of the views held on the subject from antiquity to Copernicus, Galileo, Gassendi, Kepler and Descartes.[34] However, it was probably not a work in prose that most vividly brought Newton's name before Russian readers but a poem, whose author (wrote the translator) has "penetrated into the greatest subtleties of Metaphysics".

He was referring to Pope's *Essay on Man*, which, if Warburton's words are to be believed, was "not a system of *Naturalism*, on the philosophy of Bolingbroke, but a system of Natural Religion, on the philosophy of Newton."[35] This fine distinction did not impress the Holy Synod, which blamed the author for not taking anything "from Holy Writ and the dogma of our Orthodox Church" to support his arguments. Worse, Pope had based his views, "edinstvenno . . . na

31. Cf. *Ezhemesiachnyia sochineniia* (January 1757), p. 456:
«И мне кажется... нельзя удалятся без причины от преизряднаго Невтонианскаго правила, которое учит, что ежели в натуральных вещах какія перемены случаться, оныя между собою сходны, то и причины, от которых бывают, между собою сходны суть.»
This principle (from the *Principia*) is often cited in the second half of the eighteenth century, perhaps in part owing to Voltaire's particular fondness for quoting it.
32. "Razmyshleniia o vozvrate komet, s kratkim izvestiem o nyne iavivsheisiia komete," *Ezhemesiachnyia sochineniia* (October 1757), Part II, p. 331:
«Вообще может небесное тело, привлекаемое своею тяжестію к солнцу, описывая кривой путь, двигаться токмо тремя кривыми линеями...»
33. "O Vodnykh meteorakh ili o prikliucheniiakh ot vodianykh parov v vozdukhe byvaemykh," *ibid.*, Part II, pp. 354–355:
«Но сія притягательнаія сила, которая во всех телах находится, постоянна, она уменьшается и увеличивается в разсуждении количества материи, оныя тела поставляющей, разделенной на квадрат их разстояния.»
34. From the introduction to *Opyt o Cheloveke gospodina Popa*, translated by Nikolai Popovskii in 1754. Owing to the objections of the Holy Synod, the book was not published until three years later (by the press of Moscow University in 1757).
35. *The Works of Alexander Pope Esq., with Notes and Illustrations by himself and others. To which are added, A New Life of the Author, an estimate of his Poetical Character & writings, and occasional remarks, by William Roscoe, Esq.*, new ed., 8 vols. (London, 1847), IV, 25n.

estestvennykh i natural'nykh poniatiiakh.''[36] The poet who translated it into Russian—Nikolai Popovskii, a pupil of Lomonosov—was forced to make several changes, while one of the more literate members of the Synod interpolated his own verses into the translation, with a view to ridding the text of "anything inclining to naturalism." Nevertheless, Russian readers would have recognized the vivid portrayal in the poem of the workings of Newtonian gravitation:

> Взирай, какъ для того трудится естество,
> Чтобъ было съ существомъ въ союзѣ существо.
> Частица малая въ союзѣ со другою
> Теперь къ себѣ влечетъ, то сходится съ иною;
> Взаимной силою къ себѣ другихъ влекутъ,
> И сами въ тотъ же часъ къ другимъ пристать бѣгутъ [p. 40].[37]

It is implacable, this Newtonian universe:

> When the loose mountain trembles from on high
> Shall gravitation cease if you go by? [IV, 127–128].

For,

> Shall burning AEtna, if a sage requires,
> Forget to thunder, and recall her fires?
> On air or sea new motions be impress'd,
> O blameless Bethel! to relieve thy breast? [IV, 122–126].

Which Popovskii renders in these palpitating lines:

> Достойно-ль быть тому, чтобъ Этны нутръ горящій
> Погаснулъ, и умолкъ звукъ съ сѣрою кипящій,
> Для удовольствія ученыхъ лишь людей,
> Что любопытствуютъ знать силу дѣйствій въ ней?
> Чтобъ естество свое вода перемѣнила,
> Или бы воздуха переменилась сила,
> Лишь для того, чтобъ праведный Вефель
> Свободнѣе въ себѣ дыханіе имѣлъ?

36. An account of the trials and tribulations faced by Popovskii in attempting to have his translation published is given by N. S. Tikhonravov, "Istoriia izdaniia 'Opyta o cheloveke' v perevode Popovskogo," in his *Sochineniia* (Moscow, 1898), III, Part I, 88–89. Quotations are from *Opyt o cheloveke gospodina Popa*, tr. Popovskii (Moscow, 1802; reprint of 1757 ed.), p. 9.
37. Cf. *Works of Pope*, ed. Roscoe, III, 9–12.
> See plastic nature working to this end,
> The single atoms each to other tend,
> Attract, attracted to, the next in place
> Form'd and impell'd its neighbour to embrace.
The Russian translation, it should hastily be added, was not from Pope's original, but from Silhouette's version of it in French prose.

Чтобъ въ трусъ земной горы дрожащая вершина,
Противъ правъ тяжести и естества и чина,
Паденье и ударъ сдержала съ звукомъ свой,
За темъ, что близъ горы ты путь имѣешь той? [p. 63].

But the *Essay on Man* could not, at that time, be considered a facile poem. Even Pope, it appears, was unaware of some of the logical implications hostile critics drew from the Newtonian principles the hapless author had innocently espoused. Nor would Russian readers have gained any information from Pope-Popovskii's poem concerning Newton's life or the manner in which he made his discoveries. For such mundane facts they turned to the *Pismovnik*, a work which has been characterized as the first Russian encyclopaedia.

It was written by Nikolai Gavrilovich Kurganov, who like Soimonov began his studies at Bruce's Navigation School, graduating from there to the Naval Academy in St. Petersburg. His progress, especially in astronomy, was thought so remarkable that he was asked to teach the subject himself; eventually, in 1774, he was appointed professor at the Academy of Sciences, where he remained until his death in 1796. He compiled some useful textbooks on mathematical subjects and also did some translations into Russian, of which the most notable was his new version of Euclid.[38] This replaced Andrew Ferquharson's "Newtonian" translation of the *Elements,* but the *Pismovnik* was the only work of Kurganov's to reach a really wide audience. Indeed, according to one eminent Soviet authority, "it is difficult to find a book published in the XVIIIth century to match its popularity."[39]

This is why its allusions to Newton and his contemporaries are of such peculiar interest. The *Pismovnik* (see fig. 46) contains a copious if indiscriminate array of information on history, literature, theology, folklore, geography, philosophy (both natural and moral), and much else. It opens with a section on Russian and Slavonic grammar and ends with a commentary on the Scriptures, which is placed next to some pages devoted to "Considerations on Physics," but the most interesting part for our purposes is the "Didactic dialogues," of which

38. It was published in 1769 (*Elementy geometrii po Evklidu*).
39. Cf. G. P. Makogonenko, *Radishchev i ego vremia* (Moscow: Gosudarstvennoe izdatel'stvo khudozhestvennoi literatury, 1956), p. 102; as the author points out, he is one of the very few scholars who has paid any attention to the *Pismovnik*, of whose existence not many students of the period have been aware. Kurganov's academic career, however, is discussed briefly by V. E. Prudnikov in *Russkie pedagogi-matematiki XVIII–XIX vekov* (Moscow: Gosudarstvennoe uchebno-pedagogicheskoe izdatel'stvo Ministerstva Prosveshcheniia RSFSR, 1956), pp. 102–114.

one—"razgovor o liubomudrii"—discusses various phenomena of nature, such as meteors, comets, tides, lightning, and so on. Another provides a sketch of all the "arts and sciences." It is here that references to Newton and his discoveries are repeatedly made. Not only is he given credit for proving the existence of gravity "clearly and incontrovertibly," but its effect on tides and the orbits of planets is also described in detail.[40] In another passage, Kurganov mentions his optical discoveries: "we are indebted to Sir Isaac Newton for his theory of light, and also for his telescope with mirrors and for many other discoveries in physics . . . This Englishman of wonderful intelligence died in 1727 at the age of 86."[41] Elsewhere in the *Pismovnik,* there are also references to Bradley's "invention of aberration" in the same year, and to Dollond's solution of the problem, as well as to the achievements of Edmond Halley and of Flamsteed.[42]

The *Pismovnik* was first published in 1769; a second edition, with a changed title, appeared in 1777, and the demand was still so insistent several years later that a third edition came out in 1796.[43] By then Newton had been accorded one more testament of recognition—Samuel Horsley's edition, in five volumes, of Newton's collected works (1779–1785).[44] Though not a critical edition and incomplete, its publication was an event of some importance, and it is marked as such in the protocols of the St. Petersburg Academy, which received the final volume on August 22, 1795.[45]

By that date no reputable scientific authority in Russia questioned the validity of the theory of universal gravitation. Aleksandr Radishchev, probably its last serious critic, believed that there were exceptions to Newton's law, a view which may easily be explained by the

40. The first edition of Kurganov's work is entitled *Rossiiskaia universal'naia Grammatika ili Vseobshchee Pismoslovie* . . . (St. Petersburg, 1769). The references that follow are to the edition of 1796 (*Pismovnik, soderzhashchii v sebe nauku rossiiskago iazyka so mnogim privosokupleniem raznago uchebnago i poleznozabavnago veshchesloviia . . . piatoe izdanie . . .*), pp. 206–207. Cf. SK, no. 3374.

41. Kurganov, *Pismovnik*, p. 211.

42. *Ibid.*, pp. 206, 209, 211, 215.

43. There were several imprints to some of these editions.

44. *Isaaci Newtoni Opera Quae Exstant Omnia Commentariis illustrabat Samuel Horsley*, 5 vols., ed. Samuel Horsley, (London, 1779–1785). This is still the fullest edition; a new edition to replace it is not presently deemed feasible—see Professor E. N. da C. Andrade's introduction to the *Correpdondence*, I, xv.

45. Earlier the Academy had also received three copies of Castillon's edition of Newton's *Universal Arithmetick*, which came out in Amsterdam in 1761, but the initiative came from the editor, not the Academy. See *Protokoly zasedanii konferentsii Imperatorskoi Akademii Nauk*, I, 121.

fact that Radishchev still clung to the phlogiston theory, long since disproved by Lavoisier.[46] But the manner in which Radishchev presented his criticisms gives some idea of the veneration now everywhere accorded to Newton. Only in the St. Petersburg Academy did reservations remain regarding Newton's views on light, which, curiously enough, had been the first of his doctrines to gain recognition elsewhere in Europe.[47] The situation was now complicated, however, by the fact that even outside the Academy the ondular and corpuscular theories of light still attracted its partisans, and the conroversy regarding their merits was far from over. However, at the beginning of the nineteenth century even the Academy was forced to come to terms with the *Opticks,* and it was in that period that the first efforts were undertaken to translate the *Principia* into Russian.[48] This growing interest in Newton's work even affected his so-called "fluxionary" method, which S. E. Gur'ev, a brilliant Russian mathematician, attempted to revive.[49]

46. Cf. A. N. Radishchev, *Polnoe sobranie sochinenii,* 3 vols., ed. G. A. Gukovskii *et al.* (Moscow and Leningrad: Akademiia nauk, 1938–1952), II, 147. It was, wrote Radishchev, with "trepidation" that he dared to "disagree in any way with the invention of Newton's mind"—and in so doing, he was certain he was incurring the danger of seeming "nonsensical."

47. Cf. V. I. Chenakal, "Optika v dorevoliutsionnoi Rossii," in TIIE, I (1947), 146ff. There were three other alternatives to the Newtonian theory, the best known being that of George Friedrich Parrot (1762–1852), a former rector of Dorpat University and confidant of Alexander I, who became a member of the Academy of Sciences in 1826 and reorganized its physics laboratory along modern lines (cf. his *Grundriss der Theoretischen Physik in Gebrauche für Vorlesungen* . . . [Dorpat, 1809–1811]). His theory of light had the advantage of being comprehensive, and found much favor in Russian manuals on physics in the first half of the nineteenth century, but was then forgotten. On his work in the Academy of Sciences, see S. I. Vavilov, "Ocherk razvitiia fiziki v Akademii Nauk za 220 let," in *Sobranie sochinenii,* Vol. III (Moscow: Akademiia nauk, 1956), 541–542. The most original of the three belonged to Theodore Grotthuss, who tried to link the behavior of light to electricity. His theory and that of V. V. Petrov, who was similarly prompted by an interest in electric phenomena, are discussed by Dr. Chenakal, "Optika v dorevoliutsionnoi Rossii," p. 147ff., who notes that there was little concern with optics at the end of the eighteenth century in Russia.

48. Unfortunately, the first young man asked to embark on this onerous task at the request of A. N. Muraviev, the rector of Moscow University, apparently lost his mind in the fire of 1812. The Napoleonic invasion claimed another victim: Mikhail Ivanovich Pankevich, also a graduate of Moscow University. He also attempted a translation of the *Principia,* cf. I. A. Tiulin, "Razvitie mekhaniki v Moskovskom universitete v XVIII i XIX veke," in *Istoriko-Matematicheskie issledovaniia,* Vol. VIII (Moscow: Gosudarstvennoe izdatel'stvo tekhniko-teoreticheskoi literatury, 1955), 495. Pankevich's only other known work is his thesis on the steam engine.

49. On Gur'ev, see the excellent essay by Dr. Iushkevich in TIIE, I (1947), 219–268. Gur'ev's work belongs mainly to the nineteenth century, and therefore falls outside the scope of the present study. It is here worth noting, however, that Gur'ev was the first to give an analysis in Russian of Newton's "fluxionary" method. This he did in a paper

describing the differential calculus, presented to the Academy of Sciences on July 16, 1813. Cf. "Kratkoe izlozhenie razlichnykh sposobov iz'iasniaia differentsial'noe ischislenie," which appeared in the *Umozritel'nye issledovaniia Imperatorskoi Akademii Nauk* (St. Petersburg, 1815), IV, 159–212. (The *Umozritel'nye issledovaniia* were a translation of the *Nova Acta Petropolitanae* and were published partly as a result of Gur'ev's insistence on the need to make its contributions accessible to Russian readers. Five volumes appeared [1809–1819]). Here Gur'ev reviews the methods used by Leibniz and Euler, as well as those of Newton, Maclaurin, and d'Alembert, but his own sympathies are revealed by the biased and one-sided exposition he gives of the Leibnizian formula. Indeed, he tries to prove that the assumptions underlying Newton's "fluxions" were not only borrowed by d'Alembert, who acknowledged the debt, but also by Euler, who did not. "I believe that our renowned Euler," he wrote, "drew on the same places for his method of differential calculus . . . for it hardly differs from Newton's exposition of it, as described in this introduction . . ." (Cf. "Kratkoe izlozhenie," IV, 175. Gur'ev is referring to Newton's introduction to the *Tractatus de quadratura curvarum* which was published, together with the *Opticks*, in 1704. An English translation of this work, with extensive notes, was published by John Stewart in 1745 (*Sir Isaac Newton's Two Treatises of the Quadrature of Curves, and Analysis by Equations of an infinite Number of Terms . . . London*). A facsimile is also included in *The Mathematical Works of Isaac Newton*, 2 vols., ed. T. Whiteside (New York: Johnson Reprint Corporation, 1964), I, 3–25. See also Guido Castelnuovo's *Le Origini del calcolo infinitesimale nell' era moderna con scritti di Newton, Leibniz, Torricelli* (Milan: Feltrinelli, 1962), pp. 127–162, which contains the most recent imprint of the 1704 treatise in Italian. (This is a reprint of the 1938 edition.)

Венец, наукой соплетеный,
Носим Невтоновой главой;
Таков, себе всегда мечтая,
На крыльях разума взлетая,
Дух бодр и тверд возможет вся.

Radischchev

Conclusion

The aim of this study has been to explore the historical beginnings of Newton's influence in Russia. The vicissitudes of his reputation in the nineteenth century—the reaction against the tradition he represented at the end of the Alexandrian era, the subsequent counterattacks by Kantians, Hegelians, and the followers of Schelling and Oken, his "restoration" with the ascendancy of the positivists in the 1860's—all this is beyond the scope of the present essay. According to Arago, every momentous scientific discovery passes through three well-defined phases. First, it's truth is denied. Then it is shown to be impossible or implausible because it is said to contradict religion or morality. In its final phase, it is dismissed as self-evident, and it has been suggested that Newton's reception in Russia probably followed the same pattern.[1] This, of course, is the type of generalization that may itself be traced to the philosophes, but, whatever its origin, the historical evidence does not quite fit.

Far from being proscribed in the opening phase of his influence in Russia, Newton during the latter part of his lifetime was regarded there by the few who knew of him with a veneration which his name had still to gain in France, Italy, and Germany, where established scientific orthodoxies provided a more formidable challenge to Newtonian doctrines. Moreover, Peter the Great's journey to England a decade after the publication of the *Principia* gave Bruce the oppor-

1. See T. P. Kravets, "N'iuton i izuchenie ego trudov v Rossii," p. 320.

231

tunity to make Newton's name known in Moscow and St. Petersburg many years before his theories were taught in other East European countries. After 1698 there was growing awareness of Newton's achievement in Russia, unchallenged till the years immediately following his death. This is curiously confirmed by the Cyrillic transformations of Newton's name—perhaps the most sensitive index of his fortunes in eighteenth-century Russia.

Owing largely to Lomonosov's transcription of it, Academician S. I. Vavilov and other Newtonian scholars in the Soviet Union have commonly held that he was known as Невтон throughout the eighteenth century, Ньютон or Нейтон being a nineteenth-century form adopted as the authentic pronunciation of the Englishman's name became familiar. Yet this is not strictly true. In the *Kosmotheoros,* our first printed reference to Newton in Russian literature, Newton appears in his modern incarnation as Ньютон. Cantemir, who only begins to mention him in his work after his arrival in England, also used this correct transliteration. As for the third spelling, this is to be found concurrently with the second in the first half of the nineteenth century, but not, it seems, later than that. Невтон happens to be the form usually used in the middle and second half of the eighteenth century (and much later in poetry) because German was then largely the medium through which Newton's name was exposed within the St. Petersburg Academy. Significantly enough, men like Gur'ev returned to Нютон or Ньютон, the form of the name first used by Bruce.

The semantic vagaries of gravitation in Russian, after the concept was introduced into the language, proved to be very similar in the sense that, at the end of the century, the term used in the Petrine era, and then abandoned, was adopted again by several writers.[2] The history of Newton's reputation also turned out to be no less idiosyncratic. Had Lomonosov and some of his contemporaries in the Academy of Sciences spent their formative years in England rather than Germany (as Prince Antiokh Cantemir had done), Newton's doctrines would have gained earlier appreciation in St. Petersburg.

If this helps to account for Lomonosov's opposition to Newtonianism, it does not explain it. "Nothing is so instructive," Peter Kapitsa has noted, "as the floundering of a genius."[3] Lomonosov's failure to

2. See Appendix II.
3. P. L. Kapitsa, "Lomonosov and World Science" in *Collected Papers of P. L. Kapitsa,*

understand Newton's doctrine of gravitation was linked to his own most startling and extraordinary success: his anticipation of the law of the conservation of energy (that is, for the conversion of kinetic energy into thermal energy). This alone would have brought Lomonosov European recognition had his work been better known outside Russia, but the philosophical concept which facilitated Lomonosov's remarkable discovery also held him back. He imparted universality to his concept in the belief, as his *Oration on the Origin of Light* makes clear, that in Nature there exists only one kind of interaction between bodies (osculation). This is why he rejected the doctrine of attraction at a distance (as he was also to deny the concept of electric interaction).

His opposition to Galileo's great discovery that the mass of a body is proportional to the force of gravity, or simply to its weight, sprang from the same source. Newton was able to demonstrate this law by a very simple experiment conducted in his room at Trinity. By suspending two pendulums of identical length in the doorway, he showed that they always oscillated isochronously independently of the substance suspended. This could only occur if the mass of the body were exactly proportional to its weight.[4]

Yet Lomonosov remained ardently opposed to this principle, and between 1748 and 1757 we find him again and again stubbornly opposing what Newton's elegant and convincing experiment had long ago confirmed. Thus, as late as 1755 Lomonosov proposed to the St. Petersburg Academy that the problem for its next prizewinning competition should be the experimental verification "of the hypothesis that the matter of bodies is proportional to weight." At that date the posing of such a problem, with its implied rejection of Newton's views on the subject, already required either obstinacy or a certain boldness. The topic was opposed within the Academy, and Euler was brought in to arbitrate. On this occasion Euler, who ordinarily supported Lomonosov, declined, and after 1757 Lomonosov never raised this problem again. That year may therefore perhaps be taken as marking the general acceptance of Newton's authority among the St. Peters-

---

Vol. III, ed. by D. ter Haar (Oxford: Pergamon Press, 1967), p. 179. This is a most illuminating and interesting essay.

4. Academician Kapitsa suggests that Lomonosov may not have known of Newton's experiments with the pendulum; indeed, as he so rightly observes, they are not actually mentioned in any of Lomonosov's published writings. But in view of the evidence presented above of his knowledge of Newton's other work (see Chapters 17, 18, and 19) this seems unlikely.

burg Academicians. This was one issue on which virtually all of them, except Lomonosov, took the same side.

It had taken almost sixty years in Russia, from the time that Newton's name was first heard there, for his doctrines to reach this ascendancy. Algarotti had sung their triumph "nel vasto Imperio delle Russie" prematurely. But when we consider the total absence in Russian society at the end of the seventeenth century of even the most rudimentary formal training in science and mathematics, the historical record must seem remarkable by any yardstick we choose to apply. In one respect it is unique. Russia was the first and only country in which the birth and rise of modern physics was almost coterminous with the development of Newtonian science. In a sense, therefore, the historical epilogue to the publication of the *Principia* in 1687 is Peter the Great's journey to England a decade later. That visit is one notable point in a trajectory that leads from Kepler's laws of planetary motion to Newton's inverse-square law and on to interplanetary flight in our own era.

The backwardness of Russian education on the eve of the Petrine reforms brought with it one dubious advantage. Had Bruce returned in 1699 to a country with a prevailing and deeply rooted scientific orthodoxy, be it Cartesian or Leibnizian, his championship of Newtonian principles would have met with immediate rebuttal. But a *tabula rasa* has at certain times in the historical process (so Leibniz believed) something to be said for it. The curious point is that in Russia opposition to Newton's doctrines was strongest not when they were first introduced in the reign of Peter the Great, but in the middle decades of the century. In part this was due to the growing influence of the Church in the reign of Elizabeth and the long-lasting hostility of the Holy Synod to the ideas associated with the new science. But the Greek Orthodox clergy was rarely interested in the metaphysical polemics launched, so Voltaire asserted, by the religious-confessional interest against the materialist implications of Newtonian philosophy. Freethinkers were proscribed in St. Petersburg and Moscow as they were elsewhere on the Continent and usually more harshly, but to the Russian clergy it would have made no sense to single out Newtonians for being, in the graphic phrase of the Marquise du Châtelet, "des vraies hérétiques".

Thus, that phase of religious opposition which Arago discerned as the stumbling block to any great scientific discovery of the past affected Newton's theory of universal gravitation no less and no more than the rise of heliocentrism and other scientific ideas toward which Russian clerics were sometimes acutely hostile and at other times strangely indifferent. But they could always count on the prejudice and antipathy of the populace to science and learning, which Lomonosov fought with such verve, wit, and courage in his verse.

The battles over the validity of Newton's theories in the St. Petersburg Academy were, as one might expect, an extension of the larger struggle being waged elsewhere by his followers on the Continent with the Cartesians and the school of Leibniz. This is another reason why Newton's triumph in the Russian Academy was long delayed, in spite of Bruce's and Andrew Farquharson's priority in introducing his ideas into the Navigation School and the Naval Academy. Nor was the influence of men such as Daniel Bernoulli, Josiah Weitbrecht, and Joseph de l'Isle sufficient to offset the authority enjoyed by Leonard Euler and Lomonosov, both of whom continued to oppose the doctrine of attraction at a distance long after it had been vindicated not only with reference to celestial bodies, but also to other phenomena such as electricity and magnetism.

Yet, as it turned out, this did not greatly affect the popularization of Newton's ideas, which was initiated apart from the Academy. Thus, Cantemir's attempt to launch "le verita di Newton" on its course "Dal Brittano Tamigi à Russi lidi" was prompted solely by the evident success of Francesco Algarotti's and Voltaire's efforts on Newton's behalf, not by the St. Petersburg Academy's attitude toward them. In other words, after the philosophes discovered Newton, some of the implications of Newtonianism were revealed to the Russian public through various genres bearing no direct relationship to the scientific controversies within the Academy in the first half of the century. Whether in the form of Popovskii's translation of the *Essay on Man*, or of Kurganov's *Pismovnik,* the popularization of Newton became part of the larger intellectual upheaval which the so-called "scientific revolution" had provoked in Western Europe.

There, however, the ground for the Newtonian synthesis had been prepared by the long-lasting controversy about the motion of the earth, by the discoveries of Kepler, the trial of Galileo, the martyrdom of Giordano Bruno—all of which passed largely unnoticed in Russia, which, when Copernicus was born, was still under the Tartar yoke. This is one reason why Russian intellectual history of the period is so rich in extraordinary contrasts. As late as 1815 a priest could write a work entitled *The Destruction of the Copernican System*,[5] attacking mathematicians and scientists for their apostasy in lending support to the heliocentric theory. Copernicus, Kepler, and Galileo fare worst, but even "the great Newton" comes in for his share of criticism. His works, writes the author, are so profound that Newton must surely have followed Voltaire's advice(!) by making his writings as obscure as possible—advice Voltaire is said to have given all aspiring philosophers.

To understand such anachronisms it is necessary to recall that the eighteenth century Enlightenment had not yet run its course in Russia, which also accounts for the fervor with which battles were fought that, in the West, had already been considered won. But what mattered, in retrospect, was the appeal of the new outlook: science promised success where Revelation was thought to have failed. This is why Newton's enormous influence rose above and beyond his renowned discoveries and inventions. He represented a method of human enquiry which became identified with the dominant tradition of modern science. "No one could contradict him," Radishchev would write in his philosophic treatise *On Immortality* (1792), "since respect and admiration for his discoveries is something we imbibe almost with our mother's milk."[6] In another ornate encomium written a few years later and introduced, significantly enough, into a textbook on physics intended for seminarians, Newton emerges almost as a secular saint:

Many years were spent by Nature preparing matter enough for Newton's swift and all encompassing mind: she created Kepler, Huygens, and Galileo, thereafter only to embody all their minds in one, and rearing Newton's. In his guise all embracing Reason manifested itself in all its power and maturity; Nature hid nothing from him, and once she trusted him, gave her favourite key to her mysteries. He opened the door of truth and his very first words changed the laws of the world. He weighed the sun and the moon, and was the first to give gravity as the explanation of universal phenomena. Thus, [he] let loose a flood of light on the system of the world, and conquered all motion by making it obey one law.[7]

5. *Razrushenie Kopernikovoi sistemy, rossiiskoe sochinenie* (Moscow, 1815). The late Dr. V. P. Zubov ascribes it to a priest by the name of Sokol'skii; cf. *Istoriografiia estestvennykh nauk v Rossii* (Moscow: Akademiia nauk, 1956), p. 195.

6. Radishchev, "O cheloveke, o ego smertnosti i bessmertii," in his *Polnoe sobranie sochinenii*, II, 82.

7. M. M. Speranskii, "Fizika vybrannaia iz luchshikh avktorov, razpolozhennaia i dopolnennaia nevskoi seminarii filosofii i fiziki uchitelem Mikhailom Speranskim 1797 goda v Sanktpiterburge," in *Chteniia v Imperatorskom Obshchestve istorii i drevnostei*

A similar image of Newton is often met in Russian literature throughout the first half of the nineteenth century, when his achievements entered the period of unquestioned acceptance. But there was an essential distinction, not of degree but of kind, between the adulation Newton received and the recognition accorded other great scientists. As Laplace put it, "on ne trouve qu'une fois un système du monde à établir."

rossiiskikh pri Moskovskom Universitete, Vol. II (Moscow, 1871), 3. Speranskii used this as the text for his own lectures in physics.

APPENDIXES
BIBLIOGRAPHIC NOTE
BIBLIOGRAPHY
INDEX

APPENDIX I

Draft of Newton's Letter to Menshikov

Newton wrote at least three versions of a reply to Menshikov, of which
the following, a fair copy, has not previously been published:

47. Rough draft of Newton's letter to Menshikov.

The variants between the first rough draft of Newton's letter to Menshikov and the version reproduced above are minor, i.e., "Czariensem Majestatem" instead of "Czariensem *suam* Majestatem."

"Excellentiam vestr*am*" instead of "Excellentia vestra."

"Affectu" instead of "affect*a*"; "nostr*am*" instead of "nostra"; "adhuc" instead of "amplius"; "coetus nostros prorogatos renovare licuit," instead of "nostros renovare liceret." "Electionem" with a capital instead of without. Also, the commas are omitted in the rough draft. The most noticeable difference is in the salutation, which is much extended in the version reproduced above; there is also a difference of date. The rough draft was written on October 21, 1714; the second copy on October 25.

There exists a third copy of Newton's letter, which in 1943 the Royal Society presented to the Academy of Sciences of the USSR with a copy of the first edition of the *Principia*. It, too, contains several deletions, and Oranienbourgh, instead of Oranienbaum, is repeated in all the three versions. The date of Copy No. 3 (the one presented to the USSR Academy) is the same as that of Copy No. 2. Presumably, there must be a fourth version, the fair copy Newton actually sent off to Menshikov. Copy No. 3 was presented by the Royal Society as "an original draft of a Letter by Sir Isaac Newton . . . to . . . Menshikov," and it is this version, never any other, that is cited in Soviet sources. Thus, S. I. Vavilov includes a much-reduced facsimile in the second edition of his biography of Newton (*Isaak N'iuton*, [1945]; the first edition was published to coincide with the tercentenary of Newton's birth in 1943. The second edition was subsequently included in Vavilov's *Sobranie sochinenii*, Vol. III).

The "Royal Society's copy" of the letter to Menshikov, which may be seen in the Archives of the USSR Academy of Sciences, was also reproduced in M. I. Radovskii's "N'iuton v russkoi nauchnoi literature," on page 101 of his *Iz istorii anglo-russkikh nauchnykh sviazei* (1961). Menshikov's letter to Newton (reproduced above), however, is never mentioned in Soviet sources. Except for matters of form (Menshikov's title and so forth), there are no substantial differences in the text of Copy II and Copy III. Though the letter appears to have been particularly intractable—hence, Newton's three drafts—it was not unusual for him to recopy and redraft his letters; indeed, this was Newton's habit also with his scientific and other writings, which are frequently to be found in many almost identical versions in Newton's own hand.

APPENDIX II

Gravitation and Language

Though the *idea* of gravity was not strictly Newton's, it was not apparently familiar to Russians before the Petrine era. Several classical writers refer to gravity in the general sense (as Newton himself points out in the *Principia*), and it is in this classical sense that the idea is defined in Fedor Polikarpov's trilingual dictionary of 1704, where the Aristotelian notion of "vim habens attrahendi" (in Greek, *elutikos*) is translated as привлачительный, and привлечение and привлечен, are rendered as *attractio* and *attractus*.[1] But in Russian literature, the idea became exclusively associated with the Newtonian enlightenment, Cantemir being the first to extend its use in this context. In annotations to his *Fourth Ode,* he translated it as взаимная сила, but he must have realized that this could also be taken to mean force as *impulse* in the Cartesian sense, and was therefore inadequate with reference to the Newtonian theory. Hence, in the *Seventh Satire,* written approximately three years after the *Fourth Ode,*[2] he solved the problem by using a Russified version of the Latin—аттракція, a neologism which never became part of the language.

Lomonosov referred to gravitation as:

притягательная, притягающая, ог привлекающая сила.

Close though this may seem to contemporary usage, none of these

1. Fedor Polikarpov, *Leksikon treiasychnyi sirech rechenii slavenskikh, ellinogrecheskikh i latinskikh, sokrovishche iz razlichnykh drevnikh i novykh knig sobranoe i po slavenskom alfavite v chine razpolozhennoe* (Moscow, 1704), no pagination.
2. See V. J. Boss, "La Quatrieme Ode de Kantemir et *l'Italia Liberata* de Gian Giorgio Trissino," in *Cahiers du Monde Russe et Soviétique,* IV (No. 1–2), 47–55.

terms established their place in the language either.[3] The translator of Voltaire's *Micromégas* used his own term:

Наш путешественик знал довольно уставы натуральной тягости, и всю силу привлекательную и отдалительную . . ."[4]

Even a quarter of a century after this translation was made (1755–1756), the terms used to correspond to gravitation and gravity were still flexible. In 1789 Alexander Radishchev speaks of:

существенность притяжательности
and
сила притяжения. Притяжательность почитают силою,
he writes

в веществе вкореннуто.[5]

And again:

говорят, что есть в телах сила скрытая и сокровенная, между ими притяжательность производящая.[6]

In French, as in English, "attraction" could be used in the figurative sense. The same is true of "Anziehen" and "Anziehung" in German,[7] and it appears that in Russian, too, either from translation or by the same indigenous process, the meaning of gravitation in the physical sense also acquired a literary connotation, which is hardly surprising when it is recalled what impact the *idea* of gravity had both on the poetic imagery of the age and on its thought. (Immanuel Kant, for example, thought of sympathy as a principle analogous in the moral world to gravitation in the physical). Radishchev, for instance, writes:—

3. See especially Lomonosov, "Rassuzhdenie o tverdosti i zhidkosti tel," *Polnoe Sobranie,* III, nos. 3, 6.

4. E.g., "Notre voyageur connaissait merveilleusement les lois de la gravitation, et toutes les forces attractives et répulsives," Cf. "Mikromegas, povest' filosofskaia," i.e., *Puteshestvie zhitelia svezdy Siriia v planetu Saturna,* in *Ezhemesiachnye sochineniia,* (January 1756), p. 31 (unsigned; the translation is probably by A. R. Vorontsov); and "Micromégas Histoire Philosophique," in Voltaire, *Oeuvres complètes,* XLIV, 153.

5. A. N. Radishchev, "Zhitie Fedora Vasil'evicha Ushakova," in A. V. Zapadov and G. P. Makogonenko, *Russkaia proza XVIII veka,* 2 vols. (Moscow and Leningrad: Gosudarstvennoe izdatel'stvo khudozhestvennoi literatury, 1950), II, 73.

6. *Ibid.,* p. 72.

7. Cf. Jakob Grimm & Wilhelm Grimm, *Deutsches Wörterbuch* (Leipzig, 1854); and *Le Cellarius françois ou méthode très facile pour apprendre . . . les mots les plus nécessaires de la langue françoise . . . ,* new ed. (Moscow, 1782).

Но как скорбь и отвращенне от зла и притяжательность веселия суть равно всеобщие.[8]

Sumarokov illustrates this use more clearly still:

Так ты червонцев то себе не воображала: какой у них вид... и какая у них притягательная сила?[9]

Привлекательная сила is one of the terms used by A. N. Krylov in his translation of the *Principia* in 1914, and we find that Karamzin was familiar with this usage a century earlier:

Он производил сей феноменон привлекательной силы.[10]

Simultaneously, however, Karamzin also used the term first used by Antiokh Cantemir:

закон привлекательности или аттракции,

which shows how fluid the scientific vocabulary remained *even as late as 1798*.[11] And it shows, too, how subtle and complex the vagaries of a scientific idea can be before it becomes a part of everyday speech.

8. Radishchev, "Zhitie Fedora Vasil'evicha Ushakova," p. 59.
9. A. P. Sumarokov, "Opekun" [written in 1765], *Sobranie dramaticheskikh delanii* (Moscow, 1765), I, 51.
10. N. M. Karamzin "Panteon inostrannoi slovesnosti," in *Perevody Karamzina*, 3rd ed. (St. Petersburg, 1835), IX, 165.
11. *Ibid.*

BIBLIOGRAPHIC NOTE[1]

Very little is known of Bruce's early life. Indeed, there is no biography in any language. A contemporary account, written before Bruce had reached his highest rung in the Tsar's service, describes him as: "ein stiller/verstaendiger und fleissiger Mann/den der Czaar wegen seiner guten treuen Dienste und Erfahrenheit in der Mathesi, Artillerie und Kriegs-Wesen/ zum G o u v e r n e u r von Novgorod/ und Praesidenten von der Puschkarskoi Pricasse, das ist von der C o n n e t - a b l e r Cantzeley, wo dasjenige/ was zur Artillerie, Geschuess/ und andern Gewehr gehoeret/ befoerdert wird/ d e n o m o n i r e t" (from *Des Grossen Herrens/Czaars und Gross Fuerstens von Moscau/ Petri Alexiewiz, des gantzen grossen/kleinen und weissen Reusslandes Selbsthallers/ etc., etc., etc.* Leben und Thaten aus besonderen Nachrichten beschrieben/ mit schœnen Kupfern gezieret/ in Fiven Theilen von F.H. v. L. [Franckfurt-Leipzig, 1710], pp. 178–179).

The book from which this passage is taken, now a bibliographical rarity, is one of the very few extant sources of the Petrine era that even attempts a description of Bruce's character. Bruce's connections with Flamsteed, Halley, and Newton have not been described before.

Of the many accounts in several languages of Peter the Great's first journey to Europe, A. M. Bogoslovskii's (originally published in 1941) is still the most comprehensive. Certain errors, such as his understandable confusion over the identity of the "Ivan Kolsun" who trained Bruce, also vitiate the good essay by A. I. Andreev on Peter's sojourn in England, which appeared in 1947. He came closest to spot-

1. When full citations are not given for works mentioned in this note, they appear in the list of references that follows or in the abbreviations listed at the beginning of the volume.

ting the connection between Peter the Great and Newton, but neglected Bruce.

"Collonel Bruce" is mentioned in Volume IV of Newton's *Correspondence,* published in 1967, but his identity is not revealed. Newton's famous allusion to being "teezed by forreigners about Mathematical things" in his puzzling letter to Flamsteed (IV, No. 601) acquires fresh overtones in light of Flamsteed's angry gloss to that letter (which refers to Colson and, by implication, to Bruce). It is interesting to recall in this context that Bruce later corresponded with Leibniz, but again, little relevant material has survived. More evidence concerning Bruce's connections with Newton, Flamsteed, and Halley will no doubt emerge with time. I strongly suspect that he knew David Gregory.

The inventory of Bruce's library, which has such a crucial bearing on the present work, has never been transcribed. The USSR Academy of Sciences intended to do so in the 1930's, but the project was either abandoned or left in abeyance. The Soviet copy of the first edition of the *Principia* cited in Chapter 3 has not been described before; to the best of my knowledge it is the only copy of that edition in the Soviet Union (apart from the copy presented by the Royal Society in 1943).

There has been no detailed study of Newton's influence in eighteenth-century Russia in any language. Indeed, much as we now know about the dissemination and impact of Newton's doctrines in Western Europe, at this point more information is readily available about their influence in Japan than any single country in Eastern Europe. This surely needs to be corrected.

The first attempt to raise the problem of Newton's influence in Russia came in 1943 with the third centenary of his birth. In a remarkable symposium organized in his honor by the USSR Academy of Sciences (S. I. Vavilov, ed., *Isaak N'iuton 1643–1727*),[2] two papers were peripherally concerned with the eighteenth century: T. I. Rai-

2. For an account of the trying circumstances surrounding the celebration of the third centenary of Newton's birthday in the Soviet Union, see M. S. Filippov, chief ed., *Istoriia Biblioteki Akademii nauk SSSR (1714–1964)* (Moscow and Leningrad: Izdatel'stvo "Nauka," 1964), p. 441. The Newtoniana exhibited at the tercentenary celebrations in Moscow in 1943 were brought from Leningrad, then under siege. The books and documents were loaded onto a barge, which traversed Lake Ladoga under enemy fire, where the valuable cargo had to be reloaded onto a train, and it eventually reached the Russian capital in time for the festivities in January (the exhibition is described in the second chapter of this most recent and most comprehensive history of the Academy Library).

nov's "Newton and Russian Science"[3] and T. P. Kravets' "Newton and the Study of His Works in Russia" (pp. 312–328).

Rainov's paper (pp. 329–344) deals with three outstanding Russian physicists: Lomonosov, Mendeleev, and A. N. Krylov. (It was Krylov who first successfully translated the *Principia* into Russian). Only the four-page section on Lomonosov falls within the chronological scope of the present work, however, and its usefulness ended with the publication of Lomonosov's collected works (beginning in 1950), for they revealed that Lomonosov was far more familiar with Newton's writings than Rainov, and others, had been led to suppose.

The short survey by Kravets, who later became a corresponding member of the USSR Academy of Sciences, begins with an allusion to the discussion of the problem taken from the *Principia* at the opening session of the St. Petersburg Academy and then leaps to the middle of the eighteenth century and the dispute over Newton's lunar theory (resolved when Clairaut's treatise was crowned in St. Petersburg). After referring briefly to Euler and Lomonosov, Kravets then turns to the nineteenth century, repeating some of the points made by Rainov.

His essay was criticized for its omissions by A. P. Iushkevich, himself a distinguished Newtonian scholar (see "Obzor sovetskoi iubileinoi literatury o N'iutone," in TIIE, I (1947), 440–455, but there was no attempt to correct this until 1957, when the late M. I. Radovskii published an article on "Newton and Russia," in the *Vestnik istorii mirovoi kul'tury* (No. 6, 1957), pp. 96–106. Radovskii was able to profit from A. I. Andreev's essay on Peter the Great's journey to England ("Petr v Anglii v 1698g."), which appeared a decade earlier, as well as from Vavilov's brief discussion of Newton's epistle to Menshikov (see Appendix I)—a copy of which was presented to the USSR Academy of Sciences by the Royal Society on the occasion of the tercentenary. Radovskii then extended the scope of his article in his book on Anglo-Russian connections (*Iz istorii anglo-russkikh nauchnykh sviazei*, pp. 61–113), but this, too, is largely devoted to the nineteenth and twentieth centuries.

Before his death T. P. Kravets returned to the problem he first posed, in however shadowy a form, in 1943. An essay carrying the same

---

3. I.e., "N'iuton i russkoe estestvoznanie" (rather curiously translated in the bilingual table of contents as "Newton and the Natural History in Russia" [*sic*]).

title as the paper he gave that year opens his recently published collection of shorter pieces on physics, *Ot N'iutona do Vavilova,* which appeared in 1967 (pp. 1–30). According to the editor of this book (p. 377, n. 1), this essay is a reprint of the paper given during the war, but it is in fact a more polished and amended version, though the factual contents remain unchanged. The following extract may interest readers of the present volume:

Блеск имен Галилея, Гильберта, Ферма, Паскаля, Ньютона, — дошли ли его лучи до Москвы? — Увы, нет! Только в 1703 г. вышла в свет первая на русском языке энциклопедия школьных знаний по математике — знаменитая «Арифметика» Магницкого. А это было через 16 лет после выхода НАЧАЛ...

И тем не менее невозможное и невероятное осуществилось: в России заговорили о «Невтоне» и его творениях, стали спорить об их значении, сомневаться в них и в конце концов убеждаться в их всепобеждающей мудрости. Произошло это потому, что в 1725 г. по мановению Петра Великого в Санкт-Петербурге открылась Академия наук.[4]

In reality, as we have seen, the early discussions in the Academy were not conducive to the early acceptance of Newton's doctrines. The tradition initiated by Bruce found its real successor in Prince Antiokh Cantemir, rather than the Academy proper. Cantemir is not an obscure figure. Apart from Grasshoff's recent monograph (and an earlier French biography), his life and poetical works have been examined by Russian, Soviet, and other scholars. His six and a half years in England, however, have eluded close scrutiny, and his activity as a popularizer of Newtonian science and thought has, therefore, been wholly neglected.

Andrew Ferquharson or 'Farvarson' is better known by repute in Russian sources than through any study of his writings. Until more are recovered, it will be difficult to disentangle satisfactorily the legend from his real achievements in the Naval Academy. Mr. W. F. Ryan has recently suggested that a Russian manuscript on navigation of 1703, presently in the British Museum, originated with Ferquharson. (See the forthcoming *Trudy* of the XIII International Congress of the

---

4. Translation: "Galileo, Gilbert, Fermat, Pascal, Newton—did the brilliant rays emanating from these names reach Moscow? Alas, no! Only in 1703 did the first compendium of elementary mathematics appear in Russian—the celebrated *Arithmetic* of Magnitskii. And that was sixteen years after the publication of the *Principia* . . .

And yet the impossible and unlikely occurred: in Russia they began to speak of "Nevton" and his works, to discuss them, debate them, and question them. And at last, to be convinced by their incontestable truth. This happened because Peter the Great had willed that an Academy of Sciences be opened in St. Petersburg."

History of Science in Moscow. The paper was delivered in Section VI, B, August 22, 1971). The tantalizing connections of Ferquharson's work with Newton's name have hitherto passed unnoticed.

Similarly, the controversies relating to Newton in the early years of the St. Petersburg Academy have so far escaped attention. Some of Bilfinger's writings, essential for their understanding, are not, oddly enough, to be found in the Library of the USSR Academy of Sciences in Leningrad. A paper given by L. S. Minchenko at the XIII International Congress (Section VI, B, August 23, 1971) covered some of the ground, but its scope was necessarily narrower.[5] On the other hand, the sources for the celebrated dispute over the configuration of the earth, described in Todhunter's invaluable work, are more easily accessible. The same is true, with some reservations, of the disputes over electricity and "vis viva." A paper on the earlier and "non-Russian" phase of the latter dispute (involving William 's Gravesande and John Bernoulli) was given by Dr. Carolyn Iltis, also at the XIII International Congress (Section VI, B, August 19, 1971).

G. W. Richmann's work in St. Petersburg has not received the attention it deserves in languages other than Russian. The *Trudy po Fizike* contains Russian translations of several manuscripts he left unpublished when his life was so tragically and prematurely cut short by a scientific experiment. F. U. T. Aepinus, who later settled in Russia, is of course familiar to historians of science, but not the Russian period of his life. A new edition of his *Tentamen Theoriae Electricitatis et Magnetismi,* already available in a Soviet translation, is presently being prepared by Professor R. W. Home of the University of Melbourne.

On Lomonosov, though he is still perhaps not as well known as he should be outside the Soviet Union, the sheer quantity of Russian books and other items relating to him is copious. However, the only article to be specifically devoted to Lomonosov and Newton is P. S. Kudriavtsev's "Lomonosov i N'iuton," in which the author attempts to establish Lomonosov's materialism. Allusions to Newton are secondary. Some of the most interesting writings on Lomonosov belong to the late president of the USSR Academy of Sciences, S. I. Vavilov. His works on the history of science are listed in TIIE, IV (1952), 18–30, which Vavilov helped found.

5. "N'iutonianskie i kartesianskie idei v rabotakh po fizike peterburgskikh akademikov (1726–1746)."

Besides writing a biography of Newton and translating his *Opticks* and *Lectiones Opticae* into Russian, Vavilov also edited the *Polnoe sobranie sochinenii* of Lomonosov, Vols. III–X. Hence it became possible for the first time to form some accurate estimate of Lomonosov's total achievement. His more important writings on physics and chemistry have recently been translated by Professor Henry M. Leicester and published by Harvard University Press (1970).[6] The scientific work Lomonosov wrote in Latin is also available in the original in the *Polnoe sobranie sochinenii*. This also contains all the relevant material pertaining to Lomonosov's Newtonian telescope and the "Tubus Nyctopticus modo Lom.N." Vavilov's claim that Lomonosov anticipated the principle heretofore associated with William Herschel has not, to the best of my knowledge, been examined in other than Soviet writings.

The subject of the last chapter (Newton's reputation in the Catherinian era) has been untouched. Clearly, the picture of Newton's achievements and of his place in history formed at this time in Russian literature owed much to the image cultivated by Voltaire and the "philosophes," but this could not be described here with any of the complexity the historical record suggests without going far beyond the scope of the present study.

Finally, a few words on more general sources. There is no historical bibliography of eighteenth-century Russian science. In monographic literature (such as P. N. Pekarskii's *Nauka i literatura v Rossii pri Petre Velikom*). Newton is hardly mentioned. The second volume of that erudite work (*Opisanie slaviano-russkikh knig i tipografii . . .*) has largely been superseded by the *Opisanie izdanii grazhdanskoi pechati . . .* and the *Opisanie izdanii napechatannykh kirillitsei . . .* , both edited by the late P. N. Berkov for the USSR Academy of Sciences. These excellent reference works, together with the *Svodnyi katalog* (SK) have become indispensable to students of the period, but, since Newton's scientific writings did not actually appear in print in eighteenth-century Russia, they are not of primary use.

The late V. P. Zubov's *Istoriografiia estestvennykh nauk v Rossii* deals with the eighteenth century only in part, and seldom mentions Newton in that period. The same may be said of the first volume of Alexander Vucinich's *Science in Russian Culture*, which has the en-

6. I.e., *Mikhail Vasil'evich Lomonosov on the Corpuscular Theory.*

viable distinction of being the first full-length book in English on the history of science in Russia. A. P. Iushkevich's excellent *Istoriia matematiki v Rossii do 1917 goda* (Moscow, 1968), which is certainly the most detailed historical study of its kind, devotes less than a page to Bruce, two to Ferquharson, and none to the influence of Newtonian mathematics in Russia before Leonard Euler.

Institutional histories, such as that by Pekarskii or Sukhomlinov's study of the St. Petersburg Academy, are more useful in this respect, though only in a general sense. Even the new *Istoriia Akademii nauk SSSR*, Vol. I, *1724–1803*, edited by K. V. Ostrovitianov, is curiously reticent about some of the early scientific disputes in St. Petersburg. Serious discussion of Newton, as in Iushkevich's study, only begins with Leonard Euler's adaptation of the binomial theorem.

Any study of post-Newtonian philosophers quickly reveals that they were philosophizing quite definitely in the light of his achievements. One might, therefore, expect to find some treatment of Newton's ideas in comprehensive histories of Russian thought. This is not the case either in Soviet works on the subject or in those written abroad (such as Zenkovsky's *Istoriia russkoi mysli* [2 vols.; Paris: YMCA Press, 1950] or Boris Jakovenko's *Dějny Ruské Filosofie* [Prague: Orbis, 1939]). Nor, despite much painstaking study of the eighteenth century by Soviet scholars, are there as yet any books dealing with the evolution of scientific or secular ideas in that period—the notable exception being B. E. Raikov's history of heliocentrism (*Ocherki po istorii geliotsentricheskogo mirovozzreniia v Rossii . . .*). Until these appear, the histories of individual disciplines, will continue to be useful. The same may be said of the specialized studies and bibliographies brought out by the Soviet Institute of the History of Science (TIIE; see its collection of essays entitled *Istoriia estestvoznaniia v Rossii*). But the more interesting intellectual problems in the first half of the eighteenth century are more likely to be overcome by individual effort and good fortune rather than collective studies.

To list all the works consulted in an inquiry such as this would clearly be cumbersome without necessarily being helpful. In the bibliography that follows I have included books and articles mentioned in the text, with some additional items—articles and bibliographical works dealing with the St. Petersburg Academy, for example —which, while not actually cited, proved particularly useful. Some items, such as the material on Menshikov, were not of much value for the present study, but they may indicate the terrain covered. Works of a different kind—John Kelly's *Life of John Dollond* or A. N. Krylov's essay on Leonard Euler—while not in fact cited, were interesting for reasons that should be self-evident from the text. Similarly, I have not included in the bibliography items from Bruce's inventory now in the Library of the USSR Academy of Sciences in Leningrad, unless they are described at length in the text.

The same is true of the catalogues, articles, and inventories published under the auspices of the Archives section of the USSR Academy of Sciences. On the other hand, encyclopedias and more familiar works of reference (from Poggendorff, Haan, Murhard, Brockhaus-Efron to the Biographie Universelle and the DNB) have been omitted. Nor did there seem any point in including specialized bibliographies such as those of Riccardi (Euclid), Houzeau and Lancaster (astronomy), Mottelay (electricity), Smith (arithmetic), Fletcher (tables) and so on, though a few (such as De Morgan, Zeitlinger, and E. G. R. Taylor) are included because they are cited specifically in the text. Also excluded are several individual works mentioned in the text that can be found in the author's collected writings (e.g., *La Figure de la Terre déterminée par les Observations . . . au cercle polaire* [Paris, 1738] by Maupertuis, Clairaut, Le Monnier, and others, which is already included in the *Oeuvres de Maupertuis;* Euler's *Nova theoria lucis et colorum* is already contained in his *Opuscula varii argumenti).* Nor have I included articles dealing with the main subjects of the book that can be found in the scientific and learned journals of the eighteenth century, notably the *Commentarii* (and its sequel) in St. Petersburg, the *Philosophical Transactions* of the Royal Society, and the annual volumes of the *Histoire* and *Mémoires* of the Académie Royale des Sciences in Paris. I have included in this bibliography all items pertaining to Newton's influence in eighteenth-century Russia.

All biographical studies of the major figures discussed have been omitted unless they were specifically cited in the text. And I have not included historical works on science, from William Whewell's *History of the Inductive Sciences* (with its analysis of Newton's laws of motion) to more recent histories of the development of dynamics, calculus, hydraulics, and other areas in which Newton's discoveries played so crucial a part. I have also deliberately omitted all works of critical Newtonian scholarship published after 1917 unless specifically cited or directly relevant to the themes discussed. There has been considerable interest in Newton in the Soviet Union, and a comprehensive bibliography of contributions by Soviet scholars and scientists will be included in the sequel to the present volume.

Entries are arranged alphabetically under the name of the author, not the editor; in cases where several institutions have collaborated in a publication, it has been listed under the institution that appears first on the title page.

BIBLIOGRAPHY

BOOKS AND ARTICLES

Aepinus, F. U. T. "Razmyshlenie o vozvrate komet, s kratkim izvestiem o nyne iavivsheisia komete . . . ," *Ezhemesiachnyia sochineniia k pol'ze i uveseleniiu sluzhashchiia,* II (October 1757), 329–348.

————*Tentamen Theoriae Electricitatis et Magnetismi: Accedunt Dissertationes duae quarum prior, phaenomenon quoddam electricum, altera, magneticum, explicat* (St. Petersburg, [1759]).

Akademiia nauk, Biblioteka. *Istoricheskii ocherk i obzor fondov rukopisnogo otdela biblioteki Akademii nauk.* Vol. I, *XVIII vek,* by M. N. Murzanova, E. I. Bobrova, and V. A. Petrov (Moscow and Leningrad: Akademiia nauk, 1956).

————*Istoricheskii ocherk i obzor fondov rukopisnogo otdela biblioteki Akademii nauk: karty, plany, chertezhi, risunki i graviury sobraniia Petra I,* by M. N. Murzanova, V. F. Pokrovskaia, and E. I. Bobrova (Moscow and Leningrad: Akademiia nauk, 1961).

————*Opisanie izdanii grazhdanskoi pechati 1708–ianvar' 1725,* ed. P. N. Berkov (Moscow and Leningrad: Akademiia nauk, 1955).

————*Opisanie izdanii napechatannykh kirillitsei 1689–ianvar' 1725,* ed. P. N. Berkov (Moscow and Leningrad: Akademiia nauk, 1958).

Akademiia nauk, Leningradskoe otdelenie Instituta Istorii. *Putevoditel' po arkhivu Leningradskogo otdeleniia instituta istorii,* ed. A. I. Andreev (Moscow and Leningrad: Akademiia nauk, 1958).

Aleksandrenko, V. N. *K biografii Kn. A. D. Kantemira* (Warsaw, 1896).

————*Russkie diplomaticheskie agenty v Londone v XVIII veke,* Vol. I (Warsaw, 1897).

Alekseev, M. P. "Didro i russkie pisateli ego vremeni," in *XVIII vek, Sbornik 3* (Moscow and Leningrad: Akademiia nauk, 1958), pp. 416–431.

Alembert, Jean Le Rond d'. *Traité de Dynamique, dans lequel les loix de l'équilibre et du mouvement des corps sont réduites au plus petit nombre possible* (Paris, 1743).

Algarotti, Conte Francesco. *Opere varie,* latest ed., 17 vols. (Venice, 1791–1794).

——*Opere inedite . . . nuovamente raccolte per servire di supplemento all'edizioni di Livorno e di Cremona,* Vols. V–VII (Venice, 1796).

——*Il Newtonianismo per le Dame, ovvero Dialoghi sopra la Luce, i Colori, e l'Attrazione,* rev. ed. (Naples, 1739).

——*Le Newtonianisme pour les Dames, ou Entretiens sur la lumière, sur les couleurs, et sur l'attraction,* tr. Du Perron de Castera (Amsterdam, 1741).

——"Viaggi di Russia," in *Opere,* Vol. V (Livorno, 1764).

——*Saggio di Lettere sopra La Russia,* 2nd rev. ed. (Paris, 1763).

——*Viaggi di Russia,* 2nd rev. ed., ed. Pietro Paolo Trompeo ([Turin]: Einaudi, 1961).

Amburger, Erik. "Die nichtrussischen Schüler des Akademischen Gymnasiums in St. Petersburg in den Jahren 1726–1750," *Beiträge zur Geschichte der deutsch- russischen Kulturellen Beziehungen* (Giessen, 1961), pp. 183–213.

Anderson, P. J., ed. *Lists of Officers. University and King's College Aberdeen (1495–1860)* ([Aberdeen], 1893).

——, ed. *Roll of Alumni in Arts of the University and King's College Aberdeen (1596–1860)* (Aberdeen, 1900).

Andreev, A. I., ed. *Petr Velikii[,] Sbornik statei,* Vol. I (Moscow and Leningrad: Akademiia Nauk, 1947).

Anon. *Historische nachricht von dem ehemaligen grossen russischen Staats-Ministro, Alexandro Danielowiz, Fuerst von Menzikof, nebst dessen abwechslenden curieusen Fatalitaeten,* (n.p., 1728).

——*Kartina zhizni i voennykh deianii Rossiisko-imperatorskago generalissima, kniazia Aleksandra Danilovicha Menshikova, Favorita Petra Velikago,* Vol. I (Moscow, 1809).

——"Kratkaia bibliografiia po istorii otechestvennoi fiziki," in *Ocherki po istorii fiziki v Rossii,* ed. A. K. Timiriazev (Moscow: Gosudarstvennoe Uchebno-Pedagogicheskoe Izdatel'stvo Ministerstva Proveshcheniia RSFSR, 1949), 326–341.

——*Leben und todt des erst hoch erhabenen Aber desto tieffer wiederum gestuerzteten Fuerst Menzikoff, worinnen die ganze notable Geschicht [sic] dieses grossen Gluecks-Favoriten, nebst der Beschaffenheit des Koenigreichs Siberien, in welchem er sein Leben kuemmerlich beschlossen kurz und gut beschrieben ist* (Frankfurt, 1730).

(Archimedes). *Arkhimedovy teoremy Andreem Takkvetom ezuitom vybrannye i Georgiem Petrom Domkio sokrashchennye i iz latinskogo na rossiiskii iazyk chirurgiusom Ivan Satarovym prelozhennye . . .* (St. Petersburg, 1745).

Babson Institute Library. *A Descriptive Catalogue of the Grace K. Babson Collection of the Works of Sir Isaac Newton and the Material Relating to Him in the Babson Institute Library, Babson Park, Mass.* (New York: Herbert Reichner, 1950).

——*A Supplement to the Catalogue of the Grace K. Babson Collection of the Works of Sir Isaac Newton and Related Material in the Babson Institute, Babson Park, Massachusetts,* comp. Henry P. Macomber (Babson Park, Mass.: Babson Institute, 1955).

Baily, Francis. *An Account of the Revd. John Flamsteed, the first Astronomer*

Royal compiled from his own manuscripts, and other authentic documents, never before published, & Supplement to the Account of the Revd. John Flamsteed, with an author index (London, 1835; reprinted by Dawsons of Pall Mall, London, 1966).

Bakmeister, I.-K. *Opyt o Biblioteke i Kabinete redkostei i istorii natural'noi Sanktpeterburgskoi imperatorskoi Akademii nauk, izdannoi na frantsuzskom iazyke Ioganom Bakmeisterom, podbibliotekarem Akademii nauk, a na rossiiskii iazyk perevedennoi Vasil'em Kostygovym* ([St. Petersburg], 1779).

Ball, W. W. Rouse. *An Essay on Newton's Principia* (London and New York, 1893).

Banks, John. *The History of the Life and Reign of the Czar Peter the Great, Emperor of all Russia, And Father of his Country*, 2nd ed. (London, 1740).

Bantysh-Kamenskii, D. N. *Slovar' dostopamiatnykh liudei russkoi zemli soderzhashchii v sebe zhisn' i deianiia znamenitykh polkovodtsev, ministrov i muzhei gosudarstvennykh, velikikh ierarkhov pravoslavnoi tserkvi, otlichnykh litteratorov i uchennykh, izvestnykh po uchastiiu v sobytiiakh otechestvennoi istorii . . .* In five parts, ed. A. Shiriaev (Moscow, 1836).

Barrow, Sir John Bart. *A Memoir of the Life of Peter the Great* (London, 1832).

Baumgart, K. K. *Tri memuary po mekhanike* (Moscow: Akademiia nauk, 1951).

Bell, A. E. *Christian Huygens and the Development of Science in the 17th Century* (London: E. Arnold, 1947).

Bergholtz, F. V. *Dnevnik kamer-iunkera F. V. Berkhgol'tsa 1721–1725*, rev. ed., tr. I. F. Ammon (Moscow: Universitetskaia tipografiia, 1902–1903).

Berkeley, George, Bishop of Cloyne. *The analyst: or, A discourse addressed to an infidel mathematician. Wherein it is examined whether the object, principles, and inferences of the modern analysis are more distinctly conceived or more evidently deduced, than religious mysteries and points of faith.* By the author of *The minute philosopher . . .* (London, 1734).

Naturforschende Gesellschaft in Basel. *Der Briefwechsel von Johann Bernoulli*, Vol. I (Basel: Birkhäuser Verlag, 1955).

Bernoulli, Daniel. "Dissertatio de actione fluidorum in corpora solida et motu solidorum in fluido," *Commentarii* (St. Petersburg), Part II (1729), 304–342.

———*Exercitationes quaedam mathematice* (Venice, 1724).

———*Hydrodynamica, sive De viribus et motibus fluidorum commentarii.* (Strassburg: sumptibus J. R. Dulseckeri, 1738).

———*Hydrodynamics . . .: and Hydraulics, by Johann Bernoulli*, tr. Thomas Connody and Helmut Kobus under the auspices of the Iowa Institute of hydraulic research (New York: Dover Publications [1968]).

———*Nouveaux principes de mechanique et de physique tendans à expliquer la nature & les propriétés de l'aiman* (Paris, 1748).

Bialov, V. V., *et al.* "Knigi i zhurnal'nye stat'i po istorii estestvoznaniia, opublikovannye v SSSR za period s 1939 po 1944 g. vkl. na russkom i drugikh iazykakh," *TIIE*, I (1947), 459–533.

Bienemann, Friedrich. *Der Dorpater Professor Georg Friedrich Parrot und Kaiser Alexander I* (Reval: Kluge, 1902).

Bil'basov, V. A. *Didro v Peterburge 1773–74,* (St. Petersburg, 1884).

Bilfinger, G. -B. *De causa gravitatis physica generali disquisitio experimentalis quae praemium a Regia Scientiarum Academia promulgatum, retulit: anno 1728* (Paris, 1728).

————*De harmonia animi et corporis humani maxime praestabilita, ex mente illustris Leibnitii commentatio hypothetica. Accedunt solutiones difficultatum a Foucherio, Baylio, Lamio, Tourneminio, Newtono, Clarkio . . .* (Tübingen, 1734; 3rd ed., 1741).

————*Dilucidationes philosophicae de Deo' anima humana, mundo, et generalibus rerum affectionibus* (Tübingen, 1725).

————*. . . Varia in fasciculos collecta* (Stuttgart, 1743).

————*et al. Raspolozhenie uchenii Ego Imperatorskogo Velichestva Petra Vtorago Imperatora i Samoderzhtsa Vserossiiskago . . .* (St. Petersburg, 1728).

Biliarskii, P. S. *Materialy dlia biografii Lomonosova sozdanny ekstraordinarnym akademikom Biliarskim* (St. Petersburg, 1865).

Biot, J. B. *Mélanges scientifiques et littéraires*, 3 vols. (Paris, 1858).

Birch, Thomas. *The History of the Royal Society of London for Improving of Natural Knowledge, from its first rise, in which the most considerable of those papers communicated to the Society, which have hitherto not been published, are inserted in their proper order, as a supplement to the Philosophical Transactions*, 4 vols. (London, 1756–1757).

Bloch, Léon. *La Philosophie de Newton* (Paris: F. Alcan, 1908).

Bludov, D. N. "Dnevnye zapiski kniazia Menshikova," in *Sochineniia E. P. Kovalevskago, Graf Bludov i ego vremia* (St. Petersburg, 1871), pp. 219–233.

Bluntschli, J. C. *Geschichte des Allgemein Staatsrechts und der Politik seit dem sechzehnten Jahrhundert bis zur Gegenwart*, Vol. I (Munich, 1864).

Bobrov, Evgenii. "A. N. Radishchev kak filosof," in his *Filosofiia v Rossii, materialy, izsledovaniia, i zametki*, Vol. I (Kazan, 1889), 359–560.

Bobrova, E. I. "Obzor inostrannykh pechatnykh knig Sobraniia Petra I," in *Istoricheskii ocherk i obzor fondov rukopisnogo otdela biblioteki Akademii nauk*. Vol. I, *XVIII vek* (Moscow and Leningrad: Akademiia nauk, 1956), pp. 143–170.

Bobynin, V. V. "Rumovskii, Stepan Iakovlevich," *Russkii biograficheskii slovar'*, XVII (1918), 441–450.

————"Magnitskii (Leontii Filipovich, 1669–1739)," *Brockhauss-Efron*, XVIII (1896), 327–328.

————"Georg-Vil'gel'm Rikhman," *Russkii biograficheskii slovar'*, XVI (1913), 233–240.

————*Russkaia fiziko-matematicheskaia bibliografiia*, 2 vols., 6 pts. (Moscow, 1886–1892).

Bogoslovskii, M. M. "Pervoe zagranichnoe puteshestvie," in *Petr I[:] Materialy dlia Biografii*, ed. V. Lebedev), Vol. II, Parts I and II, ([Moscow]: Ogiz, 1941; reprinted by Mouton in cooperation with Europe printing, Vaduz, at the Hague, 1969).

————*Petr Velikii i ego reforma* (Moscow: Izdanie tsentral'nogo tovarishchestva "Kooperativnoe izdatel'stvo," 1920).

Boss, V. J. "La Quatrième Ode de Kantemir et *l'Italia Liberata* de Gian Giorgio Trissino," *Cahiers du Monde Russe et Soviétique*, IV (No. 1–2), 47–55.

————"Kantemir and Rolli-Milton's *Il Paradiso Perduto*," *Slavic Review*, XXI, (September 1962), 441–455.

————"Russia's First Newtonian: Newton and J. D. Bruce," *Archives internationales d'Histoire des Sciences* (Nos. 60–61, 1962), pp. 233–265.

Brasch, F. E. "Newton's First Critical Disciple in the American Colonies—John Winthrop," in *Sir Isaac Newton 1727–1927. A Bicentenary Evaluation of His Work* (Baltimore: Williams & Wilkins Company, 1928), pp. 301–340.

Brewster, Sir David. *Memoirs of the Life, Writings, and Discoveries of Sir Isaac Newton,* 2 vols. (Edinburgh, 1855).

Bruce, J. D. (Brius, Ia.V.). *Priemy tsirkulia i lineiki ili izbranneishoe nachalo vo matematicheskikh iskustvakh* ... (Moscow, 1709; tr. from a German work by Burckhard von Pürckenstein).

Bruce, P. H. *Memoirs of Peter Henry Bruce Esq., A military Officer in the services of Prussia, Russia, and Great Britain, containing an Account of his Travels in Germany, Russia, Tartary, Turkey, the West Indies, &c. as also several very interesting private Anecdotes of the Czar, Peter I of Russia* (London, 1782).

Brunet, Pierre. *L'Introduction des théories de Newton en France au XVIIIᵉ siècle. Avant 1738.* (Paris: A. Blanchard, 1931).

————*Maupertuis, étude biographique,* 2 vols. (Paris: A. Blanchard, 1929).

————*Les Physiciens hollandais et la méthode expérimentale en France au XVIII siècle.* (Paris: A. Blanchard, 1926).

————"La vie et l'oeuvre de Clairaut," *Revue d'Histoire des Sciences* (Paris), IV (1951), 13–40, 109–153; V (1952), 334–349; VI (1953), 1–7.

Buchanan, David. *Observations on the Subjects treated of in Dr. Smith's Inquiry into the Nature and Causes of the Wealth of Nations.* (New York: A. M. Kelley, 1966).

Burckhardt, Fritz. *Die Basler Mathematiker Daniel Bernoulli und Leonhard Euler* (Basel, 1884).

Burtt, E. A. *The Metaphysical Foundations of Modern Physical Science* (New York: Doubleday, 1955: reissue of 2nd rev. ed. that appeared in 1932).

Butterfield, Herbert. *The Origins of Modern Science, 1300–1800* (London: Bell, 1957).

Cantemir, Antiokh, s.v. Kantemir, A. D.

Carré, J. R. *La Philosophie de Fontenelle ou le sourire de la raison* (Paris: F. Alcan, 1932).

Casper, Max. *Kepler,* tr. and ed. by C. D. Hellman (London: Abelard-Schuman, 1959).

Castelnuovo, Guido, ed. *Le Origini del calcolo infinitesimale nell' era moderna con scritti di Newton, Leibniz, Torricelli* (Milan: Feltrinelli, 1962).

A Catalogue of the Portsmouth Collection of Books and Papers written by or belonging to Sir Isaac Newton, the Scientific Portion of which has been presented by the Earl of Portsmouth to the University of Cambridge (Cambridge, Eng., 1888).

Celsius, Andreas. *De observationibus pro figurâ telluris determinandâ in Gallia habitis, disquisitio* (Uppsala, 1738).

Châtelet, G. E. D. du. *Lettres de la Mse Du Châtelet réunies pour la première fois ... par Eugène Asse* (Paris, 1878).

————*Réponse de Mme . . . à la lettre que M. de Mairan lui a écrite sur la question des forces vives* (Paris, 1741).

————*Institutions de Physique* (Paris, 1741).

Chebotarev, A. I. *Biblioteki Akademii nauk SSSR,* comp. by chief eds. V. I. Abramov, G. A. Chebotarev, V. I. Shunkov (Moscow: Akademiia nauk, 1959).

Chenakal, V. L. "Nauchnye sviazi russkikh i angliiskikh astronomov v XVIII v," in *Voprosy istorii fiziko-matematicheskikh nauk,* ed. K. A. Rybnikov *et al.* (Moscow: Gosudarstvennoe Izdatel'stvo "Vysshaia Shkola," 1963), pp. 486–489.

————"*Ocherki po istorii russkoi astronomii, nabliudatel'naia astronomiia v Rossii XVII i nachala XVIII v.* (Moscow and Leningrad: Akademiia nauk, 1951).

————"Optika v dorevoliutsionnoi Rossii," TIIE, I (1947), 121–168.

————ed., "Pis'ma Iakova Vilimovicha Briusa k Iogannu-Georgu Leitmanu," in *Nauchnoe nasledstvo,* Vol. II, ed. S. I. Vavilov (Moscow: Akademiia nauk, 1951), pp. 1083–1101.

————"Priroda sveta v vozzreniiakh russkikh estestvoispytatelei XVIII i nachala XIX veka," TIIE, III (1949) 173–199.

————*Russkie priborostroiteli pervoi poloviny XVIII veka* (Leningrad: Akademiia nauk, 1953.

Chistiakov, M. V. "Narodnoe predanie o Briuse," *Russkaia Starina* (St. Petersburg), IV (August 1871), 167–170.

Chistovich, I. A. *Feofan Prokopovich i ego vremia* (St. Petersburg, 1867).

Chuchmarev, V. I. "G. V. Leibnits i russkaia kul'tura 18 stoletiia," *Vestnik istorii mirovoi kul'tury* (No. 4, 1957), pp. 120–132.

Clairaut, Aléxis Claude. "Nouvelle Theorie de la figure de la terre, ou l'on concilie les mésures actuelles avec les principes de la gravitation universelle," in *Académie des Sciences, Inscriptions et Belle Lettres de Toulouse, Pièces qui ont remporté le prix . . . depuis l'année 1747, jusqu'en 1750* (Toulouse, 1758).

————*Theorie de la lune déduite du seul principe de l'attraction reciproquement proportionelle aux quarrés des distances.* [Pièce qui a remportée le prix de l'Académie impériale des Sciences de St. Petersbourg proposé en MDCCL sur la question, Si toutes les inégalités, qu'on a observées dans le mouvement de la lune s'accordent avec la théorie Newtonienne ou non? etc.] (St. Petersburg, 1752).

————*Theorie du mouvement des comètes, dans laquelle on a égard aux altérations que leurs orbites éprouvent par l'action des planètes* [Avec l'application de cette théorie à la comète qui a été observée dans les années 1531, 1607, 1682, 1759] (Paris, 1760).

Clarke, John D. D., Dean of Sarum. *A Demonstration of some of the Principal Sections of Sir Isaac Newton's Principles of Natural Philosophy. In Which His peculiar Method of treating that useful Subject, is explained, and applied to some of the chief phenomena of the System of the World* (London, 1730).

Clarke, Samuel. "On the present controversy among mathematicians concerning the proportion of velocity and forces of bodies in motion," *Royal Society of London, Philosophical Transactions,* XXXV (1727–1728), 381–388.

———*A Collection of Papers, which passed between the late learned Mr. Leibniz and Dr. Clarke, in the years 1715 &1716, relating to the Principles of Natural Philosophy and Religion* . . . (London, 1717).

Cohen, I. Bernard. *Franklin and Newton* (Philadelphia: The American Philosophical Society, 1956).

———*Introduction to Newton's 'Principia',* (Cambridge University Press, 1971).

Collins, John. *Commercium Epistolicum D. Johannis Collins, et aliorum de Analysi Promota* (London, 1712).

Craig, Sir John H. M. *History of the London Mint from* A.D. *287 to 1948* (Cambridge, Eng.: the University Press, 1953).

———*Newton and the Mint* (Cambridge, Eng.: the University Press, 1946).

Crommelin, C. A. *Descriptive catalogue of the Physical Instruments of the 18 century, including the Collection 'S Gravesande-Musschenbroek,* in the *Rijksmuseum voor de Geschiedenis der Natuurwetenschappen at Leyden* ([Leiden] 1951).

Deborin, A. M. (ed.) *Leonard Eiler, 1707–1783. Sbornik statei i materialov k 150-letiiu so dnia smerti* (Moscow and Leningrad: Akademiia nauk, 1935).

Demkov, M. I. *Istoriia russkoi pedagogii.* Vol. I, Drevne-Russkaia pedagogiia (X–XVII vv); Vol. II, Novaia russkaia pedagogiia (XVIII vek), 2nd ed. (Moscow and St. Petersburg, 1899, 1910).

De Morgan, Augustus. *Essays on the Life and Work of Newton* (London and Chicago: Open Court Publishing Company, 1914).

———*Newton: His Friend and His Niece,* reprint (London: Dawsons of Pall Mall; earlier ed., 1885).

Denison, A. P. "Nikolai Gavrilovich Kurganov (1725–1796)," TIIE-T, XXIV (1960), 360–383.

Depman, I. Ia. "Georg Petr Domkino (o pervom izdanii 'Nachal' Evklida na russkom iazyke)," TIIE, II (1948), 573–574.

———"Leontii Magnitskii. K dvukhsotletiiu so dnia ego smerti," *Morskoi Sbornik* (January 1940), pp. 112–126.

Derham, William. *Astro-Theology: or, a Demonstration of the Being and Attributes of GOD, from a survey of the Heavens* (London, 1715).

Desaguliers, J. T. *The Newtonian System of the World, the Best Model of Government: An Allegorical Poem* [with a plain and intelligible account of the system of the world, by way of annotations . . . to which is added, Cambria's Complaint against the intercalary day in the leap year], ([London], 1728).

Descartes, René. *Œuvres* . . . , 13 vols. (Paris, 1897–1913).

———*Opera philosophica.* 4 vols. (Amsterdam, 1677[78]).

Diderot, Denis. *Œuvres complètes* [. . . revues sur les éditions originales comprenant ce qui a été publie à diverses époques et les manuscrits inédits conservés a la bibliothèque de l'Ermitage . . . par J. Assézat et M. Tourneux (with "Mémoires pour servir d'histoire de la vie et des ouvrages de Diderot" by M. A. de Vandeul)], 20 vols. (Paris, 1875–1877).

———*Sobranie sochinenii v desiati tomakh,* ed. I. K. Luppol (Moscow: Ogiz, Gosudarstvennoe Izdatel'stvo Khudozhestvennoi literatury, 1935–   ).

————*Encyclopédie ou dictionnaire raisonné des sciences, des arts et des métiers* ... *Mis en ordre par M. Diderot,* ... *et quant à la partie mathématique par M. d'Alembert* ... , 17 vols. (Paris, 1751–1763).

Ditton, Humphrey. *An Institution of Fluxions* ... *with some of the uses and applications of that method according to Sir Is. Newton* (London, 1706).

Domcke, Georg Peter. *Philosophiae Mathematicae Newtonianae Illustratae Tomi Duo* ... (London, 1730).

Droysen, Hans. "Marquise du Châtelet, Voltaire, und der Philosoph Christian Wolff," *Zeitschrift für französische Sprache und Literatur,* XXXV (1909), 226–248.

Dvigubskii, I. A. *Fizika, v pol'zu vospitannikov blagorodnogo universitetskogo pansiona izdannaia* (Moscow, 1808).

Edleston, Joseph. *Correspondence of Sir Isaac Newton and Professor Cotes, including Letters of Other Eminent Men, now first published from the originals in the Library of Trinity College, Cambridge; together with an Appendix containing other unpublished Letters and Papers by Newton; with Notes, Synoptical View of the Philosopher's Life, and a Variety of Details illustrative of his History.* (Cambridge and London, 1850).

Ehrhard, Marcelle. *Le Prince Cantemir à Paris, 1738–1744* (Paris, 1938).

Eliseev, A. A., and Murzin, A. M. "G. V. Rikhman—vydaiushchiisia russkii fizik 18 veka," *Izvestiia Akademii Nauk SSSR—otdelenie tekhnicheskikh nauk* (No. 8, 1953), pp. 1166–1174.

Ellis, Brian D. "Newton's Concept of Motive Force," *Journal of the History of Ideas,* XXIII (1962), 273–278.

Esipov, G. V. "Zhisneopisanie kniazia A. D. Menshikova, po novootkrytym bumagam," *Russkii Arkhiv,* XIII (No. 7, 1875), 233–247; (No. 9, 1875), 47–74.

————"Ssylka Menshikova v Berezov", *Otechestvennyia Zapiski,* CXXXI (1860), 379–426; CXXXIV (1861), 55–70.

Euclid. *Elementy geometrii po Evklidu,* tr. by N. G. Kurganov (Moscow, 1769).

Euler, J. A. (written by Leonard Euler). *Disquisitio de causa physica Electricitatis ab Academia Scientiarum Im. Petropolitanae praemio coronata* (St. Petersburg, 1755).

Euler, Leonard. *Opera omnia sub auspiciis Societatis Scientiarum Naturalium Helveticae* ... , Ser. 2, *Opera mechanica et astronomica* (Leipzig and Berlin: B. G. Teubner, 1911–   ; imprint varies).

————*Opuscula varii argumenti,* 3 vols. (Berlin, 1746–1751).

————*Letters of Euler* ... *to a German Princess, on Different Subjects in Physics & Philosophy,* 2 vols., tr. Henry Hunter (London, 1795; 2nd ed., London, 1802).

————*Trudy Arkhiva,* No. 17, ed. M. V. Krutikova *et al.* Vol. I, *Rukopisnye materialy L. Eilera v arkhive Akademii nauk SSSR,* by Iu. Kh. Kopelevich, M. V. Krutikova, G. K. Mikhailov, and N. M. Raskin (Moscow and Leningrad: Akademiia nauk, 1962).

————*Trudy Arkhiva,* No. 20, ed. G. K. Mikhailov. Vol. II, *Trudy po mekhanike,* tr. I. A. Perel'muter (Moscow and Leningrad: Izdatel'stvo "Nauka," 1965).

————*Pis'ma k uchenym,* ed. V. I. Smirnov (Moscow and Leningrad, Akademiia nauk, 1963).

————*Défense de la révélation contre les objections des esprits forts: suivie des pensées de cet auteur sur la religion, supprimées dans la dernière édition de ses lettres à une princesse d'Allemagne* (Paris, 1805).

Fel', S. E. "Petrovskaia geometriia," TIIE, IV (1952), 140–155.

Feofan Prokopovich. *Slova i Rechi pouchitel'nyia, pokhval'nyia i pozdravitel'nyia sobrannyia* . . . (St. Petersburg, 1760–1774), chaps. 1–4.

Ferquharson, [Andrei]. *Evklidovy Elementy iz dvenatzati neftonovykh knig vybrannyia, i v osm' knig chrez professora Andreia Farkhvarsona sokrashchennye* . . . (St. Petersburg, 1739).

————, with Stephen Gwyn and Leontii Magnitskii. *Tablitsy sinusov, tangensov i sekansov . . . pod nazreniem gospodina general felizeikhmeistera ikavalera [sic] Iakova Vilimovicha Briusa* . . . (n.p., 1716; reprint of earlier ed., Moscow, 1703).

Filippov, M. S., ed. *Istoriia Biblioteki Akademii nauk SSSR* (Moscow and Leningrad: Izdatel'stvo "Nauka," 1964).

————*et al.,* eds. *Biblioteka Akademii Nauk SSSR (1714–1964), Bibliograficheskii ukazatel'* (Leningrad; [Biblioteka Akademii Nauk SSSR], 1964).

Flamsteed, John (see also s.v. Horrocks, Jeremiah.) *Atlas Cœlestis* (London, 1729).

————*Historiae Cœlestis Britannicae Volumen Primum. Completens Stellarum Fixarum Nec non Planetarum Omnium Observationes Sextante, Micrometro, & peractas* (London, 1725).

Floridov, A. A. *Imperatorskaia Publichnaia Biblioteka za sto let 1814–1914,* ed. D. F. Kobeko, (St. Petersburg, 1914).

Florovskii, A. V. "Latinskie shkoly v Rossii v epokhu Petra I," in *XVIII Vek, Sbornik 5* (Moscow and Leningrad: Akademiia nauk, 1962).

Fontenelle, Bernard Le Bovier de. *Oeuvres de M. de Fontenelle,* new ed., 12 vols. (Amsterdam, 1764), Vols. I–IV.

————*An Account of the Life and Writings of Sir I. Newton,* 2nd ed., from the *éloge* of M. Fontenelle, (London, 1728).

————*Entretiens sur la Pluralité des Mondes* (London, 1707).

————*Razgovory o mnozhestve mirov gospodina Fontenella parizhskoi akademii nauk sekretaria,* tr. Antiokh Cantemir (St. Petersburg, 1740).

Freind, John. *Chymical lectures: in which almost all the operations of chymistry are reduced to their true principles, and the laws of nature* (London, 1712).

Fuss, P. H., ed., *Correspondance Mathématique et Physique de quelques célèbres géomètres du XVIIIème Siècle, précédée d'une notice sur les travaux de Léonard Euler, tant imprimés qu'inédits et publiée sous les auspices de l'Académie Impériale des Sciences de Saint-Pétersbourg* . . . Vol. XI (St. Petersburg, 1843).

Gaedeke, Arnold. "Peter der Grosse in England i.j. 1698, aus Archivalien, *Im Neuen Reich,* I (1872), reprint of pages 217–224.

Gennadi, G. N. *Ukazatel' bibliotek v Rossii* (St. Petersburg, 1864).

————*Russkie knizhnye redkosti* (St. Petersburg, 1872).

Gerhardt, C. I. *Briefwechsel zwischen Leibniz und Christian Wolff aus den Handschriften der Koeniglichen Bibliothek zu Hannover* . . . (Halle, 1860).

Goldbach, Christian, s.v. Iushkevich, A. P. (and Winter, Eduard J.)

Gordon, Alexander. *History of Peter the Great, Emperor of Russia,* 2 vols. (Aberdeen, 1755).

Gosudarstvennaia Ordena Lenina Biblioteka SSSR imeni V. I. Lenina, *et al. Svodnyi katalog russkoi knigi grazhdanskoi pechati XVIII veka 1725–1800,* 5 vols., ed. I. P. Kondakov *et al.* (Moscow: Izdanie Gosudarstvennoi Biblioteki SSSR imeni V. I. Lenina, 1962–1967).

Gottsched, J. Ch. *Historische Lobschrift des weiland hoch- und wohlgebohrnen Herrn Herrn Christians, Freiherrn von Wolf . . . Nebst des hoch seligen Freyherrn Kupferbilde* (Halle, 1755).

Grabo, Carl. *Newton among Poets: Shelley's Use of Science in Prometheus Unbound* (Chapel Hill: Unive:rsity of North Carolina Press, 1930).

Graf, Arturo. *L'Anglomania e l'influsso inglese in Italia nel secolo XVIII* (Turin: E. Loescher, 1911).

Grasshoff, Helmut. "Kantemir und Fénélon," *Zeitschrift für Slawistik,* III (No. 2–4, 1958), 369–383.

———*Antioch Dmitrievič Kantemir und Westeuropa. Ein russischer Schriftsteller des 18 Jahrhunderts und seine Beziehungen sur westeuropaischen Literatur und Kunst* (Berlin: Akademie-Verlag, 1968).

's Gravesande, W. James. *Mathematical Elements of Natural Philosophy confirm'd by Experiments: Or, an introduction to Sir Isaac Newton's Philosophy,* 2 vols., 6th ed., tr. J. T. Desaguliers (London, 1747).

———*Philosophiae Newtonianae Institutiones, In Usus Academicos,* 2nd ed. (Leiden and Amsterdam, 1728).

Gray, George J. *A Bibliography of the Works of Sir Isaac Newton, Together with a List of Books Illustrating His Works,* 2nd ed. (Cambridge, Eng.: Bowes & Bowes, 1907).

Gregory, David. *Astronomiae Physicae et Geometricae Elementa* (Oxford, Eng., 1702).

Grey, Ian. "Peter the Great in England," *History Today,* VI (April 1956), 225–234.

Grigorian, A. T., and Iushkevich, A. P. (with T. N. Klado and Iu. Kh. Kopelevich). *Russko-frantsuzskie nauchnye sviazi—Relations scientifiques Russo-françaises* (Leningrad: Izdatel'stvo "Nauka," Leningradskoe otdelenie, 1968).

Gross, Friedrich. *Institutiones philosophiae rationalis seu logicae; conscriptae in gratiam celsissimi Principis Antiochi Cantemir* (St. Petersburg, 1726).

Grot, Ia. [K.], ed. *Pis'ma Imperatritsy Ekateriny II k Grimmu (1774–1796),* 2nd ed., *Sbornik Imperatorskago Russkogo Istoricheskago Obshchestva,* Vol. XLIV (St. Petersburg, 1885).

Guasco, l'Abbé Count Octavien de. *Satyres du Prince Cantemir* (London: Jean Nourse, 1750).

Gudzii, N. K. "Feofan Prokopovich," in *Istoriia russkoi literatury,* Vol. III (Moscow and Leningrad: Akademiia nauk), chap. 6, 157–175.

Guerlac, Henry. "Newton's Changing Reputation in the Eighteenth Century," in *Carl Becker's Heavenly City Revisited,* ed. Raymond O. Rockwood (Ithaca, N.Y.: Cornell University Press, 1958), pp. 3–26.

Guerrier, W. I. *Relations de Leibniz avec la Russie et Pierre le Grand d'après des papiers inédits de la Bibliothèque de Hanovre.* (St. Petersburg, 1871).

Guhrauer, G. E. *Gottfried Wilhelm, Freiherr von Leibnitz: Eine Biographie.* Vol. I (Breslau, 1846).

Gur'ev, S. E. "Kratkoe izlozhenie razlichnykh sposobov iz'iasniaia differentsial'-noe ischislenie," in *Umozritel'nyia izsledovaniia Imperatorskoi Akademii Nauk,* Vol. IV (St. Petersburg, 1815), 159–212.

Hamburger, G. E. *Elementa physicae methodo mathematico in usum auditorum conscripta,* 2nd ed. (Jena, 1734).

Hans, Nicholas. "The Moscow School of Mathematics and Navigation (1701)," *The Slavonic and East European Review,* XXIX (No. 73, 1951), 532–536.

Hartsoeker, Nicolas. *Recueil des plusieurs pièces de physique, òu l'on fait princi-palement voir l'invalidité du Système de Mr. Newton, et où se trouve . . . une dissertation sur la peste, etc.* (Utrecht, 1722).

Hermann, Jakob. "De mensura virium corporum," *Commentarii* (St. Petersburg), I (1726, pub. 1728), 1–42.

———*Phonoromia, sive De viribus et motibus corporum solidorum et fluidorum libri duo* (Amsterdam, 1716).

———*Responsio ad cl. Nieuwenteyt. Considerationes secundes circa differentialis principia* (Basel, 1700).

Hevelius, Johannes. . . . *Selenographia: Sive, Lunae descriptio; atque accurata, tam macularum ejus, quam motuum diversorum, aliarumque omnium vicisitu-dinum, phasiumque . . . delineatio . . . (*Gdansk, 1647).

Hinz, Walther. *Peters des Grossen Anteil an der wissenschaftlichen und künst-lerischen Kultur seiner Zeit.* (Breslau: F. W. Jungfer, 1933).

Hoffmann, P. "Zur Verbindung Eulers mit der Petersburger Akademie während seiner Berliner Zeit," in *Die deutsch-russische Begegnung und Leonhard Euler: Beiträge zu den Beziehungen zwischen der deutschen und der russischen Wissenschaft und Kultur im 18 Jahrhundert,* ed. E. J. Winter (Berlin: Akademie-Verlag, 1958), pp. 150–156.

Hopkins, M. R. *Married to Mercury. A Sketch of Lord Bolingbroke and His Wives* (London: Constable, 1936). :

Horrocks, Jeremiah. *Jeremiae Horrocii . . . Opera Posthuma, viz. Astronomia Kepleriana, defensa & promota . . . Excerpta ex epistolis ad Crabtræum suum. Observationum Cœlestium Catalogus. Lunae theoria nova . . . Adjiciuntur J. Flamstedii de Temporis Æquatione Diatriba. Numeri ad Lunae Horroccianam,* ed. John Wallis (London, 1672–1673).

Huygens, Christian. *Oeuvres Complètes,* 22 vols. (The Hague: Martin Nijhoff, 1888–1950).

———*Celestial Worlds Discover'd: or, Conjectures concerning the Inhabitants, Plants, and Productions of the Worlds in the Planets . . .* (London, 1698).

———*Kniga mirozreniia ili mnenie o nebesnozemnykh globusakh i ikh ukrashe-niiakh . . .* (Moscow, 1724).

———*K o s m o t h e o r o s, sive de terris coelestibus earumque ornatu conjecturae* (The Hague, 1696).

Iazykov, D. D. *Vol'ter v russkoi literature* (St. Petersburg, 1879).

Idel'son, N. I. "Zakon vsemirnogo tiagoteniia i teoriia dvizheniia luny," in *Isaak N'iuton 1643–1727: Sbornik statei k trekhsotletiiu rozhdeniia,* ed. S. I. Vavilov (Moscow and Leningrad: Akademiia nauk, 1943), pp. 161–210.

Ikonnikov, V. S. *Graf N. S. Mordvinov. Istoricheskaia monografiia* (St. Petersburg, 1873).

———*Opyt Russkoi Istoriografii*, 2 vols. (Kiev, 1891–1908).

———"Russkie universitety v sviazi s khodom obshchestvennogo obrazovaniia," *Vestnik Evropy*, V (No. 9, 1876), 161–206; (No. 10, 1876), 492–550; VI (No. 11, 1876), 73–132.

Iltis, Carolyn. "D'Alembert and the 'Vis Viva' Controversy," *Studies in History and Philosophy of Science*, I (No. 2, 1970), 135–144.

Imperatorskaia Akademiia Nauk. *Materialy dlia istorii Imperatorskoi Akademii Nauk 1716–1750*, 10 vols. (St. Petersburg, 1885–1901).

———*Protokoly zasedanii konferentsii Imperatorskoi Akademii Nauk s 1725 po 1803 goda . . .*, Vols. I–III, ed. K. S. Veselovskii (St. Petersburg, 1897–1900).

———*Torzhestvo Akademii nauk . . . prazdnovannoe publichnym sobraniem sentiabria 6 dnia 1749 goda* (St. Petersburg, 1749).

Iovskii, A. A. "O vazhnosti khimicheskikh issledovanii v krugu nauk i iskusstv," in *Izbrannye prozvedeniia russkikh estestvoispytatelei pervoi poloviny XIX veka*, ed. G. S. Vasetskii and S. R. Mikulinskii (Moscow: Izdatel'stvo sotsial'no-ekonomicheskoi literatury, 1959), pp. 309–320.

Iushkevich, A. P. "Akademik S. E. Gur'ev. i ego rol' v razvitii russkoi nauki," TIIE, I (1947), 219–268.

———"Eiler i russkaia matematika v 18m veke," TIIE, III (1949), 45–116.

———*Istoriia matematiki v Rossii do 1917 goda* (Moscow: Izdatel'stvo "Nauka," Glavnaia redaktsiia fiziko-matematicheskoi literatury, 1968).

———"O pervom russkom izdanii trudov Evklida i Arkhimeda," TIIE, II (1948), 567–572.

———and Winter, Eduard J., eds. *Leonhard Euler und Christian Goldbach: Briefwechsel 1729–1764* (Berlin: Abhandlungen der Deutschen Akademie der Wissenschaften, 1965).

———and Winter, Eduard J., eds. *Die Berliner und die Petersburger Akademie der Wissenschaften im Briefkwechsel Leonhard Eulers*, Vols. I and II (Berlin: Akademie-Verlag, 1959–1961).

Ivask, U. G. "Chastnye biblioteki v Rossii: opyt bibliograficheskogo ukazaniia," *Russkii Bibliofil* (St. Petersburg), (No. 3, 1911), pp. 55–74.

Jakovenko, Boris. *Dějiny Ruské Filosofie* (Prague: Orbis, 1939).

Jones, H. S., Sir. "Halley as an Astronomer," *Notes and Records of the Royal Society*, XII (No. 2, 1957), 175–192.

Jurin, James. *De conservatione virium vivarum dissertatio* (London, 1744).

———*Dissertationes physico-mathematicae, partim antea in Actis philosophicis Londinensibus . . . partim nunc primum impressae* (London, 1732).

———*Jacobo Jurini Regiae scientiarum Londinensi Societati A secretio Dissertationis de Motu Aquarum fluentium contra nonullas Petri Antoni Michelotti Animadversiones Defensio. Accedit ejusdem Michelotti Ad Illustris atque Excellentiss. Virum Antonium De Comitibus Patritium Venetum Eruditissimum Epistola in qua illi ipsi Jurinianae Defensioni respondetur* (Venice, 1724).

———"De motu aquarum fluentium," *Philosophical Transactions*, XXX (No.

355, 1719–1720), 748–766; "Disquisitiones physicae de tubulis capillaribus a Jacobo Jurino ad Academiam transmissae ut ejusdem Commentariis inserentur, una cum notis a Georgio Bernhardo Bilfingero, ad quem id negotium pertinuit, adjectis," *Commentarii* (St. Petersburg) III (1728, pub. 1732), 221–292.

Kafengauz, B. B., and Pavlenko, N. I., eds. *Ocherki i istorii SSSR. Rossiia v pervoi chetverti XVIII veka* (Moscow: Akademiia nauk, 1954).

Kaminer, L. V. "Iz istorii otkrytiia zakona sokhraneniia veshchestva M. V. Lomonosovym," TIIE, IV (Moscow: Akademiia nauk, 1952), 306–311.

Kant, Im. *Gedanken von der wahren Schätzung der lebendigen Kräfte und Beurteilung der Beweise* (1747), in *Werke,* ed. Ernst Cassirer, Vol. I (Berlin: B. Cassirer, 1912).

Kantemir, A. D. *Satiry i drugie stikhotvorcheskie sochineniia kniazia Antiokha Kantemira, s istoricheskimi primechaniiami i s kratkim opisaniem ego zhisni* (St. Petersburg, 1762).

——*Sochineniia,* ed. Aleksandr Smirdin (St. Petersburg, 1847).

——*Sochineniia, Pis'ma i izbrannye perevody s portretom avtora, so stat'eiu o Kantemire i s primechaniiami. V. Ia. Stoiunina,* Vols. I and II, ed. P. A. Efremov (St. Petersburg, 1867–1868).

——*Sobranie stikhotvorenii,* ed. P. N. Berkov, introd. by F. Ia. Priima, text and notes prepared by Z. I. Gershkovich (Leningrad: Sovetskii Pisatel', 1956).

P. L. Kapitsa. *Collected Papers of P. L. Kapitsa,* ed. D. ter Haar, Vol. III (Oxford, Eng.: Pergamon Press, 1967).

Karamzin, N. M. *Izbrannye sochineniia,* 2 vols. (Moscow and Leningrad: Izdatel'stvo "Khudozhestvennaia literatura", 1964).

Keill, John. *Introductio ad veram astronomiam* (Oxford, Eng., 1718).

——*Introductio ad veram physicam . . .* (Oxford, Eng., 1702).

Kelly, John. *The Life of John Dollond F. R. S., Inventor of the Achromatic Telescope. With a copious appendix of papers referred to,* 3rd ed. with additions (London, 1808).

Khmyrov, P. D. "Vtoroi general fel'dzekhmeister Iakov Vilimovich Brius," *Artileriiskii Zhurnal* (1866), pp. 86–136, 153–199, 249–291.

Khrgian, A. Kh. *M. F. Spasskii* (Moscow: Izdatel'stvo Moskovskogo universiteta 1955).

Kircher, Athanasius, *Arco noë in tres libros digesta . . .* (Amsterdam, 1675).

——*Itinerarium exstaticum, quo mundi opificium, id est coelestis expansi, siderumque . . . compositio et structura in nova hypothesi exposuitur . . . interlocutoribus Cosmiele et Theodidacto* (Rome, 1656).

——*Magnes; sive de arte magnetica opus tripartitum . . .* (Rome, 1641, 3rd ed., 1654).

——. . . *Mundus subterraneus, in XII libros digestus, quo divinum subterrestris mundi opificium . . .* 2 vols. (Amsterdam, 1665).

Kirpicheva, I. K. *Bibliografiia v pomoshch nauchnoi rabote.* Leningrad: Akademiia nauk, 1958).

Kniazev, G. A., ed. Trudy Arkhiva, No. 1. Vol. I, *Arkhiv Akademii Nauk SSSR, Obozrenie arkhivnykh materialov,* Part I (Leningrad: Akademiia nauk, 1933).

————and Modzalevskii, L. B. Trudy Arkhiva, No. 5. Vol. II, *Arkhiv Akademii Nauk SSSR, Obozrenie arkhivnykh materialov,* Part II (Moscow and Leningrad: Akademiia nauk, 1946).

————*et al.* Trudy Arkhiva, No. 9. Vol. III, *Arkhiv Akademii Nauk SSSR, Obozrenie arkhivnykh materialov, Part III* (Moscow and Leningrad: Akademiia nauk, 1950).

————*et al.* Trudy Arkhiva, No. 16. Vol. IV, *Arkhiv Akademii Nauk SSSR, Obozrenie arkhivnykh materialov,* Part IV (Moscow and Leningrad: Akademiia nauk, 1959).

————*et al.,* eds. Trudy Arkhiva, No. 19. *Arkhiv Akademii Nauk SSSR, Obozrenie arkhivnykh materialov,* Part V (Moscow and Leningrad: Akademiia nauk 1963).

————and Shafronovski, K. I. "Istoriia Biblioteki Akademii nauk I. Bakmeistera 1776 goda," *Trudy Biblioteki Akademii nauk i fundamental'noi biblioteki obshchestvennykh nauk Akademii nauk SSSR,* Vol. VI (Moscow and Leningrad: Akademiia nauk, 1962), 251–264.

Kogan, I. I. "Rukopisi F. I. Soimonova, v sobranii Gosudarstvennogo Istoricheskogo Muzeiia," in *Problemy Istochnikovedeniia,* Vol. VII (Moscow: Akademiia Nauk, 1959), 228–238.

Kononkov, A. F. "Pervye magisterskie dissertatsii po istorii fiziki v Moskovskom universitete", in *Voprosy istorii fiziko-matematicheskikh Nauk* ed. K. A. Rybnikov *et al.* (Moscow:Gosudarstvennoe Izdatel'stvo "Vysshaia Shkola," 1963), pp. 248–251.

Kopelevich. Iu. Kh. "Perepiska L. Eilera i Ia. V. Briusa," *Istoriko-matematicheskie Issledovaniia* (Moscow) X (1957), 95–116.

Korf, M. A. "Iz zapisok barona (v posledstvii grafa) M. A. Korfa," *Russkaia Starina,* CII (No. 18, 1900), 27–50.

Korovin, G. M. *Biblioteka Lomonosova—materialy dlia kharakteristiki literatury, izpol' zovannoi Lomonosovym v ego trudakh, i katalog ego lichnoi biblioteki* (Moscow and Leningrad: Akademiia nauk, 1961).

Kotel'nikov, S. K. *Slovo o pol'ze uprazhneniia v chistykh matematicheskikh razsuzhdeniiakh, predlozhennoe v publichnom sobranii Imp. Akademii Nauk, Sentiabria 6 dnia 1761 goda professorom Semenom Kotel'nikovym* (V Sanktpeterburge pechatano pri imp. Akademii nauk [1761]).

Krafft, G. W. "De Calore et Frigore Experimenta Varia," *Commentarii* (St. Petersburg), XIV (1751), 218–239.

Krasotkina, T. A. "Iz istorii nauchno-prosvetitel'nykh nachinanii Peterburgskoi Akademii nauk v 18om veke," TIIE-T, XXXI (1960), 364–389.

Kravets, T. P. "276 Zametok M. V. Lomonosova po fizike i korpuskuliarnoi filosofii," TIIE-T, XXII (1959), 106–113.

————"N'iuton i izuchenie ego trudov v Rossii," in *Isaak N'iuton 1643–1727, Sbornik statei k trekhsotletiiu so dnia rozhdeniia,* ed. S. I. Vavilov (Moscow and Leningrad: Akademiia nauk, 1943), pp. 312–328.

Krutikova, M. V., and Chernikov, A. M. "Didro v Akademii Nauk," *Vestnik Akademii Nauk SSSR* (No. 6, 1947), pp. 64–73.

Krylov, A. N. "Leonard Eiler," in *Leonard Eiler, 1707–1783. Sbornik Statei i materialov k 150-letiiu so dnia smerti,* ed. A. M. Deborin (Moscow and Leningrad: Akademiia nauk, 1953), pp. 1–28.

——*Sobranie trudov,* Vol. VII (Moscow and Leningrad: Akademiia nauk, 1936).

Kudriavtsev, P. S. "Lomonosov i N'iuton," TIIE-T, IV (1955), 32–51.

Kurganov, N. G. *Rossiiskaia universal'naia Grammatika, ili Vseobshchee Pismoslovie, predlagaiushchee legchaishchii sposob osnovatel'nago ucheniia ruskomu [sic] iazyku s sedm'iu prisovokupleniiami raznykh uchebnykh i poleznozabavnykh veshchei* (St. Petersburg, 1769).

Kuznetsov, B. G. *Tvorcheskii put' Lomonosova* (Moscow: Akademiia nauk, 1961).

Kuz'min, S. "Zabytaia rukopis' Didro (Besedy Didro s Ekaterinoi II)," *Literaturnoe Nasledstvo* (Moscow), LVIII (1952), 927–948.

Lalande, Joseph le Français de Jérôme. *Bibliographie Astronomique avec l'Histoire de l'Astronomie, depuis 1781 jusqu'à 1802* (Paris, 1803).

Lamanskii, V. I. *Mikhail Vasil'evich Lomonosov* (St. Petersburg, 1883).

Lambert, J. H. *Beyträge zum Gebrauche der Mathematick, und deren Anwendung,* Vol. III (Berlin, 1772).

Lapshin, I. I. "Filosofskiia vozzreniia Radishcheva," in *Sobranie sochinenii Radishcheva,* Vol. XI, ed. A. K. Borozdin *et al.* (Moscow: Izdanie M. I. Akinfieva, 1907), vii–xxxii.

Lebedev, D. M. *Geografiia v Rossii petrovskogo vremeni* (Moscow and Leningrad: Akademiia nauk, 1950).

Leibniz, G. W. *A Collection of Papers which passed between the late learned Mr. Leibnitz and Dr. Clarke in the Years 1715 and 1716 . . .* (London, 1717).

——*Virorum celeberr. Got. Guil. Leibnitii et Iohannis Bernullis, Commercium philosophicum et mathematicum,* 2 vols. (Lausanne and Geneva, 1745).

Leutmann, Johann Georg. *Anmerkungen zum Glasschleifen* (Wittenberg, 1729).

Levin, Iu. D. "Angliiskaia prosvetitel'skaia zhurnalistika v russkoi literature XVIII veka," in *Epokha prosveshcheniia,* ed. M. P. Alekseev (Leningrad: Nauka, Leningradskoe otdelenie, 1967), pp. 3–109.

L'Isle, Joseph De. *Predlozhenie o merianii zemli v Rossii, chtennoe v konferentsii sanktpeterburgskiia imperatorskiia [sic] Akademii nauk, genvaria 21 dnia, 1737 goda chrez gospodina de l'Ilia, pervogo professora astronomii.* (St. Petersburg, 1737).

——with Daniel Bernoulli. *Discours lu dans l'Assemblée publique de l'Académie des Sciences le 2 mars 1728 par Mr. De l'Isle, avec la Réponse de Mr. Bernulli [sic]* (St. Petersburg, 1728).

Liubimenko, I. I. "Ob osnovanii Rossiiskoi Akademii," *Arkhiv istorii nauki i tekhniki* (No. 6, 1935), pp. 97–116.

——*Trudy Arkhiva,* No. 2. comp. and ed. D. S. Rozhdestvenskii *et al. Uchenaia korrespondentsiia Akademii Nauk XVIII veka, Nauchnoe opisanie 1766–1782* (Moscow and Leningrad: Akademiia nauk, 1937).

Liubimov, N[ikolai]. *Zhisn' i trudy Lomonosova,* Part I (Moscow: Universitetskaia tipografiia, 1872).

Locke, John. *Some Thoughts Concerning Education,* (Cambridge, Eng., 1899).

Loginova, A. I., and Firsova, I. N., eds. *Katalog arkhivovedcheskoi literatury 1917–1959 gg.* (Moscow: [Glavnoe arkhivnoe upravlenie], 1961).

Lomonosov, M. V. *Polnoe sobranie sochinenii,* Vols. I–X (Moscow and Leningrad: Akademiia nauk, 1950–1957).

———*Mikhail Vasil'evich Lomonosov on the Corpuscular Theory,* tr., with an introduction by Henry M. Leicester (Cambridge: Harvard University Press, 1970).

———*Trudy M. V. Lomonosova po fizike i khimii,* ed. B. N. Menshutkin (Leningrad: Akademiia nauk, 1936).

Luppol, I. K. "Pol' Gol'bakh—russkii akademik (k 150-letiiu so dnia smerti)" *Vestnik Akademii nauk SSSR,* (Nos. 4–5, 1939), pp. 163–167.

Luppov, S. P. "Istoriia Biblioteki Akademii Nauk SSSR, i ee knizhnykh fondov," in *250 let Biblioteke Akademii Nauk SSSR, Sbornik dokladov nauchnoi konferentsii 25–26 Noiabria 1961,* cd. M. S. Filippov *et al.* (Moscow and Leningrad: Izdatel'stvo "Nauka," 1965), pp. 268–284.

Luttrell, Narcissus. *A Brief Historical Relation of State Affairs from September 1678 to April 1714,* 6 vols. (Oxford, Eng., 1857).

Mabilleau, Léopold. *Histoire de la philosophie atomistique . . .* (Paris, 1895).

Macaulay, Thomas, *The History of England from the Accession of James the Second,* 5 vols. (London, 1849–1861).

MacLaurin, Colin. *An Account of Sir Isaac Newton's Philosophical Discoveries . . .* in Four Books. (London, 1748).

MacPike, E. F. *Hevelius, Flamsteed, and Halley, Three Contemporary Astronomers and Their Mutual Relations.* (London: Taylor and Francis, 1937).

———*The Correspondence and Papers of Edmond Halley . . .* (London: Taylor and Francis, 1937).

Magnitskii, Leontii. *Arifmetika, sirech nauka chislitel'naia . . .* (Moscow, 1703).

Maikov, C. N. "Materialy dlia biografii kn. A. D. Kantemira," in *Sbornik otdeleniia russkogo iazyka i slovesnosti Imperatorskoi Akademii Nauk* (St. Petersburg), LXXIII (1903), 1–330.

Mairan, D'Ortous de. *Lettre à Mme. . . . sur la question des forces vives, en réponse aux objections qu'elle lui fait sur ce sujet dans ses institutions de physique* (Paris, 1741).

Makogonenko, G. P. *Nikolai Novikov i russkoe Prosveshchenie XVIII veka* (Moscow and Leningrad: Gosudarstvennoe izdatel'stvo khudozhestvennoi literatury, 1951).

Malin, M. "Anglo-russkie kul'turnye i nauchnye sviazi (do osnovaniia Peterburgskoi Akademii nauk), *Vestnik Istorii Mirovoi Kul'tury* (No. 3, 1957), pp. 98–107.

Martynov, A. A. *Russkiia dostopamiatnosti.* Vol. I, *Sukhareva bashnia v Moskve* (Moscow, 1825).

Masanov, Iu.F. *Slovar' psevdonimov russkikh pisatelei i obshchestvennykh deiatelei,* 4 vols., ed. B. L. Koz'min (Moscow: Izdatel'stvo vsesoiuznoi knizhnoi palaty, 1956–1960).

Maty, P. H. *General Index to the Philosophical Transactions from the first to the end of the Seventieth Volume.* (London, 1787).

Maupertuis, Pierre Louis Moreau de. *Oeuvres . . .* rev. ed., 4 vols. (Lyons, 1756).

Maury, Louis-Ferdinand Alfred. *Les Académies d'autrefois. L'Ancienne Académie des Sciences,* 2nd ed. (Paris, 1864).

McKie, Douglas. "Bernard le Bovier de Fontenelle 1657–1757, F.R.S.," *Notes and Records of the Royal Society of London,* XII (No. 2, 1957), 193–200.

Menshutkin, B. N. *Zhisneopisanie Mikhaila Vasil'evicha Lomonosova, s dopolneniiami P. N. Berkova, S. I. Vavilova i L. B. Modzalevskogo,* 3rd ed. (Moscow and Leningrad: Akademiia nauk, 1957).

Metzger, Hélène. *Newton, Stahl, Boerhaave, et la Doctrine Chimique,* (Paris: F. Alcan, 1930).

———*Attraction Universelle et Religion Naturelle chez quelques Commentateurs Anglais de Newton.* (Paris: Hermann & Cie, Editeurs, 1938).

Michell, John. *A treatise of Artificial Magnets; in which is shewn an easy and expeditious method of making them superior to the best natural ones . . .* (Cambridge, Eng., 1751).

Michelotti, P. A., s.v.Jurin, James.

Miller, D. C. "Newton and Optics," in *Sir Isaac Newton, 1727–1927. A Bicentenary Evaluation of His Work.* (Baltimore: Williams & Wilkins Company, 1928), pp. 15–50.

Mitchell, Mairin, F.R.C.S. *The Maritime History of Russia 848–1948* (London: Sidgwick and Jackson, 1949).

Modzalevskii, L. B. "Literaturnaia polemika Lomonosova i Trediakovskogo v 'Ezhemesiachnykh Sochineniiakh' 1755 goda." *XVIII Vek. Sbornik 4,* ed. P. N. Berkov (Moscow and Leningrad: Akademiia nauk, 1959), pp. 45–65.

Moiseeva, G. N. "Ekzempliar 'Kamernogo kataloga' prinadlezhavshchii M. V. Lomonosovu," *Trudy Biblioteki Akademii nauk SSSR i fundamental'noi biblioteki obshchestvennykh nauk Akademii nauk SSSR,* Vol. VII (Moscow and Leningrad: Akademiia Nauk, 1963), 108–116.

Montucla, J. F. *Histoire des Mathématiques,* Vol. III, new ed. (Paris, 1802).

More, Louis Trenchard. *Isaac Newton: A Biography* (New York: C. Scribner's Sons, 1934).

Morozov, A. A. *M. V. Lomonosov—put' k zrelosti, 1711–1741.* (Moscow and Leningrad: Akademiia nauk, 1962).

Motley, J. L. *Peter the Great.* New York, 1877.

Mouy, Paul. *Le Développement de la Physique Cartésienne,* 1646–1712, . . . (Paris: J. Vrin, 1934).

Navrotskii, N. N. "Khronologicheskaia i sinkhronicheskaia tablitsa izvestneishikh i primechatel'neishikh matematikov vo vsekh vekakh, p. 1800 g.," *Zhurnal ministerstva narodnogo prosveshcheniia,* V (No. 2, 1846), 61–70.

Neustroev, A. N. *Istoricheskoe rozyskanie o russkikh povremennykh izdaniiakh i sbornikakh za 1703–1812 gg., bibliograficheski i v khronologicheskom poriadke opisannykh . . .* (St. Petersburg, 1875).

Newton, Sir Isaac. *Isaaci Newtoni Opera Quae Exstant Omnia Commentariis Illustrabat Samuel Horsley,* 5 vols., ed. Samuel Horsley (London, 1779–1785.

———*Abrégé de la chronologie de M. le chevalier Isaac Newton,* tr. Nicolas Fréret (Paris, 1725).

———*Arithmetica Universalis . . .* (London, 1707).

———*Universal Arithmetick: or, A Treatise of Arithmetical Composition and*

Resolution ... tr. Mr. Joseph Raphson, rev. by Mr. Samuel Cunn (London, 1720; 1728).

——*The Correspondence of Isaac Newton,* ed. H. W. Turnbull *et al.,* Vols. I–IV (Cambridge, Eng.: the University Press for the Royal Society, 1959–   ).

——*The Mathematical Papers of Isaac Newton,* ed. D. T. Whiteside, assisted by M. A. Hoskin and A. Prag, Vols. I–III (Cambridge, Eng.: the University Press, 1967–1969). More volumes forthcoming.

——*The Mathematical Works of Isaac Newton,* assembled with an introduction by D. T. Whiteside, Vols. I–II (New York: Johnson Reprint Corporation, 1964–1967).

——*The Method of Fluxions and Infinite Series; with Its Application to the Geometry of Curve-lines* ... (London, 1736).

——*Optical Lectures Read in the Publick Schools of the University of Cambridge, Anno Domini, 1669. By the Late Sir Isaac Newton* (London, 1728).

——*Lektsii po optike,* 2nd ed., tr. S. I. Vavilov (Moscow and Leningrad: Akademiia nauk, 1946; 1st Russian ed., 1927).

——*Opticks: Or, a Treatise of the Reflexions, Refractions, Inflexions and Colours of Light. Also Two Treatises of the Species and Magnitude of Curvilinear Figures* (London, 1704, 1718, 1721, 1730; Batavian tr., 1749; French tr. by M. Coste, Paris, 1722, 1739, 1749; German tr. by W. Abendroth from the Latin of John Colson, London, 1736, 1737; 2nd French tr. by G. L. Le Clerc, Paris, 1740).

——*Philosophiae Naturalis Principia Mathematica* (ed. Roger Cotes, London, 1687, and Cambridge, Eng., 1713; ed. Henry Pemberton, London, 1726; ed. T. Le Sieur and F. Jacquier, Geneva, 1739–1742; Cologne, 1700; Eng. ed. by A. Motte, 2 vols., London, 1729 and 1803; same ed., New York, 1848; ed. R. Thorp, London, 1777 and 1802; Fr. ed. by Madame la Marquise Chastellet, 2 vols., Paris, 1759).

——Various items in the *Philosophical Transactions,* including (s.v. Maty, P.H.):

Answer to Mr. Linus' letter, animadverting on the theory of light and colours, Royal Society of London, *Philosophical Transactions,* IX (1674), 219; abridged ed. I (1716), 161.

Considerations on Mr. Linus' reply, X (1675), 501.

Another letter on the same argument, X (1675), 503.

A particular answer to Mr. Linus' letter, XI (1676), 556; I (1716), 163.

Answer to Mr. Pardie's letter on Newton's theory of light, VII (1672), 5014; I (1716), 142.

Answer to some considerations on Newton's doctrine of light and colours, VII (1672), 5084; I (1716), 202.

Answer, further explaining his theory of light and colours, VIII (1673), 6087; I (1716), 158.

Hopes of perfecting telescopes by reflections rather than refractions, VIII (1673), 6090; I (1716), 160.

On the number of colours, VIII (1673), 6108; I (1716), 157.

Experiments proposed in relation to Mr. Newton's theory of light: With observations, VII (1672), 4059; I (1716), 135.

Farther suggestions about his reflecting telescope: Together with his table of apertures and charges for the several lengths of that instrument, VII (1672), 4032; I (1716), 200.

Answer to some objections by an ingenious French philosopher, to the new reflecting telescope, VII (1672), 4034; I (1716), 201.

New Theory about *Light* and *Colours,* VI (1671), 3075.

Scala graduum caloris et frigoris, XXII (1701), 824.

Some considerations upon part of the letter of M. de Berce concerning the catadioptrical telescope pretended to be improved and refined, by M. Cassegrain, VII (1672), 4056; I (1716), 204.

————*A Treatise of the System of the World* (London, 1728–1731).

————*Unpublished scientific papers of Isaac Newton; a selection from the Portsmouth collection in the University Library, Cambridge,* ed. and tr. A. Rupert Hall and Marie Boas Hall (Cambridge, Eng.: the University Press, 1962).

Nicolson, M. H. *Newton demands the Muse: Newton's Opticks and the Eighteenth Century Poets* (Princeton, N. J.: Princeton University Press, 1946).

Nikol'skii, N. K. "K voprosu ob istoricheskom znachenii i pervonachal'nom knizhnom fonde Biblioteki Akademii Nauk SSSR. Publikatsiia i posleslovie G. A. Kniazeva i K. I. Shafranovskogo," in *Trudy Biblioteki Akademii nauk SSSR i fundamental'noi biblioteki obshchestvennykh nauk Akademii nauk SSSR,* Vol. VII (Moscow and Leningrad: Akademiia nauk, 1963), 70–86.

Novikov, N. I. *Truten . . . 1769–1770* (St. Petersburg, 1865).

————*Izbrannye sochineniia,* ed. by G. P. Makogonenko (Moscow: Gosudarstvennoe izdatel'stvo khudozhestvennoi literatury, 1951).

Ostrovitianov, K. V., ed. *Istoriia Akademii Nauk SSSR,* Vol. I (1724–1803) (Moscow and Leningrad: Akademiia nauk, 1958).

Ovsianiko-Kulikovskii, D. N. *Istoriia russkoi intelligentsii—itogi russkoi khudozhestvennoi literatury XIX veka* (Moscow: V. M. Sablin, 1906).

Pabst, E. "Nicolaus Bulow, Astronom, Dolmetsch und Leibarzt beim Grossfürsten in Russland," in *Beiträge zur Kunde Ehst-, Liv- und Kurlands,* Vol. I (Reval, 1873), 83–86.

Parrot, G. F. *Grundriss der Theoretischen Physik im Gebrauche für Vorlesungen,* Part I (Dorpat, 1809).

Pavlova, G. E. *M. V. Lomonosov v vospominaniiakh i kharakteristikakh sovremennikov* (Moscow and Leningrad: Akademiia nauk, 1962).

Pavlova, T. E. "Laland i Peterburgskaia Akademiia Nauk," in *Voprosy istorii fiziko-matematicheskikh nauk,* ed. K. A. Rybnikov *et al.* (Moscow: Gosudarstvennoe Izdatel'stvo "Vysshaia Shkola," 1963), pp. 489–494.

Pekarskii, P. P. "Ekaterina i Eiler," *Zapiski Imperatorskoi Akademii Nauk,* VI (1865), 59–92.

————*Istoriia Imperatorskoi Akademii Nauk v Peterburge,* 2 vols. (St. Petersburg, 1870–1873).

————*Nauka i literatura v Rossi pri Petre Velikom.* Vol. I, *Vvedenie v istoriiu prosveshcheniia v Rossii XVIII stoletiia;* Vol. II, *Opisanie slaviano-russkikh knig i tipografii 1698–1725 godov* (St. Petersburg, 1862).

Pemberton, [H. A.] *A View of Sir Isaac Newton's Philosophy,* (Dublin, 1728).

Perry, John. *The State of Russia under the Present Czar . . .* (London, 1716).

Peter I. *Journal de Pierre le Grand depuis l'année 1698. Jusqu'à la conclusion de la paix de Neustadt . . . imprimé d'après les Mss corrigés de la propre main de sa Majésté Impériale . . .*, ed. Mikhail Shcherbatov (Berlin, 1773).

———*Pis'ma i bumagi imperatora Petra Velikogo,* 11 vols. (St. Petersburg or Moscow: Gosudarstvennaia tipografiia [and others], 1887–1964).

Philalethes Cantabrigiensis (pseud. for James Jurin). *Geometry no Friend of Infidelity: or, a Defence of Sir Isaac Newton and the British Mathematicians in a Letter to the author of the Analyst . . .* (London, 1734).

———*The Minute Mathematician or the Free Thinker no Just-Thinker . . . a second letter to the author of the Analyst; containing a defence of Sir Isaac Newton and the British Mathematicians, against a late pamphlet entituled, A Defence of Free-Thinking in Mathematicks* (London, 1735).

Plekhanov, G. V., *Istoriia russkoi obshchestvennoi mysli,* in *Sochineniia,* ed. D. B. Riazanov, Vol. XXI (Moscow and Leningrad: Gosudarstvennoe Izdatel'stvo, 1925).

Pliushkin, L. N., and Geidrich, E. K., comp. *Katalog imeiushchikhsia v prodazhe izdanii Akademii nauk 1769–1935[:] knigi na russkom iazyke,* ed. N. K. Kuz'min (Moscow and Leningrad: Akademiia nauk, 1936).

Pod'iapol'skaia, E. P. "Shifrovannaia perepiska v Rossii v pervoi chetverti XVIII veka," *Problemy istochnikovedeniia,* Vol. VIII (Moscow: Akademiia Nauk, 1959), 314–342.

Podozerskaia, G. F. "Iz istorii katalogov Biblioteki Akademii nauk (1714–1850 gg.)," *Trudy Biblioteki Akademii nauk SSSR i fundamental'noi biblioteki obshchestvennykh nauk Akademii nauk, SSSR,* Vol. VII (Moscow and Leningrad: Akademiia nauk, 1963), 117–137.

Polikarpov, Fedor. *Leksikon treiazychnyi; sirech rechenii slavenskikh Ellindgrecheskii i latinskii sokrovishche iz razlichnykh drevnikh i novykh knig sobranoe i po slavenskom alfavite v chine razpolozhenoe* (Moscow, 1704).

Poludenskii, M. P. "Petr Velikii v Parizhe," *Russkii Arkhiv* (No. 6, 1865), pp. 675–702.

Pope, Alexander. *The Works of Alexander Pope, Esq., with Notes and Illustrations by Himself and Others. To Which Are Added a New Life of the Author, an Estimate of His Poetical Character and Writings, and Occasional Remarks, by William Roscoe, Esq.,* new ed., 8 vols. (London, 1847).

———*The Works of Alexander Pope,* ed. Whitewell Elwin and W. J. Courthope, 10 vols. (London, 1871–1889).

———*The Essay on Man . . .,* notes by William Warburton (London, 1745).

———*Essais sur l'Homme . . . Traduit de l'Anglois en François, par M. D. S.[ilhouette]* (London, and Amsterdam, 1736).

———*Les Oeuvres de M. Pope traduites en François: Les Principes de la Morale et du Goût. En deux Poèmes,* new ed., tr. M. Du Resnel (Paris, 1745).

———*Opyt o Cheloveke gospodina Pope,* tr. Nikolai Popovskii (Moscow, 1802; reprint of 1757 ed.).

Popov, Nil [A.]. *V. N. Tatishchev i ego vremia* (Moscow, 1861).

Pososhkov, I. T. *Zaveshchanie otecheskoe*—[*sochinenie I. T. Pososhkovo*] *Novoe izdanie, dopolnennoe vnov' otkrytoiu vtoroiu polovinoiu "Zaveshchaniia,"* ed. E. M. Prilezhaev (St. Petersburg, 1893).

Posselt, Moritz. *Der General und Admiral Lefort—Sein Leben und seine Zeit; Ein beitrag zur geschichte Peter's des grossen,* 2 vols. (Frankfurt am Main, 1866).

Priima, F. Ia. "Antiokh Kantemir i ego frantsuzskie literaturnye sviazi," in *Trudy otdela novoi russkoi literatury, Institut russkoi literatury,* Vol. I (Moscow and Leningrad: Akademiia nauk, 1957).

Prudnikov, V. E. *Russkie pedagogiki-matematiki XVIII–XIX vekov* (Moscow: Gosudarstvennoe uchebno-pedagogicheskoe izdatel'stvo Ministerstva prosveshcheniia RSFSR, 1956).

Pypin, A. N. "Dopetrovskoe predanie v XVIII veke," *Vestnik Evropy* (No. 7, 1886), pp. 306–345.

——*Russkoe masonstsvo,* XVIII i pervaia chetvert XIX v., ed. G. V. Vernadskii (Petrograd: Ogni, 1916).

Rabiqueau, Charles. *Lettre èlèctrique sur la mort de M. Richmann* (Paris, 1754).

Radishchev, A. N. *Polnoe sobranie sochinenii* 3 vols. (Moscow and Leningrad: Akademiia nauk, 1938–1952).

Radovskii, M. I. "Angliiskii naturalist XVII veka Gans Sloan i ego nauchnye sviazi s Peterburgskoi Akademiei Nauk," TIIE-T, XXIV (No. 5, 1958).

——*Iz istorii anglo-russkikh nauchnykh sviazei,* (Moscow and Leningrad: Akademiia nauk, 1961).

——*Antiokh Kantemir i Peterburgskaia Akademiia Nauk* (Moscow and Leningrad: Akademiia nauk, 1959).

——*M. V. Lomonosov i Peterburgskaia Akademiia Nauk* (Moscow and Leningrad: Akademiia nauk, 1961).

Raikov, B. E. *Ocherki po istorii geliotsentricheskogo mirovozzreniia v Rossii, iz proshlogo russkogo estestvoznaniia,* 2nd ed. (Moscow and Leningrad: Akademiia nauk, 1947).

Rainov, T. I. "Daniel Bernulii i ego rabota v Peterburgskoi Akademiia Nauk (k 200 letiiu 'Gidrodinamiki')," *Vestnik Akademii Nauk SSSR,* (No. 7–8, 1938), pp. 84–93.

——*Nauka v Rossii XI–XVII vekov; ocherki po istorii do nauchnykh i estestvenno-nauchnykh vozzrenii na prirodu.* (Moscow and Leningrad: Akademiia nauk, 1940).

——"N'iuton i russkoe estestvoznanie," in *Isaak N'iuton 1643–1727, sbornik statei k trekhstoletiiu so dnia rozhdeniia,* ed. S. I. Vavilov (Moscow and Leningrad: Akademiia nauk, 1943), 329–344.

Ranft, Michael. *Das merkwurdige Leben des beruehmten Fuerstens Menshikow welches mit vielen Anekdoten ans Licht stellt ein Liebhaber der Wahrheit* (Leipzig, 1774).

Raphson, Joseph. *The History of Fluxions, shewing in a compendious manner the first Rise and various Improvements made in that Incomparable Method* (London, 1715).

Ravier, Emile *Bibliographie des oeuvres de Leibniz* (Paris: Librairie Félix Alcan, 1937).

Réau, Louis ed. *Correspondance de Falconet avec Catherine II, 1769–1778, Publiée avec une introduction et des notes par Louis Réau* (Paris: E. Champion, 1921).

The Record of the Royal Society of London for the Promotion of Natural Knowledge, 4th ed. (London: The Royal Society, 1940).

Richmann, G. W., s.v. Rikhman, G. V.

Richter, Liselotte. *Leibniz und sein Russlandbild.* (Berlin: Akademie-Verlag, 1946).

Ridgway, R. S. *La Propaganda philosophique dans les tragédies de Voltaire.* Vol. XV in Th. Besterman's *"Studies on Voltaire and the Eighteenth Century"* (Geneva: Institut et Musée Voltaire, 1961).

Rigaud, S. J. *Defence of Halley against the Charge of Religious Infidelity* (Oxford, 1844).

Rigaud, Stephen Peter. *Historical Essay on the First publication of Sir Isaac Newton's Principia* (Oxford, Eng., 1838).

———*Correspondence of Scientific Men of the Seventeenth Century . . .* , 2 vols. (Oxford, Eng., 1841).

Rihs, Charles. *Voltaire: Recherches sur les origines du matérialisme historique* (Geneva: Droz, 1962).

Rikhman, G. V. *Trudy po fizike,* ed. A. T. Grigor'ian (Moscow: Akademiia nauk, Institut Istorii Estestvoznaniia i Tekhniki, 1956).

Robinson, Sir Robert, PRS. "An Address of Welcome to the Delegates," the introduction to *The Royal Society Newton Tercentenary Celebration 15–16 July, 1946* (Cambridge, Eng.: the University Press, 1947).

Rohault, Jacques. *System of Natural Philosophy,* Vol. I (London, 1735).

Rousset de Missy, Jean. *Mémoires du Règne de Pierre le Grand Empereur de Russie, père de la Patrie* (Amsterdam, 1730).

Rovinskii, D. A. *Russkie narodnyia kartinki . . .* , 5 vols. ([St. Petersburg], 1881).

———*Podrobnyi slovar' russkikh gravirovannykh portretov . . .* , 4 vols. (St. Petersburg, 1886–1889).

Rumovskii, S. Ia. *Rech o nachale i prirashchenii optiki, do nyneshnikh vremen* (St. Petersburg, 1763).

Russell, Bertrand. *A Critical Exposition of the Philosophy of Leibniz with an Appendix of Leading Passages* (London: G. Allen and Unwin, 1937).

Ryan, W. F. "Rathbone's Surveyor (1616/1625): The first Russian translation from English," *Oxford Slavonic Papers,* XI (1964), 1–7.

Sakharov, I. P. *Zapiski russkikh liudei (grafa Matveeva, Krekshina, Zheliabuzhskago, i Medvedeva) Sobytiia vremen Petra Velikago,* Vol. (St. Petersburg, 1841).

Scott, Wilson. "The Significance of 'Hard Bodies' in the History of Scientific Thought," *Isis,* L (1959), 199–210.

Shafronovskii, K. I., and Fedotova, Z. S. "Bibliografiia pechatnykh rabot Biblioteki Akademii nauk za 1714–1916 gg.," *Trudy Biblioteki Akademii nauk SSSR i fundamental'noi biblioteki obshchestvennykh nauk Akademii nauk SSSR,* Vol. VI (Moscow and Leningrad: Akademiia nauk, 1962), pp. 332–353.

Shevyrev, Stepan. *Istoriia imperatorskago Moskovskago universiteta napisannaia k stoletnemu iubileiu, 1755–1855* (Moscow, 1855).

Shishkin, I. "Mikhail Avramov, odin iz protivnikov petrovskoi reformy," *Nevskii Sbornik*, Vol. I (St. Petersburg, 1867), 375–429.

Shmurlo, E. F. *Mitropolit Evgenii kak uchenyi. Rannie gody zhizni 1767–1804* (St. Petersburg, 1888).

Shpet, Gustav. *Ocherk razvitiia russkoi filosofii*, Vol. I (Petrograd, 1922).

Sipovskii, V. V. *Ocherki iz istorii russkago romana i povesti (materialy po bibliografii, istorii i teorii russkago romana)*, Vol. I, *XVIII vek*, 2nd ed. (St. Petersburg: Otdeleniie Imperatorskoi Akademii nauk, 1903).

Sk. Al. "O nachale morskago korpusa," *Zapiski gidrograficheskago departamenta morskago ministerstva*, IV (1846), Section II, pp. 312–319.

Slavenas, P. V. "Rasprostranenie ucheniia Kopernika i astronomicheskikh znanii v Litve," in K. A. Rybnikov *et al. Voprosy istorii fiziko-matematicheskikh nauk* (Moscow: Gosudarstvennoe Izdatel'stvo "Vysshaia Shkola," 1963), pp. 515–516.

Slovar' professorov i prepodavatelei Moskovskogo Universiteta, Vol. I (Moscow, 1855).

Smirnov, V. I., and Iushkevich, A. P., eds. *Leonard Eiler. Perepiska-annotirovannyi ukazatel'* (Leningrad: Izdatel'stvo "Nauka," Leningradskoe Otdelenie, 1967).

Sobol', S. L. *Istoriia mikroskopa i mikroskopicheskikh issledovanii v Rossii v XVIII veke*, (Moscow and Leningrad: Akademiia nauk, 1949).

Soimonov, F. I. *Svetil'nik morskoi, to est' opisanie vostochnago ili variazhskago moria* (St. Petersburg, 1738).

———*Kratkoe is'iasnenie o astronomii v kotorom pokazany velichiny i razstoianiia nebesnykh tel kupno s poriadkom v ikh raspolozhenii i dvizhenii po raznym sistemam i o velichiny i dvizhenii zemnago globusa vypisano iz astronomicheskikh i fizicheskikh avtorov*. (St. Petersburg, 1765).

Sokolov, Al. "Andrei Danilovich Farvarson," *Morskoi Sbornik*, Nos. 14 and 15 (November and December 1856).

———*Russkaia morskaia biblioteka 1701–1851. Ischislenie i opisanie knig, rukopisei i statei po morskomu delu za 150 let*, 2nd ed., ed. V. K. Shul'ts (St. Petersburg, 1883).

[Sokol'skii]. *Razrushenie Kopernikovoi sistemy, rossiiskoe sochinenie* (Moscow, 1815).

Sotheby & Co. *Catalogue of the Newton Papers sold by Order of the Viscount Lymington to whom they have descended from Catherine Conduitt, Viscountess Lymington, Great Niece of Sir Isaac Newton* ([London:] H. Davy, 1936).

Sopikov, V. S. *Ukazatel' k opytu russkoi bibliografii . . .* , rev. ed. (St. Petersburg: A. S. Suvorin, 1908)

Speranskii, M. M. "Fizika vybrannaia iz luchshikh avktorov, razpolozhennaia i dopolnennaia nevskoi seminarii filosofii i fiziki uchitelem Mikhailom Speranskim, 1797 goda. V. Sanktpiterburge," ed. O. M. Bodianskii, *Chteniia v Imperatorskom Obshchestve istorii i drevnostei rossiiskikh pri Moskovskom Universitete*, III (July–September 1871), Part II, 1–56.

Spiess, Otto. *Leonhard Euler; ein Beitrag zur Geistesgeschichte des XVIII Jahrhunderts* (Frauenfeld: Huber & Co., [1929]).

Stählin, Karl. *Originalanekdoten von Peter dem Grossen* (Leipzig, 1875; Strassburg, 1877).

Staniukovich, T. V. *Kunstkamera Peterburgskoi Akademii Nauk,* ed. V. L. Chenakal (Moscow and Leningrad: Akademii nauk, 1953).

Stäckel, Paul. *Johann Albrecht Euler.* Supplement to *Vierteljahrsschrift der Naturforschenden Gesellschaft* (Zurich), LV (1910).

Steklov, V. A. *Mikhailo Vasil'evich Lomonosov* (Berlin: Z. I. Grzhebin, 1921).

Steuart, A. F. *Scottish Influences in Russian History from the end of the 16th century to the beginning of the 19th century* (Glasgow: J. Maclehose & Sons, 1913).

Stoiunin, V. Ia. "Kniaz' Antiokh Kantemir v Londone," *Vestnik Evropy* (March 1867), pp. 224–273; (June 1867), pp. 97–139.

———"Kniaz' Antiokh Kantemir v Parizhe," *Vestnik Evropy* (August 1880), pp. 577–620; (September 1880), pp. 173–222.

Stukeley, William. *Memoirs of Sir Isaac Newton's Life . . .* , ed. A. Hastings White (London: Taylor and Francis, 1936).

Sukhomlinov, M. I. *Issledovaniia i stat'i po russkoi literature i prosveshcheniiu,* Vols. I and II (St. Petersburg, 1889).

———*Istoriia Rossiiskoi Akademii,* 8 vols. (St. Petersburg, 1874–1888). Individual volumes appended as supplements to *Zapiski Imperatorskoi Akademii Nauk,* XXIV (1874), XXVII (1876), XXIX (1876–1877), XXXII (1878–1879), XXXVIII (1880–1881), XLII (1882), XLIX (1884), LVIII (1888).

———"Lomonosov—student Marburgskogo universiteta," *Russkii Vestnik,* (No. 1, 1861).

Sushkevich, A. K. "Materialy dlia istorii algebry v Rossii," in *Istoriko-matematicheskie issledovaniia,* Vol. IV (Moscow: Gosudarstvennoe izdatel'stvo tekhniko-teoreticheskoi literatury, 1951), 237–545.

Sutkevich, A. "Leonard Eiler (V dvukhsotniiu godovshchinu dnia ego rozhdeniia)," *Russkaia Starina,* CXXXII (1907), 467–506.

Sviatskii, D. O. "Chudesnost' i estestvennost' v nebesnykh iavleniiakh po predstavlenniiam nashikh predkov," *Mirovedenie,* XVI (No. 2, 1927), 91–102.

Tacquet, Cl. A. *Elementa Euclidea Geometriæ planæ ac solidæ, et selecta ex Archimede theoremata,* 1st Cambridge ed. (Cambridge, Eng., 1722).

Taylor, E. G. R. *The Mathematical Practitioners of Tudor and Stuart England,* (Cambridge, Eng.: the University Press, 1954).

Thomas, C. G. (afterward Thomas-Stanford, Sir C., Bart). *Early editions of Euclid's Elements* (London: Bibliographical Society, 1926).

Tikhonravov, N. S. "Istoriia izdaniia 'Opyta o cheloveke' v perevode Popovskogo," *Russkii Arkhiv,* (No. 7–8, 1872), 1311–1322; also contained in his *Sochineniia,* III, Pt. I, (Moscow, 1898).

———, ed. *Letopisi russkoi literatury i drevnosti,* Vol. I (Moscow, 1859).

Timbs, John. *London and Westminster: City and Suburb . . .* , (London, 1868 [1867]).

Timiriasev, A. K., ed. *Ocherki po istorii fiziki v Rossii* (Moscow, 1949).

Tiulichev, D. V. "Iz istorii komplektovaniia Biblioteki Akademii nauk SSSR otechestvennymi knigami (XVIII vek)," in *Trudy Biblioteki Akademii nauk SSSR i fundamental'noi biblioteki obshchestvennykh nauk Akademii nauk SSSR,* Vol. VII (Moscow and Leningrad: Akademiia Nauk, 1963), 87–107.

Tiulin, I. A. "Razvitie mekhaniki v Moskovskom Universitete v 18 i 19 veke," *Istoriko-matematischeskie issledovaniia,* Vol. VIII (Moscow: Gosudarstvennoe izdatel'stvo tekhniko-teoreticheskoi literatury, 1955), 489–536.

Todhunter, I. F.R.S. *A History of the Mathematical Theories of Attraction and the Figure of the Earth, from the Time of Newton to That of Laplace,* 2 vols. (London, 1873).

Topchiev, A. V., *et al. Letopis' zhizni i tvorchestva M. V. Lomonosova* (Moscow and Leningrad: Akademiia nauk, 1961).

Tourneux, Maurice. *Diderot et Catherine II* (Paris, 1899).

Treat, I. F. *Un Cosmopolite du XVIIIe Siècle Francesco Algarotti,* (Trévoux, 1913).

Tukalevskii, Vladimir. "Glavnye cherty mirosozertsaniia Lomonosova (Leibnits, Lomonosov)," in *M. V. Lomonosov, 1711–1911; Sbornik statei,* ed. V. V. Sipovskii (St. Petersburg: Akademiia nauk, 1911), pp. 13–32.

Tumanskii, F. V. *Sobranie raznykh zapisok i sochinenii, sluzhashchikh k dostavleniiu polnogo svedeniia o zhisni i deiatel'nosti Petra Velikogo,* Vol. III (St. Petersburg, 1787).

Ustrialov, N. G. *Istoriia tsarstvovaniia Petra Velikago,* 4 vols. (St. Petersburg, 1858–1863).

Vanderbech, M. S. "Praesens Russiae Literariae Status," in *Acta Physico-Medica Caesarea Leopoldino-Carolina naturae curiosorum,* I (1727), Appendix, 131–149.

Varenius, Bernhard. *Geographia generalis . . . emendata . . . & illustrata ab Isaac Newton* (Cambridge, Eng., 1672).

———*Geografiia general'naia . . .* (Moscow, 1718).

Vavilov, S. I. *Sobranie sochinenii,* Vol. III (Moscow: Akademiia nauk, 1956).

Venevitinov, M. A. *Russkie v Gollandii Velikoe posol'stvo 1697–98* (Moscow, 1897).

Verenet, George. *Pierre le Grand en Hollande et à Zaandam dans les années 1697 et 1717* (Utrecht: Broese, 1865).

Vereshchagin, V. A. . . . *L'ex-libris russe—Russkii knizhnyi znak,* (St. Petersburg: Pechatnia P. Golike, 1902).

Veriuzhskii, V. *Afanasii, Arkhiepiskop Kholmogorskii; ego zhizn' i trudy v sviazi s istoriei Kholmogorskoi Eparkhii . . .* (St. Petersburg: Tipografiia T. Leont'eva, 1908).

Veselago, F. F. *Kratkaia istoriia russkago flota,* 2 vols. (St. Petersburg, 1893, 1895).

———*Ocherk istorii morskago kadetskago korpusa s prilozheniem spiska vospitannikov za sto let* (St. Petersburg, 1852).

Veselovskii, A. [N.] *Zapadnoe vliianie v novoi russkoi literature* (Moscow, 1896).

Veselovskii, K. S. *Istoricheskoe obozrenie trudov Akademii Nauk na pol'zu Rossii v proshlom i tekushchem stoletiiakh* (St. Petersburg, 1865).

Viglione, Francesco. "L'Algarotti e l'Inghilterra," *Studi di Letteratura Italiana diretti da Erasmo Percopo,* Vol. XIII (Naples: N. Jovene & Co., 1922).

Viktorov, A. E. *Opisanie zapisnykh knig i bumag starinnykh dvortsovykh prikazov 1584–1725,* 2 vols. (Moscow, 1877; 1883).

Villamil, Richard de. *Newton: The Man,* with a foreword by Albert Einstein. (London [1931]).

Voltaire, François Marie Arouet de. *Correspondance,* ed. Th. Bestermann, (Geneva: Institut et musée Voltaire, 1953–   ).

——*Œuvres complètes,* 42 vols. (Paris, 1784).

——*Œuvres,* 72 vols., ed. Adrien Beuchot (Paris, 1829–1840).

——*Elémens de la Philosophie de Neuton, mis à la portée de tout le monde* (Amsterdam, 1738).

——*Lettres Philosophiques; Edition critique, avec une introduction et une commentaire par Gustave Lanson,* 2nd ed. (Paris: Hachette, 1915).

Vorontsov-Vel'iaminov, B. A. *Ocherki istorii astronomii v Rossii* (Moscow: Gosudarstvennoe izdatel'stvo tekhniko-teoreticheskoi literatury, 1956).

Voskresenskii, N. A. *Zakonodatel'nye akty Petra I, Redaktsii i proekty zakonov, zametki, doklady, donosheniia, chelobit'ia i inostrannye istochniki,* Vol. I, ed. B. I. Syromiatnikov (Moscow and Leningrad: Akademiia nauk, 1945).

Vucinich, Alexander. *Science in Russian Culture: A History to 1860* (Stanford, Calif.: Stanford University Press, 1963).

Wade, Ira O. *Studies on Voltaire, with some unpublished Papers of Mme du Châtelet* (Princeton, N.J.: Princeton University Press, 1947).

——*Micromégas: A study on the Fusion of Science, Myth, and Art.* (Princeton, N.J.: Princeton University Press, 1950).

——*Voltaire and Mme du Châtelet: An Essay on the Intellectual Activity at Cirey* (Princeton, N.J.: Princeton University Press, 1941).

Wahl, Richard. "Professor Bilfinger's Monadologie und prästabilirte Harmonie in ihrem Verhältniss zu Leibniz und Wolff," *Zeitschrift für Philosophie und philosophische Kritik,* new ser., LXXXV (No. 1, 1884), 66–92; (No. 2, 1884), 202–223.

Waliszewski, Kazimierz. *L'Héritage de Pierre le Grand; règne des femmes, gouvernement des favoris, 1725–41* (Paris: Plon-Nourrit et cie, 1900).

Wallis, John. *Operum Mathematicorum volumen Tertium* (Oxford, Eng., 1699).

Warburton, Bishop William. *The Divine Legation of Moses demonstrated,* 2 vols. (London, 1738–1741).

Weber, F. C. *The Present State of Russia,* 2 vols. in one (London, 1723 [1722]).

Weitbrecht, Josiah. "A lustrationem Richmannianam brevis dilucidatio," in G. V. Rikhman, *Trudy po fizike* (Moscow: Akademiia nauk, 1956), pp. 493–494.

——"Cogitationum physiologicarum de circulatione sanguinis. Caput de quantitate motus sanguinis," *Commentarii* (St. Petersburg), VIII (1736), 334–340.

——"Explicatio difficiliorum experimentorum circa ascensum aquae in tubos capillares," *Commentarii* (St. Petersburg), IX (1737), 275–309.

——"Tentamen theoriae, qua ascensus aquae in tubis capillaribus explicatur," *Commentarii* (St. Petersburg), VIII (1736), 261–309.

Weld, C. R. *A History of the Royal Society, with memoirs of the presidents . . .* 2 vols. (London, 1848).

Whiston, William. *A New Theory of the Earth From its Original, to the Consummation of all Things. Wherein The Creation of the World in Six Days, the Universal Deluge, And the General Conflagration, As laid down in the Holy Scriptures, Are shewn to be perfectly agreeable to Reason and Philosophy . . .* (London, 1696).

————*Praelectiones Astronomicae Cantabrigiae in Scholis publicis Habitae . . .* (London, 1707).

————*Sir Isaac Newton's Mathematick Philosophy More easily Demonstrated: With Dr. Halley's Account of Comets Illustrated . . .* (London, 1716).

Whittaker, Sir Edmund, F.R.S. *A History of the Theories of Aether and Electricity,* Vol. I (New York: Philosophical Library, 1951).

Wilkins, John. *The Discovery of a World in the Moone, or a Discourse tending to prove that 'tis probable there may be another Habitable World in that Planet* (London, 1638).

Winter, Eduard J. "Bericht von Johann Werner Paus aus dem Jahre 1732 über seine Tätigkeit auf dem Gebiete der russischen Sprache, der Literatur und der Geschichte Russlands," *Zeitschrift für Slawistik,* III (No. 5, 1958), 744–770.

————*Halle als Ausgangspunkt der deutschen Russlandkunde im 18 Jahrhundert* (Berlin: Akademie-Verlag, 1953). See also s.v. Iushkevich, A. P.

Wolff, Christian. *Gesammelte Werke,* ed. J. Ecole *et al.,* Vols. XXIX–XXX, Part 2 (Lateinische Schriften: Matheseos Universae, Book I) (Hildesheim: Georg Olms, Verlagsbuchhandlung, 1968).

————*Anfangsgründe aller mathematischen Wissenschaften,* Part II (Halle, 1710).

————*Briefe von Christian Wolff aus den Jahren 1719–1753; ein Beitrag zur Geschichte der Kaiserlichen Academie der Wissenschaften zu St. Petersburg* (St. Petersburg and Leipzig, 1860).

————*Ch. Wolff's Eigene Lebensbeschreibung,* ed. Wolff von Heinrich Wattke (Leipzig: Weidmann'sche Buchhandlung, 1841).

————*Kurtzer Unterricht von den vornehmsten Mathematischen Schriften aufgesetzet von Christian Wolff* (Halle, 1717).

————*Vernuenfttige Gedancken von den Wuerckungen der Natur, den Liebhabern der Wahrheit mitgetheilet von Christian Wolffen,* 5th ed. (Halle im Magedeburgschen, 1746).

————"Vol'fianskaia eksperimental'naia fizika s nemetskago podlinnika na latinskom iazyke sokrashchennaia, skotorogo na rossiiskii iazyk perevel Mikhailo Lomonosov," (1746) in Lomonosov's *Polnoe sobranie sochinenii,* Vol. I (published separately in St. Petersburg, 1746).

Wolska, Wanda. *La Topographie chrétienne de Cosmas Indikopleustès-théologie et science au VIe siècle* (Paris: Presses universitaires de France, 1962).

Zabelin, I. E. "Detskie gody Petra Velikago," in his *Opyty izucheniia russkikh drevnostei i istorii,* Vol. I (Moscow, 1872), 1–50.

————"Pervoe vodvorenie v Moskve Grekolatinskoi i obshchei Evropeiskoi nauki," *Chteniia v Imp. Obshchestve istorii i drevnostei rossiiskikh,* IV (No. 1, 1886), 1–24.

Zeitlinger, H. "Newton's Library and its Discovery," in Cambridge University, Trinity College Library, *Library of Sir Isaac Newton: presentation by the Pilgrim Trust . . . 30 October 1943 . . .* (Cambridge, Eng., 1944), pp. 13–21.

———, and H.C.S., *Bibliotheca Chemico-Mathematica* (London: H. Sotheran and Co., 1921).

Zeller, Eduard. *Geschichte der deutschen Philosophie, seit Leibniz*, 2nd ed. (Munich, 1875).

Zubov, V. P. "Kalorimetricheskaia formula Rikhmana i ee predystoriia," TIIE-T V (1955), 69–93.

———*Istoriografiia estestvennykh nauk v Rossii* (Moscow: Akademiia nauk, 1956).

———"Lomonosov i Slaviano-Greko-Latinskaia Akademiia," TIIE-T, I (1954), 5–52.

MANUSCRIPTS

Babson Park, Massachusetts: The Grace K. Babson Collection of the works of Sir Isaac Newton and the material relating to him, in the Babson Institute Library.

Cambridge, England: "Catalogue of the Library of Dr. James Musgrave," in the Trinity College Library (a version of this appears in R. de Villamil, *Newton: The Man* [London (1931)], pp. 62–103); the original note from John Newton to Isaac Newton concerning Peter the Great's visit is in the University library.

Leningrad: Rukopisi Lomonosova, F. 20, in the Archives of the Academy of Sciences of the USSR; Sobranie Dashkova, in the Institute of Russian Literature (IRLI) of the Academy of Sciences; A. E. Suknovalov, "Ocherki po istorii voenno-morskogo obrazovaniia," unpublished dissertation, Leningrad University, 1947, in the Leningrad University Library.

London: Catalogue of Newton's library, known as Huggins' List (Add. MS. 25424), in the British Museum (a version of this appears in Villamil, *Newton*, pp. 62–110); Journal Book of the Royal Society, at the Royal Society; Early Letters, G.2.5, in the Royal Society Archives; Newton MSS, 5 bound volumes, with a typescript inventory by Sir John Craig (c. 900 items listed in 105 pages), in the Royal Mint.

Sussex, England: Observing book of the Mural Arc, in the Royal Greenwich Observatory (Herstmonceaux Castle, Hailsham).

NEWSPAPERS, JOURNALS (ZHURNALY), AND ALMANACS

Akademicheskiia izvestiia (St. Petersburg), 1779–1781.
Beseduiushchii grazhdanin (St. Petersburg), 1789.
Chtenie dlia ukaza, razuma, i chuvstvovaniia (Moscow), 1791–1793.
Delo ot bezdeliia (Moscow), 1792.
Ekonomicheskii magazin (Moscow), 1780–1789.
Ezhemesiachnyia sochineniia k pol'ze i uvesileniiu sluzhashchiia (St. Petersburg), 1755–1764.

The Gentleman's Magazine (London).
Kratkoe opisanie kommentariev Akademii nauk (St. Petersburg), 1728–   [for 1726].
Irtysh prevrashchennyi v Ippokrenu (Tobol'sk), 1789–1791.
Ippokrena (Moscow), 1799–1801.
Lekarstvo ot skuki (St. Petersburg), 1786–1787.
Magazin natural'noi istorii (Moscow), 1788–1790.
Mesiatseslovy (St. Petersburg), 1727–   .
Muza (St. Peterburg), 1796.
Novyia ezhemesiachnye sochineniia (St. Petersburg), 1786–1796.
Politicheskii zhurnal (St. Petersburg), 1790–1800.
Poleznoe uveselenie (Moscow), 1760–1762.
The Postboy (London).
The Postman (London).
Priiatnoe i poleznoe preprovozhdenie vremeni (Moscow), 1794–1798.
Sankt-Peterburgskii Merkurii (St. Petersburg), 1793–[1794].
Sobranie raznykh sochinenii i novostei (St. Petersburg), 1775–1776.
Sobranie sochinenii vybrannykh iz mesiatseslovov (St. Petersburg), 1785–1793.
Soderzhanie uchennykh rassuzhdenii Imperatorskoi Akademii Nauk izdannykh v pervom(chetvertom) tome Novykh kommentariev . . . , 1750–1759.
The Spectator (London).
Truten' (St. Petersburg), 1769–1770.
Uraniia (Kaluga), 1804.
Utrennie chasy (St. Petersburg), 1788–1790.
Utrennii svet (St. Petersburg), 1777–1780.
Zerkalo sveta (St. Petersburg), 1786–1787.

OTHER PERIODICALS

Acta Eruditorum (Leipzig).
Commentarii Academiae Scientiarum Petropolitanae (St. Petersburg, 1726–1746).
Novi Commentarii (St. Petersburg, 1747–1775).
Acta Académie des Sciences Impériale de St. Pétersbourg (St. Petersburg, 1777–1782).
Nova Acta Petropolitanae (St. Petersburg, 1783–1796).
Histoire [of the Académie Royale des Sciences].
Journal des Sçavans (Paris).
Philosophical Transactions of the Royal Society (London).

Index

Aberdeen, University of, 81

aberration, chromatic, 202, 206; Dollond on, 206; Euler on, 206–207; Lomonosov on, 206–207; Newton on, 201–202, 206; Oldenburg on, 202; and telescopes, 201–202

Académie des Sciences, Paris, 51, 64, 97, 101, 104, 107, 110, 116; and Aepinus-Lomonosov dispute, 208; and Daniel Bernoulli, 135; and Bilfinger, 109–110; and Cartesianism, 116; and J. Cassini, 134–135; and Cavalieri, 135; and Euler, 135–136; and French Newtonians, 136; and gravitation, 123–131, 136; and MacLaurin, 135; and Maupertuis expedition, 134–135; prizes of, 132, 135; scientific expeditions of, 132–135

achromatic lenses, and Dollond, 206

Acta Eruditorum (Leipzig), 39, 43, 140, 168, 211

Adams, George, the elder, 205

Admiralty Board, and Ferquharson, 79, 81

Aepinus, Franz Ulrich Theodor, 151; and Berlin Academy, 160; on electricity, 160–161, 163; and electrical fluids, 160–161; and Euler, 160–161, 208; and Franklin, 160, 163; and Lomonosov, 163; and Lomonosov's telescope, 208, 210; and Newtonianism, 160, 163–164; in Russia, 160; his *Tentamen Theoriae Electricitatis et Magnetismi*, 160, 163; and Wilke, 160

Aladin, Bogdan, 34–35

alchemy, and Lomonosov's theory of colors, 187

Alembert, Jean d', 140–141, 141n10; on Euler, 216; correspondence with La-grange, 216; his *Traité de dynamique*, 141

Alexander, Tsar, and Grimm's letter to Catherine II, 221n23

Algarotti, Conte Francesco, 124; and Cantemir, 124–126; and Cartesianism, 125; and Marquise du Châtelet, 125; and Fontenelle, 124–125; on Newton, 234–235; and Newtonianism, 124–125, 134; his *Neutonianismo per le Dame*, Russian translation of, 124–126; his *Saggio di Lettere sopra la Russia*, 139n3; and scientific demonstrations, 139; visit to England, 124–125; and Voltaire, 124–125

American Tomineius, 58

Analysis per quantitatum series fluxiones ac differentias, see under Newton, Isaac

Andreev, A. I., 9–10, 12, 13

Anglomania: in France, 124; in Italy, 124; and Russian nobility, 217

Anne (Anna Ioannovna), Empress, 126n30, 212; and Avramov, 62; and Krafft, 138; and Russian nobility, 119

anthropocentrism, 218

anticlericalism, 216

Antonievo-Siiskii monastery, 165

Antwerp, 107

Arago, Dominique François Jean, 231, 234

Archangel, 133

Archimedes, 82–83, 87; problems with Russian edition, 83, 87

Aristotle: and Cotes, 172; and Lomonosov, 173; and Slavo-Greco-Latin Academy, 166; and Wolff, 168

Arithmetic, see under Magnitskii, Leontii

Arithmetica Universalis, see under Newton, Isaac

Arkhimedovy teoremy . . . . See Archimedes

Astronomiae Physicae et Geometricae Elementa, see under Gregory, David

Astronomical Lectures, see under Whiston, William

astronomical observatory near Kholmogory, 165

astronomical tables, 36

astronomy: Newtonian, 222; in Russia, 133; and Soimonov, 213–214

Athanasii, Archbishop, 165

Atkinson, James, 43

Atlas Coelestis, see under Flamsteed, John

atomic theory, of Lomonosov, 186

attraction, 135, 149, *see also* Appendix II

attraction at a distance, 171–174, 176, 233; further applications of, 152–154, 161, 163; and celestial bodies, 152; Newton on, 140, 144, 146, 148–150, 171–174, 176, 224; and Richmann, 153, 154, 170

Avramov, Mikhail Petrovich: and Empress Anne, 62; attitude toward natural philosophy, 61, 62n2, 63; and J. D. Bruce, 63–66; and Bühren (Biron), 62; and Cartesianism, 61; and Empress Elizabeth, 62; and Fontenelle, 61, 63; and Huygens, 61, 63; and *Kosmotheoros,* 61–62; and Newton, 61; and Old Believers, 62; and Peter II, 62; and Peter the Great, 62, 64–66; and Petrine enlightenment, 62–63, 66; and *Pluralité des Mondes,* 62; and Theophan Prokopovich, 62; and publication of *De Officio Hominis et Civis;* and publication of *Kosmotheoros* in Russian translation, 61, 64–66; transfer to College of Mines, 65

Bacon, Francis, 218

Barrow, Isaac, 44

Barton, Catherine, 13

Basel, 107, 131

Basel University, 162

Baumgarten, Alexander Gottlieb, 108

Beigh, Ulagh, catalogue of, 28

Beliaev, I. I., 204

Bentham, Jeremy, 218; his brother, 217

Bentley, Richard, 173

Bering, Captain Vitus, 133

Berkeley, George, Bishop of Cloyne, 215; and *Analyst,* 140; and Newtonianism, 140

Berlin Academy of Sciences, 131, 160

Bernoulli, Daniel, 94n4, 106n2, 107, 107n5, 142n15; and Académie des Sciences, 135; and Bilfinger, 105–106, 109–111, 130; and Cartesianism, 107, 110–111; and conservation of energy, 142; correspondence with Leonard Euler, 135–136, 143–144; and heliocentric theory, 119; his *Hydrodynamica,* 141; and "hypothesis cartesiana," 110; and Jurin, 141; on Krafft, 138; and Joseph de L'Isle, 133; and Newton, 107, 138, 141; on Newton's method of determining the shape of the earth, 135n20; on Newton's reputation, 135–136; as physicist and professor of mathematics, 106–108, 110–111; and *Principia,* 107, 111, 141; as professor of medicine, 107; and Riccati, 107; and St. Petersburg Academy of Sciences, 112, 115; on vortices, 136, 136n21

Bernoulli, Jacob, 107; and Hermann, 131

Bernoulli, John, 107n5, 140–141; and Cartesianism, 107, 141; on dynamics, 141–142; and Leibniz, 107; and Newton, 107, 132

Bernoulli, Nicholas, 108; and Hermann, 130–131; professor of mathematics in St. Petersburg, 107

Bibliographie Astronomique avec l'Histoire de l'Astronomie, depuis 1781 jusqu'a 1802, see under Lalande, Joseph

Bilfinger, Georg-Bernhardt, 95, 106n2, 108, 108n7, 112, 115; and the Académie des Sciences, 109, 110; and Daniel Bernoulli, 105; dispute with Daniel Bernoulli, 106, 109, 110–111, 130; and Cartesianism, 106–111, 113–114, 130; and J. and D. Cassini, 129–130; and circulation of blood, 145–146; and Samuel Clarke, 109, 109nn8,9; in the *Commentarii,* 113; his *Commentatio hypothetica,* 109; his *Dilucidationes,* 109; and experiments with running water, 113; and Frederick the Great, 111; and 's Gravesande, 103; and gravitation, 109, 110, 112, 113–114; and Gross, 118; and Hermann, 130, 132; and Jurin, 111, 144; controversy with Jurin, 112–115, 141, 171–172; and Leibniz, 109; and Newton, 109, 110, 111; and St. Petersburg Academy of Sciences, 102, 104–105, 111, 114–115; and Schumacher, 114, 115; and the shape of the earth, 130, 130n4;

at Tübingen, 108; and vortices, 113; and Wolff, 105, 108–109

Blaeu, Joh., and Guiljelmus, *Theatrum orbis terrarum*, 50n1

blood, and Bilfinger, 145–146; circulation of, 58; and Jurin, 113

Blumentrost, Lavrentii, 96, 130; and controversy with Bilfinger and Daniel Bernoulli, 106; and Wolff, 104

bodies, forces, and motion, 131

Bogoslovskii, M. M., 30

Bolkhovitinov, Evgenii, Metropolitan of Kiev, 217

Bologna Academy of Sciences, 202

Borelli, Giovanni Alfonso, and gravity, 59

Bouguer, Pierre, 132

Boyle, Robert, *Paradoxa Hydrostatica,* 169

Bradley, James, 227

Brahe, Tycho, 28, 53, 55; his heirs and disputes with Kepler, 53n9

Brandenburg, Markgraf Friedrich Heinrich von, 214

Breteuil, Emily de. *See* Châtelet, Marquise du

Brewster, David, 21n9; on Wolff. 168n10

Brius, Iakov Vilimovich. *See* Bruce, Jacob Daniel

Briusov's Kalendar', see under Bruce, Jacob Daniel

Bruce, Jacob Daniel, 15–16; and alloys, 74, 76–77; his almanacs, 37, 68; interest in astronomy, 29, 68; and *Atlas Coelestis*, 27; and Avramov, 63, 64, 65, 66; and Cantemir, 117; and coinage reform, 17–18, 34; and College of Mining and Manufactures, 34, 73, 89; and John Colson, 26, 31–32, 93; and Copernican theory, 68; and sojourn in England, 17; correspondence with Leonard Euler, 42, 94; and Andrew Ferquharson, 78–79, 81; and Flamsteed, 18, 21–22, 26, 29; and Flamsteed's works, 28; and Goldbach, 96; and Greenwich Observatory, 19; and association with Halley, 22–23, 93n2; and Hamilton, 67; and *Historia Coelestis Britannica*, 26, 28; and inventory of books and scientific instruments, and problems of transcription thereof, 33, 34–35, 39–40; and *Kalendar'*, 68; and publication of *Kosmotheoros*, 64–67; and translation of *Kosmotheoros*, 63–67, 213; correspondence with Leibniz, 94; correspondence with Leutmann, 73–

74, 76–77, 94; library of, 34, 39–40, 68–73, 94, 121, 171, 223; visit to London, 68; his map of Tartary and Asia Minor, 16; his "Mathematical Letters," 44; his study of mathematics, 29; and Prince Menshikov, 67; military career, 81; visit to Royal Mint, 15; and "Monetnyi Dvor," 17; and navigation, 43; and Navigation School, 78, 212, 226; and Newton, 32, 35, 39–40, 44, 68–69, 71–72, 76–77, 88, 96, 231–232; and Newton's later writings, 45; and Newton's lunar theory, 20, 39; and Newton's mathematical writings, 42; and Newtonianism, 104, 234; notations in his books, 44n24, figs. 11, 12, 14; and *Opera Mathematica* by John Wallis, 26; and optics, 68, 71–76; visit to Oxford, 30; and Peter the Great, 15, 29, 93–94; correspondence with Peter the Great, 64–68; and primogeniture, 17; and *Principia*, 39–40; purchase of books (for Peter the Great), 40, 41, 94n3; purchase of mathematical instruments and books, 17; and St. Petersburg Academy of Sciences, 33, 94, 115; and *School of Mathematics and Navigation*, 33; and Society of Neptune, 81; studies in England, 32; and Tatishchev, 73; and telescope, 73, 74, 76–78; translation of foreign works into Russian, 64, 67; travels abroad, 68; and *Vilima Sevela iskusstvo niderlandskogo iazyka,* 67n21; and Wallis, 44; correspondence with Wolff, 94

Bruce, James, 16

Bruce, Peter Henry, 15–16

Bruce, William, 16

Bruno, Giordano, 53, 235

Buffon, 95n7

Bühren, 62, 212

Bulfinger. *See* Bilfinger, Georg-Bernhardt

Butterfield, Herbert, 51–52n5

calculus, 144, 170; *see also* fluxions

calorimetry, 146

Calvinism, 215

Cambridge University, 201; and Cartesianism, 102–103; and Jurin, 112

camera obscura, 139, 139n5

Candide, see under Voltaire

Cantemir, Prince Antiokh, 72, 120n9, 121n14, 165, 235; and Algarotti, 124–126; and Bruce, 117; and Cartesianism, 116–117, 119, 126; and Clairaut, 127,

Cantemir, Prince Antiokh (*Cont.*)
153; correspondence of, 120; death of, 140; education of, 118; activities in England, 117; emissary to England, 119; *First Satire*, 119; and the dialogues of Fontenelle, 119; on gravitation, 123; and Gross, 118; correspondence with Lord Harrington, 73, 121; library of, 120; manual on algebra, 123; and Maupertuis, 127; and Friedrich Christian Mayer, 118; correspondence with the Marquise de Monconseil, 122; and the *Neutonianismo per le Dame*, 126; and Newton, 232; and Newtonianism, 116–117, 120–126; in Paris, 127; preoccupation with natural philosophy, 123; and the *Pluralité des Mondes*, 116–117, 124–126, 213; political role, 119; and Theophan Prokopovich, 119; and Russian Church, 216; and St. Petersburg Academy of Sciences, 115, 117–119, 126–127; and Sloane, 119, 120; and Tatishchev, 73; and Thomas, 121, 125; and Voltaire, 122–124

Carter, Elizabeth, 124n21

Cartesian cosmology, 37

Cartesianism, 55, 61, 126–127, 135, 214, 234–235; and Algarotti, 125; and Daniel Bernoulli, 107, 110–111; and John Bernoulli, 107, 141; and Bilfinger, 106–111, 113–114, 130; at Cambridge, 102, 103; and Cantemir, 116–117, 119, 126; and Samuel Clarke, 103; and Cotes, 177–178; and Leonard Euler, 136, 161–162; and Fontenelle, 116; and 's Gravesande, 103; and gravitation, 128–131; and Huygens, 54–55, 57, 59–60; on light, 197; and Lomonosov, 167, 186, 194; Newtonian criticism of, and effect, 97–101; and Newtonian theory, 110; and *Principia*, 110; and Richmann, 147, 148; and St. Petersburg Academy of Sciences, 102, 104–105; and shape of the earth, 127–137; and universal gravitation, 127; and vortices theory, 37, 97–101, 109–111, 110n2, 113, 130, 132–133, 135, 154, 157, 214; and Wolff, 167–168, 175

Cartesians: dispute with Leibnizians, 140–141; and Newtonians, 215; Voltaire on, 218

Cassegrain, and Newton, 70

Cassini, D., and earth's shape, 129n3, 129–133

Cassini, J., and earth's shape, 129n3, 129–130, 132–134; and Maupertuis, 134

catadioptric telescope, 204

catadioptrical tubes, 73–74

Catherine I, and St. Petersburg Academy of Sciences, 94

Catherine II, 149n43, 205; and Anglomania, 217; and Diderot, 221, 222; and Baron Grimm, 221; and literacy, 223; and project for a university, 222–223; and Soimonov, 213; and Voltaire, 221

Cavalieri, Bonaventura, and Cartesianism, 135

Cavendish, Lord William, 201n5

Cayenne, 129

celestial mechanics, 214

Celsius, Andreas, 134–135n18; and Maupertuis expedition, 134

centrifugal action, and Huygens, 50

centripetal forces, 37, 174, 176; and Newton, 86

Charles II: appoints Flamsteed "astronomical observator," 20; and Royal Mathematical School, 79; and Royal Society, 15

Charles XII, 117

Châtelet, Marquise du (Emily de Breteuil), 140, 141, 142n15, 234; and Algarotti, 125; and Newtonianism, 134; and *Principia*, 123; and Voltaire, 104, 123, 123n9; and Wolff, 104

Chemical and Optical Notes, see under Lomonosov, Mikhail V.

Chemistry, 185, 198–199

Chenakal, V. L., 34

China, 137

Chizhov, Nikolai, 204, 205

Christ Church, Oxford, 80

Christ's Hospital, 29; Royal Mathematical School, 79, 80

Christianity, and plurality of worlds, 55

Chronology, see under Newton, Isaac

Cirey, 123

Clairaut, Aléxis Claude, 123, 127, 136n24, 153n4, 158n8; and Cantemir, 153; and Cartesianism, 132–133; and earth's shape, 136; and lunar theory, 153; and Maupertuis, 132–133; and Newtonianism, 153; and St. Petersburg Academy of Sciences, 153

Clarke, John, 37–38, 120n9; his *A Demonstration of some of the Principal Sections of Sir Isaac Newton's Principles of Natural Philosophy*, 37

Clarke, Samuel: and Bilfinger, 109, 109n-n8,9; refutation of Cartesianism, 103; editor of *Rohault's System of Natural Philosophy*, 37, 38; and Newton, 109; and Newton-Leibniz controversy, 37; as translator, 38

Coinage reform, 34

Cole, William, 31

College of Mining and Manufactures, 34, 65, 73, 89

College of Physicians, 112

Color, Lomonosov on, 187, 197, 202; Newton on, 190–192, 197–198, 201–202

Colson, Francis, 30

Colson, John, 30–31, 95n7, 120n9; and *Arithmetica Universalis*, 43, 43n20; and J. D. Bruce, 26, 31–32, 39; teacher of mathematics, 29; correspondence with Flamsteed, 26; friendship with Flamsteed, 31; Master of the Free Grammar School, Rochester, 30; and Newton, 30, 32; and controversy between Newton and Flamsteed, 22, 25; and Peter the Great, 30; correspondence with Peter the Great, 29–31, 93; teacher of mathematics, 29; treatise on calculus, 32

Colson, Nathaniel, 43, 43n22

comets, motion of, 100; and vortices, 59

Commentarii, 106, 110, 147; paper by Bilfinger, 112–113; Bilfinger-Jurin controversy, 144; and Leonard Euler, 136; paper by Jakob Hermann, 114; and Jurin, 112

Commentatio hypothetica, see under Bilfinger, Georg-Bernhardt

compass, variations of and Halley's observations, 23

compound pendulum, 108

Condorcet, Marquis de, 101, 216, 223

conservation, principle of, 142

conservation of energy, 233; and Daniel Bernoulli, 142

conservation of force, 140

Constantinople, 117

Copernican theory, Soimonov on, 214

Copernicus, 50n1, 52, 55, 56, 60, 68, 99, 139n4, 218, 224, 235–236; dissemination through translation in manuscript form, 50n1; and Russian Church, 216

Corpuscular theory, 148–149, 157, 188–190, 194–195, 202, 228

Cotes, Roger: and Cartesians, 177–178; on gravitation, 172–173, 175, 178; on Newton, 171–173, 175, 177, 178

Court of St. James, and Cantemir, 119

Court of St. Petersburg, and Fontainebleau, 133

Crabtree, William, 28

Craig, Sir John, 13n13, 14

Crosthwait, Joseph, and publication of *Atlas Coelestis*, 27

Cusanus, Cardinal, 53

Dean, John, 78

De conservatione virium vivarum dissertatio, see under Jurin, James

De Morgan, Augustus, 36; on Newton and Whiston, 36n8

De Mundi Systemate, see under Newton, Isaac, *A Treatise of the System of the World*

De Officio Hominis et Civis, see under Puffendorff, Samuel

Depman, I. Ia., 84, 85

Deptford, shipyards and Peter the Great, 12, 14, 17, 78

Desaguliers, J. T., as translator of 's Gravesande, 38, 96n11

Descartes, René, 51, 55, 59–60, 116, 184; and Algarotti, 125; continuing power of his doctrines, 135; cosmology, 214; and Huygens, 51, 54–55, 57, 59–60; on interphenomena, 178; on light, 194; measurements of force, 140, 141; and Mersenne's father, 96; Newton's attack on basic hypothesis of, 97; his *Principia Philosophiae*, 147; effect on St. Petersburg Academy of Sciences, 101; on space, 186; on tides, 224; and Voltaire, 220; theory of vortices, *see* vortices

Destruction of the Copernican System (anon.), 236, 236n5

De Temporis Æquatione Diatriba. Numeri ad Lunae Theoriam Horrocianem, see under Flamsteed, John

De Wilde, Jacob (archeological and anthropological museum), 12

Diderot, Denis, 144, 149n43, 217, 221–223; and "l'esprit géometrique," 222; unfinished memorandum on Russia, 221–222n24

diffraction, 195

Dilucidationes, see under Bilfinger, Georg-Bernhardt

Disquitio de causa physica electricitatis, see under Euler, Leonard

Dissertatio Physica de Corporum Mixtorum Differentia, see under Lomonosov, Mikhail V.

Ditton, Humphrey, and French transla-
tion of his *An Institution of Fluxions,*
120n9
Dollond, John, 205–207, 209, 210, 227
Domcke, George Peter, 83–86; and An-
drew Ferquharson, 85, 87–88; and
Newton's mathematics, 86–87; and
Newtonian philosophy, 85; his *Phi-
losophiae Mathematicae Newtonianae
illustrata Tomi Duo . . . ,* 85–87; and
William Whiston, 85–86
Domckio. *See* Domcke, George Peter
Domkino, Georg Petr. *See* Domcke, George
Peter
Domkio, Georg Petr. *See* Domcke, George
Peter
Domckius, Georgius Petrus. *See* Domcke,
George Peter
Dorofeev, Vasilii, 165, 166
Dufour, 51–52
Duising, 170, 181
Duvernois, Johann Georg, 145
Dvigubskii, Ivan, 205n21
dynamics, 140–142

earth, circumference of, 128; configuration of, 127; motion of, 235; shape
of, 97, 127–137, 130n4
education, in Russia, 33, 118, 223
Einstein, Albert, 128n1
electrical fluids, 159, 160; and Franz Ulrich Theodor Aepinus, 160, 161
electricity, 97; cause of, 155; experiments
with, 154; Franklin on, 160–161; and
glass, 160; and gravitation, 156–157;
and gravitation and magnetism, 156;
and light, 156; Lomonosov on, 155–
156, 187, 162–163, 188, 198; problems
with understanding of, 159, 160; and
Richmann, 151, 154–155
Elémens de la Philosophie de Neuton,
see under Voltaire
Elementa Geometriae, see under Tacquet,
Claude André
*Elements of Euclid, selected from New-
ton's twelve books and condensed to
eight books by Professor Andrew
Ferquharson,* see under Ferquharson,
Andrew
Elementy. See Ferquharson, Andrew, *Ele-
ments of Euclid*
Elizabeth, Empress, 202, 234; and Avra-
mov, 62; and electrical experiments,
154
Éloge of Newton, see under Fontenelle,
Bernard Le Bovier de

emission theory, 136; hypothesis (of Gilbert), 155
empirical method, 180
empiricism vs. rationalism, 176–183
encyclopaedia, 226
Encyclopedists, 221
English merchants, and Prince Alexander
Menshikov, 48–49
Essai sur la Poésie Epique, see under
Voltaire.
Essay on Man, see under Pope, Alexander
Estonia, 146
ether, 97, 173, 186–188, 190, 191; and
electricity, 158; Euler on, 158–161;
and gravitation, 158–159; and light,
158; and Lomonosov, 162–163, 186–
188; Newton on, 190
ethereal matter, Cotes on, 172–173
Euclid, 44, 82, 180, 226
Euler, Johann Albrecht: *Disquisitio de
causa physica electricitatis,* 155
Euler, Leonard, 94, 94n4, 95n7, 136n24;
and Académie des Sciences, 135, 136;
in *Acta Eruditorum,* 211; and Aepinus, 160–161; and Aepinus-Lomonosov
controversy, 208; D'Alembert on, 216;
and attraction at a distance, 153, 154;
at Basel University, 162; on Berkeley,
215; in Berlin, 136, 137; and Daniel
Bernoulli, 135–136, 143–144; and J. D.
Bruce, 42, 94; and calculus, 144; and
Cartesianism, 136, 161, 162; and
Catherine the Great, 211; on chromatic aberration, 206; in *Commentarii,* 136; and corpuscular theory,
157; and John Dollond, 206; his
*Disquisitio de causa physica elec-
tricitatis,* 159, 161; on electricity, 155–
161; and ether, 158–161; his "Examination and Refutation of Newton's system," 215; and Frederick the
Great, 137; and Goldbach, 207n25;
and gravitation, 136, 153, 158–159;
and James Jurin, 140, 143, 144; Lagrange on, 216; and Leibniz, 144; his
Letters to a German Princess, 211–
212, 214–216; and light, 136, 158, 198;
on light and electricity, 156–158; on
Locke, 214; and Lomonosov, 206–207,
209, 233; on Lomonosov, 180; and
mathematics, 198; his *Mechanica,* 211;
on motion, 158, 159; and Newton,
143; on Newton, 211–212, 214–217;
on Newton and Cartesianism, 157; on
Newton and light, 156–157; and Newtonianism, 136, 146, 159, 161–162;

religious convictions, 215–216; and Rumovskii, 207–208, 215, 217; and St. Petersburg Academy of Sciences, 136–137, 155, 215; on space, 157; his *Sur la Nature et les Proprietés du Feu,* 136; and Weitbrecht-Richmann controversy, 143–144; and Wolff, 150
"Examination and Refutation of Newton's System," see under Euler, Leonard
Ezhemesiachnyia sochineniia, 213; Micromégas in, 218; on Newtonianism, 224

Faraday, 163
Farbenlehre, 197
Farvarson, Andrei Danilovich. *See* Ferquharson, Andrew
Fel', S. E., 34
Fénélon, Archbishop, 117n3
Feodor Alekseevich, Tsar, 16
Ferquharson, Alexander, 89
Ferquharson, Andrew, 226; and J. D. Bruce, 78–79, 81; and Domcke, 85, 87–88; and *Elements of Euclid,* 82–87; *Elements of Euclid,* Ferquharson's translation of, 226; *Elements of Euclid,* problems with, 82–89; and Grice and Gwyn, 80; intellectual background, 81; library of, 89; as mathematics teacher, 78–81; and Naval Academy, 78, 80–81; and Newton's influence on Russia, 78; as a Newtonian, 81–82, 85–86, 88–89, 104; and Peter the Great, 78–79, 81; influence on Russian science, 89; and Tacquet, 82–84, 87
First Satire, by Antiokh Cantemir, 119
fixed stars, 71
Flamsteed, John, 93n2, 227; as Astronomer Royal, 19; astronomical tables, 36; publication of *Atlas Coelestis,* 27; and Bruce, 18, 20–22, 26, 29; and John Colson, 26, 31; his *De Temporis Æquatione Diatriba. Numeri ad Lunae Theoriam Horrocianem,* 28n24; his *Historia Coelestis Britannica,* 19, 26–27; lunar theory, 32; and Newton, 44; altercations with Newton, 18, 24–27; correspondence with Newton, 25; work with Newton, 20, 21; controversy with Newton and Halley, 20, 21–28; and Newton's Lucasian lectures, 20, 20n7; observations with mural arc, 28; the parallax of the polestar "discovered," 24; and Royal Mathematical School, 79; his Table of Refractions, 20, 32; and

Wallis, 44; correspondence with Wallis, 24
fluxionary method, 198, 228; of Newton, 43
fluxions, 42–43, 68, 86–87, 120; Newton's system, 113
Folkes, Martin, and Algarotti, 124; and Royal Society, 124
Fontainebleau, and Court of St. Petersburg, 133
Fontenelle, Bernard le Bovier de, 51, 116n1, 126n29; and Algarotti, 124–125; and Avramov, 61, 63; and Cartesianism, 116; dialogues, 119; *Éloge* to Académie des Sciences, 97, 110, 179, 194; and gravitation, 116; and Huygens, 52; and Newton, 110, 116; on Newton, 97, 179; and Peter the Great, 64; *Pluralité des Mondes,* 116–117, 124–126, 211, 213–214; and Voltaire, 116
force, Leibniz, 142; Newton, 142
forces, attractive, 37; centripetal, 37
Frankfurt, on the Oder, 131
Franklin, Benjamin, 154, 188; and Aepinus, 160; on electricity, 160, 161; and Leyden vial, 160; and Richmann, 154
Frederick the Great, and Bilfinger, 111; and Euler, 137; and Wolff, 104–105, 169
Freethinkers, 215, 217
Freind, John, 113, 149, 149n42; and Newtonian laws, 146; *Praelectiones Chymicae,* 170
French Academy of Science. *See* Académie des Sciences, Paris
French Revolution, and Newtonianism, 217
Fresnel, Auguste Jean, 188, 196

Galileo, 54, 140n7, 224, 233, 235–236
Garraways, or Garroways. *See* Garways
Garways, 26
Gascoigne, in *Historia Coelestis Britannica,* 28
Gassendi, Pierre, 194, 224
"geometers of Basel," 140, 144
George, Prince of Denmark, 39
Gilbert, William, 155
Gillibrand, Henry, 43
glass, use of, for speculum, 74–77
Glover, Richard, 96n11
Godin, Louis, 132
Goethe, Johann Wolfgang von, on light, 197
Goldbach, Christian, 49, 95n7, 106, 112, 115, 134–135n18, 136n24; and Blu-

Goldbach, Christian (*Cont.*)
mentrost, 96; and Bruce, 96; and Euler, 207n25; and Leibniz, 95; and Prince Menshikov, 96; and Newton, 95; correspondence with Newton, 93; and St. Petersburg Academy of Sciences, 95, 96
Gottsched, Johann Christoph, on Newton, 168; on Wolff, 168
Gouda, Hans, and Peter the Great, 11
's Gravesande, W. James, 102n2, 149; and Cartesianism, 103; light corpuscles, 103; and Newton, 90; and Newtonianism, 38, 103; his *Philosophiae Newtonianæ Institutiones, In Usus Academicos,* 102–103; his *Physices elementa mathematica experimentis confirmata: sive Introductio ad Philosophiam Newtonianam,* 38, 90, 102; and St. Petersburg Academy of Sciences, 102; and Schumacher, 90, 103; and vortices, 97
gravitation, 86, 105, 111–114, 127–131, 136, 147–148, 150, 158–159, 171–179, 194, 213, 224–225, 227–228, 233, 235; Bilfinger on, 109–110, 112–114; Antiokh Cantemir's work on, 123; and Cartesianism, 128–131; Roger Cotes on, 172–173, 175, 178; and electricity, 156–157; Euler on, 136, 153, 158–159; Fontenelle on, 116; Huygens on, 158, 213; Jurin on, 113–114; Lomonosov on, 170, 172–179, 181, 194; compared to magnetism, 56; and moon, 128, 153; Newton on, 86, 105, 111, 123, 128–131, 136, 171–178, 194, 213, 224–225, 227–228, 233; Newton's theory of, 61; Richmann on, 150; first textbook on, 120; Wolff on, 175–176, 179, 181
gravitation, universal: and Cartesianism, 97; and Leibniz, 105; and Wolff, 105
Great Northern War, 146
Greek Orthodox clergy, and Newtonianism, 234–235
Greenwich Observatory, 17, 19, 20, 28, 29, 31; and *Historia Coelestis Britannica,* 28; Peter the Great's visit to, 19; and Royal Mathematical School, 79
Gregory, David, 21, 21n9, 24, 38–39; his *Astronomiae Physicae et Geometricae Elementa,* 38–39, 120, 222; and Halley, 25; and Newton, 24–25, 39, 81; and Newtonianism, 86; and Whiston, 39
Gregory, James, 166, 200, 202

Grew, Nehemiah, 58
Grice, Richard, 79–80, 80nn6,7; and Ferquharson, 80; and Peter the Great, 80
Grimm, Baron, 221
Grishchov, A. I., 207
Gwyn, Stephen, 79–80, 80nn6,7; and Ferfinger, 118; and Cantemir, 118
Grotthuss, Theodor, 228n47
Guasco, Octavien de, 125n27
Guhrauer, G. E., on Leibniz, 170–171
Gulf of Bothnia, Maupertuis expedition, 127, 134, 135
gunpowder, and Huygens, 59n30
Gur'ev, S. E., 228
Gur'ev, Lieutenant Captain, 34–35
Gwyn, Stephen, 79–80, 80nn6, 7; and Ferquharson, 80; and Peter the Great, 80
gymnasium, of St. Petersburg Academy of Sciences, 138, 145–146, 167

Halle, University of, 104–105, 108, 146
Halley, Edmond, 93n2, 169, 227; and *Arithmetica Universalis,* 42–43; astronomical tables, 36; and Bruce, 22–23, 93n2; as commander of *Paramour Pink,* 22n11, 23; controversy with Flamsteed, 20–28; and Gregory, 25; and Newton, 20–28; on Newton, 182; and Peter the Great, 22, 93; religious beliefs, 23; and Whiston, 36
Hamburger, G. E., and gravitation, 146
Hamilton, Hugo Johann, 67n20; and Bruce, 67
Hampton Court, 17
Harrington, Lord, 73; and Cantemir, 121
Hartzoeker, Christian, and Peter the Great, 11; and invitation to Russia, 11n8
Hayes, Charles, and Newton, 32; his *Treatise on Fluxions,* 32
heart, and James Jurin, 113
heliocentric theory, 52, 213–214, 235–236; and Joseph de L'Isle and Bernoulli, 119; Lomonosov on, 139n14, 213
Helmholtz, Hermann Ludwig Ferdinand von, 142n15
Henriade, see under Voltaire
Hermann, Jakob, 95; and Berlin Academy, 131; and Bernoulli, 130–131; and Bilfinger, 130–131; in *Commentarii,* 114; and earth's shape, 130, 130n4; at Frankfurt on the Oder, 131; and gravitation, 114, 130–131; and Leibniz, 131; Montucla's judgment of, 131;

and Newton, 130–132; and Padua, 131; his *Phonoromia, sive De viribus et motibus corporum solidorum et fluidorum libri duo,* 131; and St. Petersburg Academy of Sciences, 114–115, 130, 131, 132; and Schumacher, 114–115

Hermitage Museum, Leningrad, 77

Herschel, William, 41, 200, 203–205

Hevelius, Johannes, catalogue of, 28; his *Machina Coelestis* and *Selenographia,* 28; telescope, 72

Historia Coelestis Britannica, see under Flamsteed, John

Holy Synod, 119; and *Essay on Man,* 224–225; and Newtonianism, 234; and Voltaire, 219–221

Hoogzaat, Hans Isbrandtsen, and Peter the Great, 11

Hooke, Robert, 169; on color, 197; diary, 26; on light, 189, 191–194; and Newton, 54, 68, 189, 191–193, 197

Horrocks, Jeremiah, his *Opera Posthuma,* 28n24

Horsley, Samuel, 227

House of Commons, 17

Huiens, Kristofer. *See* Huygens, Christian

Huygens, Christian, 50–60, 89, 131, 139n4, 168, 184; and Avramov, 61, 63; criticisms of Cartesianism, 54–55, 57, 59–60; and centrifugal action, 50; and Christian theology, 55; and clergy, 54; on contemporary scientific advances, 58–59; and Descartes, 51, 54–55, 57, 59–60; and Fontenelle, 52; on gravitation, 158, 213; on interphenomena, 178; and Athanasius Kircher, 55–56; his *Kosmotheoros,* 50–67, 213–214, 232, *see also Kosmotheoros;* on light, 193–194; his microscope, 58; and Newton, 50, 51, 54–57, 193; correspondence with Newton, 59; and Newton's telescope, 201; on ondular theory, 188; his pendulum clock, 58; his reaction to *Principia,* 51n3; and discovery of satellites of Saturn, 56; and "vis viva," 108

Huyssen, Baron Heinrich von, 12n11, 63

Hydrodynamica, see under Bernoulli, Daniel

Iavorskii, Stefan, 62, 108n7

Idel'son, N. I., Acad., on Newton's lunar theory, 153n7

Imperial Cadet School, 163

Ingram, Sir John, 78n1

An Institution of Fluxions, see under Ditton, Humphrey

Institutiones Philosophiae Wolffianae, see under Tümmig, L. F.

interphenomena, 178

Introductio ad veram astronomiam, see under Keill, John

Introductio ad veram physicam, see under Keill, John

"Irtysh (being) transformed into Hippocrene," 223

Iter Exstaticus, see under Kircher, Athanasius

Iushkevich, Dr. A. P., 82–85

Jena University, 146

Johnson, Samuel, 124n21

Joule, 142n15

Journal, of Peter the Great, 19; visit to Greenwich Observatory, 19; itinerary in England, 17. *See also* Peter the Great

Jupiter, 57; and Newton's criticism of vortices theory, 98

Jurin, James, 112, 113, 141n12; and Bishop Berkeley, 140; on Daniel Bernoulli, 141; on John Bernoulli, 140, 141; and John Bernoulli, 143; and Bilfinger, 111, 144; controversy with Bilfinger, 112–115, 141, 171–172; and blood, 113; and Cambridge, 112; and College of Physicians, 112; and *Commentarii,* 112; on conservation of force, 140; *De conservatione virium vivarum dissertatio,* 140–143; and Euler, 140, 143–144; and gravitation, 113–114; in Leyden, 112; and Michelotti, 113; and Newton, 111, 113, 144; and Newton-Leibniz controversy, 142, 143; and Newtonian Laws, 146; and Dr. Pemberton, 140; his "Philalethes Cantabrigiensis," 140n6; and physiology, 113; and Richmann, 143, 149; and Royal Society, 112, 113; and St. Petersburg Academy of Sciences, 140–141; and smallpox inoculation, 112; at Trinity College, 112; and Voltaire, 112; and Weitbrecht, 142–145; and Wolff, 145

Kamchatka, expedition to, 133

Kämpfer, Engelbert, 137n25

Kant, Immanuel, 104, 108n7, 140–141, 141n9; and Wolff, 104, 168

Kantemir, Demetrius, 117, 118–119n6; and Peter the Great, 117–118; as ruler of Wallachia, 117

Kapitsa, P.L., Acad., on Lomonosov, 232–233, 233n4

Karamzin, N. M., visit to England, 218; on Sir Godfrey Kneller's portrait of Peter, 10, 11n4

Keill, John, 86, 149, 149n43; his *Introductio ad veram astronomiam,* 222; his *Introductio ad veram physicam,* 222, 223; and Wolff, in *Acta Eruditorum,* 168

Kepler, Johann, 53, 55, 139n4, 168, 175, 224, 234–236; and Brahe, 53n9; and discovery of elliptic course of planetary motion, 55; his law, 99, 100; his rule, and Newton, 129

Kholmogory, 165

Kircher, Athanasius, 55, 55n17; and Huygens, 55–56; his *Iter Exstaticus,* 55

Kneller, Sir Godfrey, 10

Knutzen, Martin, 108n7

Kolsun, Ivan. *See* Colson, John

Königsberg, University of, 145

"Konston." *See* Colson, John

"Konton." *See* Colson, John

"Kontun." *See* Colson, John

Kopelevich, Iu. Kh., 34

Korff, Baron, and St. Petersburg Academy of Sciences, 167

Kornil'ev, V. D., 223

Kosmotheoros, by Christian Huygens: and Avramov, 61–62; confusion over contents, 63; and ecclesiastical view of earth, 52; in French translation, *Nouveau Traité de la Pluralité des Mondes,* 51–52; impact on reading public, 51; introduction to Russian edition, 52; and Planetarians, 53, 56–58; problems in translating into Russian, 89; relationship of earth and other planets, 52–53; Russian publication of, 61, 64–67; second Russian printing, 65; translation into Russian, 63–67. *See also* Huygens, Christian

Krafft, G. W., and Empress Anne, 138; as astrologer, 138; and calorimetry, 146; marriage of, 138; and Morin's formula, 147; and Newtonianism, 137, 139; his public lectures, 138, 139; and Richmann, 146, 147–148; and St. Petersburg Academy of Sciences, 137–140, 144; and Schumacher, 138, 139; textbooks, 138; and Taubert, 138; at Tübingen, 139

Kratkoe iz'iasnenie o astronomii, see under Soimonov, Feodor Ivanovich

Kratkoe opisanie kommentariev, see under St. Petersburg Academy of Sciences

Kurganov, N. G., 226; his *Pismovnik,* 226, 227, 235

Kuznetsov, B. G., on Lomonosov, 163

La Condamine, Charles Marie de, 132; and Voltaire, 152

Lagrange, Joseph Louis, 141n13; correspondence with d'Alembert, 216; on Leonard Euler, 216

Lalande, Joseph le Français de Jérôme, 135n18, 209; his *Bibliographie Astronomique avec l'Histoire de l'Astronomie, depuis 1781 jusqu'à 1802,* 135n18

Lambert, Johann Georg, 209

Laplace, Pierre Simon, Marquis de, 237

"Lapland Hercules," see under Maupertuis

Lapland, Maupertuis expedition, 134

Larionov, 67

Lavoisier, Antoine Laurent, 228

Lebiadnikov, Boris, and Peter the Great, 94

Leewenhoek, Anton von, and Peter the Great, 11

Lefort, François, 16

Leibniz, Gottfried Wilhelm, 108n7, 109nn8,9, 140n7, 142n14, 168; in *Acta Eruditorum,* 140; algebraic notation (calculus), 198; on attraction at a distance, 144, 171–172; and John Bernoulli, 107; and Bilfinger, 109; correspondence with Bruce, 94; and Samuel Clarke, 109; and controversy with Samuel Clarke, 142; and Descartes, 140–141; on dynamics, 142; and Euler, 144; and Goldbach, 95; and gravitation, 105, 128; Guhrauer on, 170–171; and Hermann, 131; and Lomonosov, 167; metaphysical views of, 109; monadology and action at a distance, 175n42; and Newton, 104, 114, 144; controversy with Newton, 13, 37, 104, 114, 120, 142–145, 150; and marginalia to Newton's *Principia,* 171n26; and Peter the Great, 10; correspondence with Peter the Great, 68; meeting with Peter the Great, 93, 94, 104; and rationalism, 178–179; Richmann on, 144; and Russian science, 104; and St. Petersburg Academy of Sciences, 93–94, 101; and Voltaire, 169, 220; and Wolff, 104

Leibnizians, 234–235; dispute with Cartesians, 140–141; dispute with New-

tonians (on calculus), 168; and Wolff, 167–168, 175
Leipzig University, 154
Le Monnier, 134, 222
Les Oreilles du Comte de Chesterfield, see under Voltaire
Letters to a German Princess, see under Leonard Euler
Lettres Philosophiques, see under Voltaire
Leutmann, Johann Georg, 94, 204; and J. D. Bruce, 44; correspondence with J. D. Bruce, 73–74, 76–77, 94
Leyden, James Jurin in, 112
Leyden jar, 155
Library of the USSR Academy of Sciences (Leningrad), 43
Liddle Mathematical Tutor, 81
light: and Cantemir, 123; and colors, 68, 70–75, 97; and Descartes, 194; and electricity, 156–158; emission theory of, 103; Euler on, 136, 158, 198; general law of the interference of, 195; Goethe on, 197; Huygens on, 193–194; Lomonosov on, 185–189, 192–195, 202, 210; motion of, 157–158; Newton on, 68, 70–75, 103, 123, 136, 138, 156, 158, 185–198, 227, 228; Newton's theories of, 68, 70–76, 97, 138, 139
L'Ingenu, see under Voltaire
Linus, Franciscus, and Newton, 71
L'Isle, Louis de la Croyère de, 132–133, 133n13
L'Isle, Joseph de, 126n30, 133nn14,16, 133–134, 134n17, 134–135n18; and Daniel Bernoulli, 133; and Cartesianism, 133; and heliocentric theory, 119; and maps, 133; in *Philosophical Transactions,* 133–134; and St. Petersburg Academy of Sciences, 72, 133–134; and Todhunter, 133
Litken, 185n1
Liubimov, N. A., 180, 181
Locke, John, 123, 214; association with Newton at the Mint, 10; and fine fluids, 159; and Voltaire, 123
Lomonosov, M. V., 94, 140–141, 151; and Academy of Sciences in Bologna, 202; and Aepinus, 163, 208, 210; and atomic theory, 186; and attraction at a distance, 163, 170–171, 174–176; and John Bernoulli, 174; and Cartesianism, 167, 186, 194; his *Chemical and Optical Notes,* 209; and chemistry, 185, 198–199; on color, 187, 202; and Dollond, 209; and Duising, 181; edu-

cation of, 166–168, 170, 175, 181; on electrical fluids, 163; on electricity, 155–156, 162–163, 187–188, 198; on ether, 162–163, 186–188; and Euler, 206, 207, 209, 233; and Galileo, 233; and glass factory, 202; on gravitation, 170, 172–179, 194; on heliocentric theory, 213, 139n4; his influence in Russia, 198–199; his *Institutiones philosophiae Wolffianae,* 146; and instrument makers, 203, 204; in *Irtysh (being) transformed into Hippocrene,* 223; and Kapitsa, 232–233; and laws of refraction, 210; and Leibniz, 167; his *Letters on the Usefulness of Glass,* 139n4; life, early, 165–166; on light, 185–189, 192–195, 202, 210; on luminescence and electricity, 146; and Marburg University, 165, 169, 170, 175; and Mariotte, 187; on mass and weight, 182–183; his mathematical conception of nature, 183–184; and mathematics, 180–182; on matter, 177; and Menshutkin, 199; and mosaics, 202; in Moscow, 166–167; and "nebesnye orudiia," 202; on Newton, 231; and Newton's telescope, 200, 202–203; and Newtonianism, 162–164, 169–184, 186, 188–189, 193–195, 197, 198–199, 211, 232–234; his *Observations on the density and fluidity of bodies,* 174; his "On the Improvement of Telescopes," 205; his *127 Observations on the theories of light and electricity,* 189; and optics, 185, 197; his *Oration on light,* 194, 197; his *Oration on the Origin of Light,* 193, 233; and Pope's lines on Newton, 9; and Popovskii, 225; on primary matter, 187; his *Problema physica de tubo nyctoptico,* 208; his *Prolegomena to Natural Philosophy,* 179; as a rationalist, 179–180, 182–183; and Richmann, 146, 148; and Rumovskii, 197, 210; and the Russian Church, 216; and the St. Petersburg Academy of Sciences, 151, 155, 171, 185, 197, 233; and Prince Shuvalov, 208–210; and correspondence with Soimonov, 213; and telescopes, 200, 202–210; his *Tentamen Theoriae de Particulis Insensilibus,* 183–184; and Wolff, 104, 162, 164, 167–170, 175–181, 198
London, 10, 127; Cantemir in, 117
Louis XV, 154
Lowthorp, John, 71

Lucas, Anthony, and Newton, 71
Lull, Raymond, 55n17
luminescence and electricity, 146
lunar theory, 22, 28, 153; and J. D. Bruce, 20; of Clairaut, 153; of Flamsteed, 20, 21, 32; and Halley, 21, 22; and Horrocks, 28n24; and Newton, 20–23, 26, 39
Luttrell, Narcissus, and Peter's visit to the Mint, 13–14

Machina Coelestis, see under Hevelius, Johannes
MacLaurin, Colin, and Académie des Sciences, 135
magnetism: and electricity, 156; and John Michell, 153–154
Magnitskii, Leontii, 80n8; his *Arithmetic,* 166
Mairan, d'Ortous de, 140–141n8
Makarov, Aleksei Vasil'evich, 67
Makogonenko, G. P., on Kurganov's *Pismovnik,* 226, 226n39
Malus, Etienne Louis, 188
Manteuffel, Ernst Christoph, Count, 169n14
Marburg, University of, 104, 162, 168–170, 175, 181, 185
Marie Clare, Viscountess Bolingbroke, and Voltaire, 123–124
Mariotte, Edme C., 187; *De la Nature des Couleurs,* 187n5
Marishal College, University of Aberdeen, 81
Mars, altitude of, 129
Martini, Christian, and St. Petersburg Academy of Sciences, 105; and Wolff, 105
mass, and weight, 182–183
Materials for the History of the Imperial St. Petersburg Academy of Sciences, inventory of Bruce's books and scientific instruments, 33
mathematical analysis of phenomena, 198; and conception of nature, 183–184
The Mathematical Principles of Natural Philosophy. See *Principia*
Mathematical School in Halle, 79n5
matter, divisibility of, 177
Maupertuis, Pierre Louis Moreau de, 127, 132n12, 135n19, 142n15, 152, 219; and Cantemir, 127; and Cartesianism, 132–133; and J. Cassini, 134; and Clairaut, 127, 132–133; expedition, 132–135; and Newtonian theory, 136; and Voltaire, 135

Maxwell, James Clark, 188; on electricity and light, 158
Mayer, Friedrich Christian, 142n15; and Cantemir, 118
Mechanica, see under Euler, Leonard
medical theory, and Newtonian laws, 146
medicine, and Newtonian ideas, 113
Meier, Georg Friedrich, and Wolff, 108
Mendeleev, Dimitrii Ivanovich, 223
Menshikov, Prince Alexander, 17, 48, 48n6, 49, 67; and Bruce, 67; embezzlement of government revenue, 49; and Goldbach, 96; visit to London, 48; correspondence with Newton, 46–49; and Peter the Great, 48; power of, 49; and Royal Society, 46–49; science, interest in, 49; and Society of Neptune, 81
Menshutkin, B. N., 188–189; and Lomonosov, 199
Mersenne, father of, 96
metal, use of, for speculum, 74–77
metallic mirrors, 75
Method of Fluxions, French translation of, 95n7
Metropolitan of Kiev, 217
Michelessi, Domenico, 125n27
Michell, John, on attraction at a distance, 153–154; and magnetism, 153–154
Michelotti, P. A., and Jurin, 113
Michelson, Albert Abraham, and terrestrial tide effects, 41
Micrographia, see under Hooke, Robert
Micromégas, see under Voltaire
Milne Bursar and Liddell Mathematical Tutor (Ferquharson), 81
Milton, John, 119n7, 201n5
Mitchell, Sir David, 79
Moldavia, Hospodar of, 117
monadology, 147; of Leibniz, 109; of Wolff, 176
monetary problems, 17
Monetnyi Dvor, 17
Monconseil, Marquise de, and Cantemir, 122
Montague, Charles, Lord Chancellor of the Exchequer, 10, 13; meeting with Peter the Great, 14; president of Royal Society, 14; and Sommers, 10
Montesquieu, 217
moon, irregularities of, 21; irregularities in motions of, 173n34; Halley's observations of, 22; Flamsteed's computed places of, 22–24; synopsis of, 152; observed places, 22

Mordvinov, Count, correspondence with Jeremy Bentham's brother, 217–218

More, Louis Trenchard, 54

Morin's formula, 147

Morozov, A. A., on Lomonosov, 169

Moscow University, and Wolff, 108

motion: Euler on, 158–159; laws of, 37; lunar, 37; Newton on, 158–159; Newton's laws of, 86; solar, 37; theory of Newton, 113

Motte, Andrew, 40

mountain crystal, 77

Müller, G. V., 106; correspondence with Rumovskii, 214

Musschenbroek, Pieter van, 154, 223

natural law, Voltaire on, 218–219

Naval Academy, St. Petersburg, 78, 80–81, 83, 88–89, 226; and *Introductio ad Philosophiam Newtonianam*, 90; overshadowed by St. Petersburg Academy of Sciences, 90

Naval Department, move to St. Petersburg, 80

navigation, and Bruce, 43

Navigation School, in Moscow, 34, 78, 80, 212, 226

Neutonianismo per Le Dame, see under Algarotti, Francesco

New Theory of the Earth, see under Whiston, William

Newton, Isaac, 95n7, 96n11, 109nn8, 9, 112, 115–116, 125; in *Acta Eruditorum*, 43; and Algarotti, 124–125; alloys, use of, 76; and the apple, 128n1; his *Arithmetica Universalis*, 36, 42, 86, 120n9; his *Arithmetica Universalis* in Cantemir's library, 120; and preface to *Astronomiae Physicae et Geometricae Elementa*, 120; on astronomy, 222; and attraction at a distance, 140, 144, 146, 148–150, 171–174, 176, 224, 233; and attraction at a distance, further applications, 152–154, 161, 164; and Avramov, 61; and Daniel Bernoulli, 107, 138, 141; and John Bernoulli, 107; and Bilfinger, 109–111; and Lord Bolingbroke, 224; and Bruce, 32, 44–45, 68–69, 71–72, 76–77, 96, 231–232; and calculus, 144; and Cantemir, 232; Cantemir's library, Newton's works in, 120; and Cartesianism, 97–101, 110; and Cassegrain, 70; influence under Catherine II, 221–222; on celestial mechanics, 214; and the Marquise du Châtelet, 123n19; and chromatic aberration, 201–202, 206; his *Chronology of the Ancient Kingdoms Amended*, 120, 120–121n11, 137, 137n25; and Samuel Clarke, 109; on color, 190–192, 197–198, 201, 202; and John Colson, 30, 32; on comets, motion of, 100; on conservation, principle of, 142; and Copernican hypothesis, 99; cosmology of, 50; Cyrillic transformation of name, 231–232; death of, 96; dynamics of, 140–142, 175; and earth's circumference, 128; and earth's diameter, 129n2; and earth's shape, 127–136, 130n4; and *Elements of Euclid* in Russian, 82, 85; and *Éloge*, by Fontenelle, 194; his emission theory, 136; as empiricist, 177–178; Encyclopedists on, 221; on equations, 166; and *Essay on Man*, by Pope, 224–226, 235; on ether, 190; and Euclid, 82; and Euler, 143; Euler on, 211–212, 214–217; and Andrew Ferquharson, 82; and Flamsteed, 44; collaboration with Flamsteed, 20–21; correspondence with Flamsteed, 25; and "fluxionary" method, 43, 198, 228; and fluxions, 86–87, 113; and Fontenelle, 110, 116; Fontenelle on, 97, 179; on force, 142; and Galileo, 233; and *Geographia Universalis*, see under Varenius, Bernhard; and Goldbach, 93–95; Gottsched on, 168; and 's Gravesande, 90; on gravitation, 86, 105, 111–114, 123, 127–131, 136, 147–148, 150, 158–159, 171–179, 194, 213, 224, 225, 227–228, 233, 235; and gravitational theory, 55–56, 59, 61; and Gregory, 24–25, 39, 81; and Halley, 20–28; and Charles Hayes, 32; on heat, 148; and Hermann, 130–132; and Hooke, 54, 68, 189, 191–193, 197; Horsley's edition of, 227; and Huygens, 50–51, 54–57, 193; and hypotheses, 150, 150n47; on hypotheses, 191–192; and the inductive method, 150, 218; on interphenomena, 178; observations of Jupiter, 98; and Jurin, 111, 113, 144; and Kepler, 99–100, 129; laws of motion, gravity, and centripetal force, 86; controversy with Leibniz, 13, 37, 104, 114, 120, 142–145, 150; on light, 123, 136, 138, 156, 158, 185–198, 227–228; light, his theory of, 68, 70–75, 103, 123, 136; light and color, his theories of, 68, 70–75, 97; and Linus, 71; London,

Newton, Isaac (*Cont.*)
move to, 10; and Lucas, 71; Lucasian lectures at Cambridge, 20, 20n7; Lucasian professor of mathematics, 69; lunar theory, 20, 23, 26, 39; on mass, 233–234; Master of the Mint, 10n2; mathematical conception of nature, 183–184; on matter, 177; and Maupertuis, 127; and Prince Menshikov, 46, 48–49; correspondence with Prince Menshikov, 46–49; his *Method of Fluxions and Infinite Series,* first French translation of, 120n9; monuments to, 218; Count Mordvinov on, 218; biography by L. T. More, 54; on motion, 113, 158–159; and Natural Religion, 224; and naturalism, 224–225; letter from John Newton, 9, 14; his *Observations upon the Prophecies of Daniel and the Apocalypse of St. John,* 120–121n11; correspondence with Oldenburg, 76, 201–202; his *Optical Lectures Read in the Publick Schools of the University of Cambridge* and S. I. Vavilov's Russian translation of, 68–69, 69n26; optical theories, 68, 71–76, 139; his *Opticks: or a Treatise of the Reflexions, Refractions, Inflexions and Colours of Light,* 40, 68–69, 74, 88, 189, 191–192, 197, 219, 223, see also *Opticks;* and optics, 68, 71–76, 137, 197; optics, lectures on, 201; and Pardies, 70; and Peter the Great, 9, 13, 30, 45; Peter the Great's second copy of *Principia,* 45–46; phenomenalism, 55; and philosophes, 217; and "philosophie anglaise," 218; in *Philosophical Transactions,* 70–71, 74, 76; on physics, 222–223; and Jean Picard, 129; on planets, orbits of, 99–100, 227; popularization of, 223; his *Principia,* 18, 20, 28, 49–51, 88, 94, 143, 178, 197, 223, 231, 234, 236, see also *Principia;* prism, use of, 74, 139; first publication, 201n8; Radishchev on, 231, 236; reflector, Newtonian, see Newton, Isaac, telescope; reform of English coinage, 18; refraction, laws of, 210; Royal Mint, papers at, 13; and Royal Society, 95–96, 119; acceptance in Russia, 231, 233–237; influence in Russia, 89, 231–237; and St. Petersburg Academy of Sciences, 94–96, 104, 112; and scientific method, 221; *Sir Isaac Newton's Mathematick Philosophy,* see under

Whiston, William; Soimonov on, 212; Speranskii on, 236–237n7; his "System of the World," 86; his telescope, 70–72, 74–76, 78, 200–204, 209–210, 227; and construction of reflecting telescope, 74–76; and discovery of reflecting telescope, 74; and terrestrial tide effects, 41; on tides, 224, 227; his treatise of "Series and Fluxions," 68; *Treatise of the System of the World,* 41; his *Universal Arithmetick,* and Castillon's edition of, 171, 227n45; and Uranus, 41; and de Vercé, 70; and "verita di Newton," 235; and Voltaire, 121–123, 152, 218–221, 235–236; and Wallis, 44; Warden of the Mint, 10; on wave theory, 196; and Whiston, 36, 42–43; and white light, decomposition of, 74; work mistakenly credited to him, 78

Newton, John, letter to Isaac Newton, 9, 14
Newtonian laws, and medicine, 146
Newtonian phenomenalism, 55
Newtonian philosophy, 102, 102n2; explanations of, 85, 86
Newtonian physics, and St. Petersburg Academy of Sciences, 102–105
Newtonian synthesis, 235
Newtonianism, acceptance on the continent, 212; and Académie des Sciences, 136; and Algarotti, 134; and Bolkhovitinov, 217; and Bruce, 104, 234; and Cantemir, 116–117, 120–126; and Clairaut, 153; and electrical fluids, 160; and Euler, 159, 161–162, 211, 214–217; and fine fluids, 159; and French Revolution, 217; and Krafft, 137, 139; and Lomonosov, 162–164, 169–184, 186, 188–189, 193–195, 197–199, 211, 232–234; and medicine, 113; and modern physics in Russia, 234; problems of acceptance of, 152, 159; and Richmann, 146–150; and St. Petersburg Academy of Sciences, 90, 136–137, 156–157, 211; and Voltaire, 134; and Wolff, 168–169, 175–176
Newtonians, and Cartesians, 215; and Leibnizians (on calculus), 168
night telescope, of Lomonosov, 207–210
Nollet, Abbé, 154
Norwood, 43, 129n2
Nouveau Traité de la Pluralité des Mondes, see under *Kosmotheoros*

Observations on the density and fluidity

of bodies, see under Lomonosov, Mikhail V.
Ocherk istorii morskago kadetskago korpusa, see under Veselago, F. F.
Old Believers, 52, 165; and Avramov, 62
Oldenburg, Henry, and Newton, 76, 201–202
"On the Improvement of Telescopes," see under Lomonosov, Mikhail V.
ondular theory, 188, 228
127 Observations on the theories of light and electricity, see under Lomonosov, Mikhail V.
On Immortality, see under Radishchev, Aleksandr
Opera Mathematica, see under Wallis, John
Optical Lectures Read in the Publick Schools of the University of Cambridge, see under Newton, Isaac
Opticks . . ., by Isaac Newton: in Cantemir's library, 120; and electricity, 155; French translation of, 120n9; and gravitation, 171, 176, 177; and Richmann, 149–151; and St. Petersburg Academy of Sciences, 228. *See also* Newton, Isaac
optics: and Bruce, 68, 71–76; and Lomonosov, 185, 197; and Newton, 68, 71–76, 137, 197; Newton's theories on, 68, 71–76, 139; at St. Petersburg Academy of Sciences, 137
Oration on Light, see under Lomonosov, Mikhail V.
Oration on the Origin of Light, see under Lomonosov, Mikhail V.
Osborne, Vice Admiral Peregrine, Marquis of Carmarthen, and Peter the Great, 79
Ottomans, and Russia, 117
Ovid, 61–62n5, 62, 66
Oxford, 17, 30, 170; and John Freind, 113; and Halley, 23; and Keill, 86; Peter's visit to, 12, 12n11; Postnikov's visit to, 79

Padua, 131
palace revolution of *1762,* 205
Pankevich, M. I., 228n48
Paradoxa Hydrostatica, see under Boyle, Robert
Paramour Pink, 22n11, 23
Pardies, Ignace Gaston, 70, 191
Paris Academy. *See* Académie des Sciences, Paris
Parrot, G. F., 228n47

Paus, Johann Werner, 63
Pemberton, and Jurin, 140
pendulums, 37, 132–133
Perkins, 43
Pernau, Estonia, 146
Perry, Captain John, 45, 78
Peru, expedition to, 132
Peter II, 49; and Avramov, 62
Peter III, assassination of, 205
Peter the Great, 121n14; acquisition of books in England, 41; acquisition of books and scientific instruments, 94, 103; astronomy, interest in, 29, 68; and Avramov, 62, 64–66; and Bruce, 15, 29, 93–94; correspondence with Bruce, 64–68; and Charles XII, 117; and John Colson, 29–31, 93; correspondence with John Colson, 30; death of, 49; in Deptford, 12; and Dutch scientists, 11; invitation to England, 11; visit to England, 9, 17, 78, 104, 231, 234; his European debut in Koppenburg, 10; first European journey, 17; second European journey, 64; and Ferquharson, 78–79, 81; and Fontenelle, 64; and Gouda, 11; visit to Greenwich Observatory, 19–20, 31; and Grice and Gwyn, 80; and Halley, 22, 93; and Hartzoeker, 11, 11n8; and Holland, "scientific attractions" in, 12; and Hoogzaat, 11; his *Journal,* 13–15, see also *Journal;* and Demetrius Kantemir, 117, 118; and publication of *Kosmotheoros,* 63–66; and translation of *Kosmotheoros,* 50–51; and *Kunstkammer,* 49, 94; and Lebiadnikov, 94; and Leewenhoek's microscope, 11; and Leibniz, 10, 104; correspondence with Leibniz, 68; meeting with Leibniz, 93–94; his visit to London, 48; and Prince Menshikov, 48; his meeting with Charles Montague, 14; arrival in the Netherlands, 10; meetings with Newton, 9, 13, 30; and Vice Admiral Peregrine Osborne, Marquis of Carmarthen, 79; visit to Oxford, 12, 12n11, 30; and *Principia,* 39; acquisition of second copy of *Principia,* 45, 46; and Dirk Raven, 11; reform of Russian coinage, 17, 18; reforms, 234; later reforms, anticipation of, 17; visit to Royal Mint, 13; visit to Royal Society, 13–15, 93; and St. Petersburg Academy of Sciences, 93–94; and Schumacher, 89, 94, 103; science, con-

Peter the Great (*Cont.*)
tinuing interest in, 45, 47; and secularization, 62; shipbuilding, interest in, 11; and Society of Neptune, 81; and Soimonov, 212; translation of foreign works into Russian, 66–67; and translation, problems with, 89; and his Turkish campaign, 16; and Van Coehorn, 11; war against Sweden, 146; war against Sweden and Turkey, 80; and William III, 11–12; and Wolff, 104

Petrine enlightenment, and Avramov, 62, 63, 66

Petrov, V. V., 228n47

"Philalethes Cantabrigiensis," see under Jurin, James

philosophes, 116, 126, 217

Philosophiae Mathematicae Newtonianae Illustratae, see under Domcke, George Peter

Philosophiae Newtonianæ Institutiones, in Usus Academicos, see under 's Gravesande, W. James

Philosophical Transactions: in Bruce's possession, 39–40; and Joseph de L'Isle, 133, 134; Newton in, 148; Newton's telescope, 70–71, 74, 76; Newton's theory of light and color, 70; Newton's use of alloys in, 76; Newton's writings on light in, 71. *See also* Royal Society

phlogiston theory, 198, 228

Phonoromia, sive de viribus et motibus corporum solidorum et fluoridorum, see under Hermann, Jakob

Physices elementa mathematica experimentis confirmata: sive Introductio ad Philosophiam Newtonianam, see under 's Gravesande, W. James

physics, modern, and Newtonianism in Russia, 234

Physics, see under Rohault, Jacques

physiology, 113

Picard, Jean, 129, 132

pietists, and Wolff, 105

Pismovnik, see under Kurganov, N.G.

Planck, Max, 189

planetarians, 57–60

planets, orbits of, 129, 227; orbits around sun, 99, 100

Plekhanov, G. V., 116

Plenist vs. Vacuist, 147

plenum, 147

Pluralité des Mondes, see under Fontenelle, Bernard le Bovier de

polarization, 196

polestar, Flamsteed's "discovery" of its parallax, 24

Poltava, Battle of, 117

Pope, Alexander, *Essay on Man:* 217, 224–226, 235; in French translation, 217n13, 225n37; in Russian translation, by Bolkhovitinov and by Popovskii, 217, 225–226; and Warburton, 224

Popov, N. I., 207

Popovskii, Nikolai, translation of Pope's *Essay on Man*, 225–226, 235

Postnikov, Dr. Pëtr, and recruitment of Gwyn and Grice, 79, 80, 80n6

Prévost d'Exiles, Antoine François, 217

primary matter, 187

primogeniture, English laws of, 17

Principia, by Isaac Newton: Amsterdam edition of, 120n9; and Daniel Bernoulli, 107, 111, 141; third book of, intended, 41; and Bruce, 35, 39–40; in Cantemir's library, 120; and Cartesianism, 110; criticism of Cartesianism, 97–101; and Marquise du Châtelet, 123; John Clarke's commentary on, 37–38; controversies over, 68; Domcke's commentary on, 86–87; and earth's shape, 128–130; first edition, in USSR Academy of Sciences, 40n15, 41n16; English commentaries on, 121, 122; first English translation of, 40; and ether, 173; and Newton's controversy with Flamsteed, 27; observations received from Flamsteed, 28; French translation of, 104, 123; General Scholium, 99–100; Geneva edition of, 120n9; commentary by 's Gravesande, 38; influence on 's Gravesande, 103; and gravitation, 114, 171, 176, 179, 182; and David Gregory, 39; and Lomonosov, 170–171, 173–175, 182–183; on mass and weight, 182–183; and Peter the Great, 39; Peter the Great's acquisition of second copy, 45, 46; and *Pluralité des Mondes*, 116; and Richmann, 150–151; second edition in Russia of, 49; translation into Russian of, 228; and proposed new Russian university, 222; and effect in St. Petersburg, 97, 101; and St. Petersburg Academy of Sciences, 94, 101; Voltaire on, 219; and Wolff, 168. *See also* Newton, Isaac

Principia Philosophique, see under Descartes, René

prism, 74, 77, 139
Prokopovich, Theophan, 62; and Cantemir, 119
Pruth, 117
Ptolemy, catalogue, 28
Puffendorf, Samuel, 66
Pukhort, Ivan, responsibilities for Bruce's library, 34

Quaker meetings, 17
quantum theory, 189

Rainov, T. I., 189
Radishchev, Aleksandr, 198, 227–228; on Newton, 231, 236; his *On Immortality*, 236; and Russian Church, 216
Raven, Dirk, and Peter the Great, 11
Ray, John, 169
Razgovory o svete. See Algarotti, Francesco, *Neutonianismo per le Dame*
Razumovskii, Count K. G., 153n3, 153n8, 163
rationalism vs. empiricism, 176–183
refracting telescope, and Dollond, 209–210
refraction, laws of, 210
Reichenbach, E. H., 178
Riccati, Giordano, Count, 107
Richmann, Georg Wilhelm, 94, 146, 188; and attraction, 149; and attraction at a distance, 153, 154, 170; and calorimetry, 146; and Cartesianism, 147–148; and corpuscular philosophy, 148–149; and electricity, 151, 154–155; on electricity and lightning, 154; electrocution of, 151, 154–155; and Franklin, 154; and Freind, 149; and 's Gravesande, 149; on gravitation, 150; and Halle University, 146; on heat and electricity, 154; and Jena University, 146; and Jurin, 143, 149; and Keill, 149; and Krafft, 146; controversy with Krafft, 147–148; on Leibniz, 144; and light, 149; and Lomonosov, 146, 148; on luminescence and electricity, 146; and magnetism, 154n9; and Newtonianism, 146–150; and *Opticks*, 149–151; in *Primechaniia na vedomosti*, 146–147; and *Principia*, 150–151; and St. Petersburg Academy of Sciences, 146, 155; and Schumacher, 146; his *Tentamen stabiliendi leges cohaesiones*, 146, 148–150; and Tümmig, 146; and Josiah Weitbrecht, 142–143; controversy with Weitbrecht, 149; and Wolff, 150; his work, later, 147
Rohault, Jacques, his *Physics*, 103; his *System of Natural Philosophy*, 37–38

Rolli, Paolo, 119n7
Ronayne, Philip, 44n24
Rostock, University of, 160
Rousseau, Jean-Jacques, 215, 217
Royal Mathematical School, Christ's Hospital, 79, 80; as a model for Navigation School, 79
Royal Mint, 15, 17. *See also* Newton, Isaac; Peter the Great
Royal Observatory. *See* Greenwich Observatory
Royal Society, 17, 58, 71, 96n10, 119; visit by Charles II, 15; and Folkes, 124; and *Historia Coelestis Britannica*, 27; and Jurin, 112–113; and light, refrangibility of, views of Newton and Hooke juxtaposed, 192; and Prince Menshikov, 46–49, *see also* Appendix I; Newton as President of, 95–96; and Newton's telescope, 74, 201–202; visit by Peter the Great, 13–15, 93; *Philosophical Transactions*, 58, 69–71, 148, see *Philosophical Transactions;* and St. Petersburg Academy of Sciences, 94–96; and Sloane, 119
Rumovskii, S. Ia., 197; correspondence with Euler, 207–208; translation of *Letters to a German Princess*, 215, 217; on Lomonosov's telescope, 210; correspondence with G. V. Müller, 214
Russell, Bertrand, on Leibniz and Newtonian gravitation, 175n42
Russia: and interest in English culture, 217; and French Revolution, 217; maps, 133; mathematics in, 198; and Ottomans, 117; scientific revolution in, 184; war with Sweden, 80, 146; war with Turkey, 80, 117
Russian Academy. *See* St. Petersburg Academy of Sciences
Russian Enlightenment, 116
Russian language, development of, 89, *see also* Appendix II
Russian nobility, and Anne of Courland, 119
Russian Synod. See under Holy Synod
Russian University, Catherinian project for, 222–223

Saardam, and Peter's trip to Holland, 10
Sage of Ferney. *See* Voltaire
St. Petersburg Academy of Sciences, 33, 49, 89, 108–109; and Aepinus, 151; and Bilfinger, 102, 104–105, 111, 114–115; Blumentrost as president of,

Saint Petersburg Academy (*Cont.*)
96; and Cantemir, 115, 117–119, 126–127; and Cartesianism, 102, 104–105; and *Commentarii*, 147; decline of, 115; and earth's shape, 129–134; and electricity, 155; and Leonard Euler, 136, 137, 155, 215; *Ezhemesiachnyia sochineniia*, 213, 218; *Ezhemesiachnyia sochineniia* on Newtonianism, 224; founding of, 93–94, 104; and Goldbach, 95–96; and 's Gravesande, 102; gymnasium of, 118, 138, 145–146, 167; and Jakob Hermann, 114–115, 130–132; and *Kratkoe opisanie kommentariev*, 213; and Kurganov, 226; and Leutmann, 204; library of, 94; on light, 228; and Joseph de l'Isle, 72, 133–134; and Lomonosov, 151, 155, 171, 185, 197, 233; and Lomonosov's telescopes, 204–205, 207–210; and Maupertuis expedition, 134; and Newton, 232–235; correspondence with Newton, 94–96; and Horsley's edition of Newton, 227; and Newton-Leibniz controversy, 145; and Newton's optics, 137; and Newtonianism, 90, 136–137, 152–154, 211; opening of, 94; and the *Opticks*, 228; and Peter the Great, 93–94; discussion of the *Principia*, 101; public competition, 152, 153; public lectures, 138–139; and Count K. G. Razumovskii, 163; and Richmann, 146, 155; and Richmann-Weitbrecht dispute, 143; and Royal Society, 94–96; science, popularization of, 139, 146–147, 213, 218, 224; and Soimonov, 213; and *Universal Arithmetick*, 227n45; and Weitbrecht, 145; and Wolff, 105

St. Petersburg Naval Academy, and Ferquharson's works, 80

St. Petersburg topography, 64

Saltykov, Count (responsible for Bruce's library), 33, 34

Satarov, Ivan, 82–83, 87

satellites of Jupiter and Saturn, discovery of, 52

Saturn, 57

scholastic education, 166

scholasticism, 221

School of St. Jean Baptiste de la Salle, 79n5

schools of Russian and ciphering, 80

Schumacher, J. D., 139; and Bilfinger and Jakob Hermann, 114–115; and 's Gravesande, 90; and 's Gravesande's book, 103–104; and Krafft, 138–139; and Peter the Great, 89, 94, 103; correspondence with Richmann, 154; scientific books and instruments, acquisition of, 89; and Taubert, 138; and Wolff, 145

scientific education, in Russia, 234–235

scientific instruments, in Bruce's library, 72–74, 76–77

scientific revolution, 184, 198

secularization, under Peter the Great, 62

Selenographia, see under Hevelius, Johannes

Sharp, Abraham, 19; his mural arc, 27

Short, James, 205

Shuvalov, Prince I., 208–210

Siberia, Bering's expedition to, 133

Slavic Grammar, see under Smotritskii, Ivan

Slavo-Greco-Latin Academy, 166–167

Sloane, Sir Hans, and Cantemir, 119, 120

smallpox inoculation, 112

Smith, Adam, 218

Smith, Robert, 120n9

Smotritskii, Ivan, *Slavic Grammar*, 166

Sobol', S. L., 34

Society of Jesus, 137

Society of Neptune, 16, 81

Soimonov, Feodor Ivanovich, 212–213; on astronomy, 213–214; and Catherine II, 213; on Copernican theory, 214; his *Kratkoe iz'iasnenie o astronomii . . .*, 213–214; on Newton, 212; and Peter the Great, 212; and Rumovskii, 214; and St. Petersburg Academy of Sciences, 213; in Siberia, 212–213; his *Svetil'nik morskoi, to est' opisanie vostochnago ili variazhskago moria*, 212; and Volynskii, 212

Sommers, Lord Keeper of the Mint, 10

Sophie Friederike Charlotte Leopoldine Luise, Princess, 214

Sopikov, Vasilii Stepanovich, 63

space, Leonard Euler on, 157

Spasskii School. See Slavo-Greco-Latin Academy

speculum, suitable materials for, 75–77

Speiser, Andreas, on Leonard Euler, 216n9

Speranskii, M. M., and Newton, 236, 236–237n7

Spinoza, Benedictus de, 184, 219n18

Staehlin, Carl, 126n30

Steklov, V. A., 166

stenthrophonic tube, 58

Strassburg, University of, 89

Street, astronomical tables of, 36

Sukharev Tower, 80
Sumarokov, Pankratii Platonovich, 223
sun, distance from earth, 129; and New-
ton's criticism of vortices theory, 98
Sur la Nature et les Proprietés du Feu,
see under Euler, Leonard
Sweden, war with Russia, 80, 146
Swedish Academy, 224
*Syndesmologia, sive historia ligamentorum
corporis humani,* see under Weit-
brecht, Josiah
System of Natural Philosophy, see under
Rohault, Jacques
System of the World, see under Newton,
Isaac

Table of Refractions, 32
tables of sines introduced into Russia, 58
Tacquet, Claude André, and *Elementa
Geometriae,* 85; abridgement of Eu-
clid, 82–87
Tames, Ivan. See under Thomas, John
Tatishchev, V. N., 73, 94, 121n14; and
Bruce, 73
Taubert, Johann Kaspar, and Krafft, 138;
and Schumacher, 138
telescopes, alloys for, 74, 76–77; arsenic,
use of, in constructing Newtonian,
76–77; of Bruce, 73–74, 76–78; of
Lomonosov, 200, 202–209, 210; of
Newton, 70–72, 74–76, 200–203, 227
Tentamen stabiliendi leges cohaesiones,
see under Richmann, Georg Wilhelm
*Tentamen Theoriae de Particulis Insensi-
bilibus,* see under Lomonosov, Mik-
hail V.
*Tentamen Theoriae Electricitatis et Mag-
netismi,* see under Aepinus, Franz
Ulrich Theodor
theomachists (Bogobortsy), 64
*Theorems of Archimedes selected by An-
drew Tacquet the Jesuit and con-
densed by Georg Petr Domkio, put
into the Russian tongue from the
Latin by Ivan Satarov, the surgeon.*
See Satarov, Ivan
Thomas, Jean. See Thomas, John
Thomas, Captain John, 72–73, 121n14;
and Cantemir, 121, 125
tides, 135, 224, 227
Tiedemann, Christoph, 34, 35
Tiriutin, F. N., 204
Tobol'sk, 223n29
Todhunter, and J. N. de l'Isle, 134
Toms. See under Thomas, John
Torgau, 93

Tornea River, Maupertuis expedition, 134
Torricelli, 168
Traité de dynamique. See under d'Alem-
bert
transverse wave, 196
Treatise on Fluxions, see under Hayes,
Charles
Treatise of the System of the World, see
under Newton, Isaac
Trinity College, Cambridge, and Jurin,
112
Trudoliubivaia Pchela, first private pe-
riodical in Russia, 220n22
Tübingen, and Bilfinger, 108; and Krafft,
139
Tübingen University, 145
"Tubus nyctopticus modo Lomonosov-
Newton," 207–209
Tümmig, L. F., *Institutiones philosophiae
Wolffianae,* 146
Turks, war with Russia, 80, 117
typography at Antonievo-Siiskii monas-
tery, 165

*Universal Arithmetick: or, A Treatise of
Arithmetical Composition and Reso-
lution,* see under Newton, Isaac
universal gravitation. *See* gravitation
universe, infinity of, 53; and Giordano
Bruno, 53n11
Uppsala University, 134
Ust Ruditsy, 202

Vacuist vs. Plenist, 147
Van Coehorn, and Peter the Great, 11
Varenius, Bernhard, 120
Vavilov, Sergei Ivanovich, 200, 205, 232
de Vercé, and Newton, 70
Verenet, George, on Peter's visit to Hol-
land, 11
*Vernuenftige Gedancken von den Wuerck-
ungen der Natur . . . ,* see under Wolff,
Christian
Veselago, F. F., 81; his *Ocherk istorii
morskogo kadetskogo korpusa,* 80
Vieta, Franciscus, 166
Villamil, Richard de, 33n1, 128n1, 129n2
Virgil, 62
"vires vivae." *See* "vis viva."
"vis viva," 97, 108, 130, 141, 143
Voltaire, 116, 126, 215, 217; and Algarotti,
124–125; his *Candide,* 169; and Prince
Cantemir, 122–124; on Cartesians,
218; and Catherine the Great, 221;
and Marquise du Châtelet, 104, 123,
123n19; correspondence with M. de la

Voltaire (*Cont.*)
Condamine, 152; and Descartes, 59–60, 220; and the earth's shape, 128; his *Elémens de la Philosophie de Newton*, 122–124, 219; and presentation of the *Elémens* to the St. Petersburg Academy of Sciences, 126n30; and English critics, 123–124; his *Essai sur la Poésie Epique*, 122; and Fontenelle, 116; his *Henriade*, 122; and Holy Synod, 220–221; in "Irtysh (being) transformed into Hippocrene," 223; and Jurin, 112; and Leibniz, 169, 220; his *Les Oreilles du Comte de Chesterfield*, 219; his *Lettres Philosophiques*, 122–123, 128, 170, 220; his *L'Ingenu*, 219; and Maupertuis, 135; and metaphysics, 220; his *Micromégas*, 169, 218–220, 223–224; and Newton, 121–123, 235–236; on Newton, 152, 218–221; and Newtonianism, 134; on Newtonianism, 234; elected to St. Petersburg Academy, 126n30; on Spinoza, 219n18; and Wolff, 104; on Wolff, 169
Volynskii, Artemii, 212
Voronezh Seminary, 217
Vorontsov-Vel'iaminov, B. A., 34
vortices, 37, 97–101, 109–111, 110n12, 113, 130, 132–133, 135, 154, 157, 214; and Bilfinger, 113

Wallachia, 117
Wallis, John, 169; correspondence with Flamsteed, 24; and Flamsteed's and Newton's involvement in preparation of his mathematical works, 44; mathematical writings of, 43; and Newton, 24; his *Opera Mathematica*, 23n12, 24, 26
Wallis, P. J., 43–44n22
Wapping, 30
Warden of the Mint, Newton as, 10
Wargentin, P., 224
wave-ether hypothesis, 191
wave theory, 188–189, 191, 196
Weitbrecht, Josiah: and experiments with blood, 145–146; and Dr. Duvernois, 145; and Jurin, 142–145; and University of Königsberg, 145; and Newtonian laws, 146; as physiologist, 145; and Richmann, 143; controversy with Richmann, 149; and St. Petersburg Academy of Sciences, 145; his *Syndesmologia, sive historia ligamentorum corporis humani*, 145

Weld, C. H., 96n10
Westminster Abbey, monument to Newton, 218
Whiston, William: his Astronomical Lectures, 36; and Domcke, 85–86; and David Gregory, 39; his *A New Theory of the Earth*, 36–37, 43; and Newton, 36, 42, 43; and Newtonian philosophy, 86; his *Praelectiones Astronomicae*, in Cantemir's library, 120; his *Sir Isaac Newton's Mathematick Philosophy*, 36, 37
White Sea, 133, 165
Wilcke, Johan, on electricity, 160–161
William III, and Peter the Great, 11–12
Williamson, Sir Joseph, 31
Winter Palace, 154
Wolff, Christian, 94; compared to Aristotle and Hasan, 104, 105n8; and Baumgarten, 108; and Bilfinger, 105, 108–109; correspondence with Blumentrost, 104; and Cartesianism, 167–168, 175; on English natural philosophers, 169; and Euler, 150; and Frederick the Great, 104–105, 169; and Frederick William of Prussia, 104; and German universities, 108; and universal gravitation, 105, 175; on gravitation, 175–176, 179, 181; at Halle, 108; flight from Halle, 104–105; on Jurin, 144–145; and Kant, 104, 168; and Keill in *Acta Eruditorum*, 168; and Leibniz, 104; and Leibnizians, 167–168, 175; limitations of, 105; and Lomonosov, 104, 162, 164, 167–170, 175–181, 198; at Marburg, 104; and mathematics, 180–181; on matter, 177; and Meier, 108; monadology of, 176; and Moscow University, 108; and Newtonianism, 168–169, 175–176; and Peter the Great, 104; and rationalists, 178–179; and Richmann, 150; and St. Petersburg Academy of Sciences, 105; correspondence with Schumacher, 145; his *Vernuenftige Gedancken von den Wuerckungen der Natur*, 105, 168, 175, 177; and Voltaire, 104, 169
Woolwich Arsenal, 17
Wren, Christopher, 19, 169

Young, Thomas, 187–188; on diffraction, 195–197; his "general law of the Interference of Light," 195–196

Zaandam. See Saardam
Zabelin, I. E., 34; inventory of Bruce's books and scientific instruments, 33

Russian Research Center Studies

1. *Public Opinion in Soviet Russia: A Study in Mass Persuasion,* by Alex Inkeles
2. *Soviet Politics—The Dilemma of Power: The Role of Ideas in Social Change,* by Barrington Moore, Jr.*
3. *Justice in the U.S.S.R.: An Interpretation of Soviet Law,* by Harold J. Berman. Revised edition, enlarged
4. *Chinese Communism and the Rise of Mao,* by Benjamin I. Schwartz
5. *Titoism and the Cominform,* by Adam B. Ulam*
6. *A Documentary History of Chinese Communism,* by Conrad Brandt, Benjamin Schwartz, and John K. Fairbank*
7. *The New Man in Soviet Psychology,* by Raymond A. Bauer
8. *Soviet Opposition to Stalin: A Case Study in World War II,* by George Fischer*
9. *Minerals: A Key to Soviet Power,* by Demitri B. Shimkin*
10. *Soviet Law in Action: The Recollected Cases of a Soviet Lawyer,* by Boris A. Konstantinovsky; edited by Harold J. Berman*
11. *How Russia Is Ruled,* by Merle Fainsod. Revised edition
12. *Terror and Progress USSR: Some Sources of Change and Stability in the Soviet Dictatorship,* by Barrington Moore, Jr.
13. *The Formation of the Soviet Union: Communism and Nationalism, 1917–1923,* by Richard Pipes. Revised edition
14. *Marxism: The Unity of Theory and Practice—A Critical Essay,* by Alfred G. Meyer. Reissued with a new introduction
15. *Soviet Industrial Production, 1928–1951,* by Donald R. Hodgman
16. *Soviet Taxation: The Fiscal and Monetary Problems of a Planned Economy,* by Franklyn D. Holzman*
17. *Soviet Military Law and Administration,* by Harold J. Berman and Miroslav Kerner*
18. *Documents on Soviet Military Law and Administration,* edited and translated by Harold J. Berman and Miroslav Kerner*
19. *The Russian Marxists and the Origins of Bolshevism,* by Leopold H. Haimson
20. *The Permanent Purge: Politics in Soviet Totalitarianism,* by Zbigniew K. Brzezinski*
21. *Belorussia: The Making of a Nation—A Case Study,* by Nicholas P. Vakar
22. *A Bibliographical Guide to Belorussia,* by Nicholas P. Vakar*
23. *The Balkans in Our Time,* by Robert Lee Wolff (also American Foreign Policy Library)

24. *How the Soviet System Works: Cultural, Psychological, and Social Themes,* by Raymond A. Bauer, Alex Inkeles, and Clyde Kluckhohn†
25. *The Economics of Soviet Steel,* by M. Gardner Clark*
26. *Leninism,* by Alfred G. Meyer*
27. *Factory and Manager in the USSR,* by Joseph S. Berliner†
28. *Soviet Transportation Policy,* by Holland Hunter
29. *Doctor and Patient in Soviet Russia,* by Mark G. Field†
30. *Russian Liberalism: From Gentry to Intelligentsia,* by George Fischer
31. *Stalin's Failure in China, 1924–1927,* by Conrad Brandt
32. *The Communist Party of Poland: An Outline of History,* by M. K. Dziewanowski
33. *Karamzin's Memoir on Ancient and Modern Russia: A Translation and Analysis,* by Richard Pipes*
34. *A Memoir on Ancient and Modern Russia,* by N. M. Karamzin, the Russian text edited by Richard Pipes*
35. *The Soviet Citizen: Daily Life in a Totalitarian Society,* by Alex Inkeles and Raymond A. Bauer†
36. *Pan-Turkism and Islam in Russia,* by Serge A. Zenkovsky
37. *The Soviet Bloc: Unity and Conflict,* by Zbigniew K. Brzezinski. Revised and enlarged edition‡
38. *National Consciousness in Eighteenth-Century Russia,* by Hans Rogger
39. *Alexander Herzen and the Birth of Russian Socialism, 1812–1855,* by Martin Malia
40. *The Conscience of the Revolution: Communist Opposition in Soviet Russia,* by Robert Vincent Daniels*
41. *The Soviet Industrialization Debate, 1924–1928,* by Alexander Erlich
42. *The Third Section: Police and Society in Russia under Nicholas I,* by Sidney Monas
43. *Dilemmas of Progress in Tsarist Russia: Legal Marxism and Legal Populism,* by Arthur P. Mendel
44. *Political Control of Literature in the USSR, 1946–1959,* by Harold Swayze
45. *Accounting in Soviet Planning and Management,* by Robert W. Campbell
46. *Social Democracy and the St. Petersburg Labor Movement, 1885–1897,* by Richard Pipes
47. *The New Face of Soviet Totalitarianism,* by Adam B. Ulam
48. *Stalin's Foreign Policy Reappraised,* by Marshall D. Shulman
49. *The Soviet Youth Program: Regimentation and Rebellion,* by Allen Kassof*
50. *Soviet Criminal Law and Procedure: The RSFSR Codes,* translated by Harold J. Berman and James W. Spindler; introduction and analysis by Harold J. Berman
51. *Poland's Politics: Idealism vs. Realism,* by Adam Bromke
52. *Managerial Power and Soviet Politics,* by Jeremy R. Azrael
53. *Danilevsky: A Russian Totalitarian Philosopher,* by Robert E. MacMaster
54. *Russia's Protectorates in Central Asia: Bukhara and Khiva, 1865–1924,* by Seymour Becker
55. *Revolutionary Russia,* edited by Richard Pipes
56. *The Family in Soviet Russia,* by H. Kent Geiger
57. *Social Change in Soviet Russia,* by Alex Inkeles
58. *The Soviet Prefects: The Local Party Organs in Industrial Decision-making,* by Jerry F. Hough
59. *Soviet-Polish Relations, 1917–1921,* by Piotr S. Wandycz

60. *One Hundred Thousand Tractors: The MTS and the Development of Controls in Soviet Agriculture,* by Robert F. Miller
61. *The Lysenko Affair,* by David Joravsky
62. *Icon and Swastika: The Russian Orthodox Church under Nazi and Soviet Control,* by Harvey Fireside
63. *A Century of Russian Agriculture: From Alexander II to Khrushchev,* by Lazar Volin
64. *Struve: Liberal on the Left, 1870–1905,* by Richard Pipes
65. *Nikolai Strakhov,* by Linda Gerstein
66. *The Kurbskii-Groznyi Apocrypha: The Seventeenth-Century Genesis of the "Correspondence" Attributed to Prince A. M. Kurbskii and Tsar Ivan IV,* by Edward L. Keenan
67. *Chernyshevskii: The Man and the Journalist,* by William F. Woehrlin
68. *European and Muscovite: Ivan Kireevsky and the Origins of Slavophilism,* by Abbott Gleason
69. *Newton and Russia: The Early Influence, 1698–1796,* by Valentin Boss

* Out of print.
† Publications of the Harvard Project on the Soviet Social System.
‡ Published jointly with the Center for International Affairs, Harvard University.